Air Quality Assessment and Management

Other books in Clay's Library of Health and the Environment:

E Coli: Environmental Health Issues of VTEC 0157 – *Sharon Parry and Stephen Palmer* 0415235952

Environmental Health and Housing – *Jill Stewart* 041525129X

Also available from Spon Press:

Air Pollution – *Jeremy Colls* 0419206507

Clay's Handbook of Environmental Health 18th Edition – *edited by W. H. Bassett* 0419229604

Decision-making in Environmental Health – *edited by C. Corvalán, D. Briggs and G. Zielhuis* 0419259406 HB, 0419259503 PB

Environmental Health Procedures 5th Edition – *W. H. Bassett* 0419229701

Groundwater Quality Monitoring – *S. Foster, P. Chilton and R. Helmer* 0419258809 HB, 0419258906 PB

Legal Competence in Environmental Health – *Terence Moran* 0419230009

Monitoring Bathing Waters – *edited by Jamie Bartram and Gareth Rees* 0419243704 HB, 0419243801 PB

Statistical Methods in Environmental Health – *J. Pearson and A. Turton* 0412484501

Toxic Cyanobacteria in Water – *edited by Ingrid Chorus and Jamie Bartram* 0419239308

Upgrading Water Treatment Plants – *Glen Wagner and Renato Pinheiro* 0419260404 HB, 0419260501 PB

Water Pollution Control – *Richard Helmer and Ivanildo Hespanhol* 0419229108

Water Quality Assessments – *Deborah Chapman* 0419215905 HB, 0419216006 PB

Water Quality Monitoring – *Jamie Bartram and Richard Balance* 0419223207 HB, 0419217304 PB

Urban Traffic Pollution – *D. Schwela and O. Zali* 0419237208

Air Quality Assessment and Management

A Practical Guide

D. Owen Harrop

London and New York

First published 2002
by Spon Press
11 New Fetter Lane, London EC4P 4EE

Simultaneously published in the USA and Canada
by Spon Press
29 West 35th Street, New York, NY 10001

Spon Press is an imprint of the Taylor & Francis Group

Printed and bound in Great Britain by
Biddles Ltd, Guildford, Surrey

Publisher's Note
This book has been produced from camera-ready copy
supplied by the author.

British Library Cataloguing in Publication Data
A catalogue record for this book is available from the British Library

Library of Congress Cataloging in Publication Data

Harrop, Owen.
Air quality assessment and management: a practical guide / Owen Harrop
p. cm. – (Clay's library of health and the environment)
Includes bibliographical references and index.
1. Air--Pollution--Measurement. 2. Air quality--Measurement. 3. Air quality management. I. Title. II. Series.

TD890.H37 2002
628.5'3—dc21 2001042024

ISBN 0-415-23411-5 (pbk)
ISBN 0-415-23410-7 (hbk)

Clay's Library of Health and the Environment

An increasing breadth and depth of knowledge is required to tackle the health threats of the environment in the 21st century, and to accommodate the increasing sophistication and globalisation of policies and practices.

Clay's Library of Health and the Environment provides a focus for the publication of leading-edge knowledge in this field, tackling broad and detailed issues. The flagship publication *Clay's Handbook of Environmental Health*, now in its 18th edition, continues to serve environmental health officers and other professionals in over thirty countries.

Contents

Preface

This book primarily describes how to undertake air quality assessment and management studies at the project level, and therefore details basic concepts, techniques, methodologies and practices.

The book reviews the principles of air quality assessment and management; primary sources of air pollution; air pollution effects on human health, flora and fauna, etc; scoping of air quality impacts; baseline monitoring; impact prediction; impact significance and pollution mitigation and control. Included are a series of case studies and exercises. The book concludes with individual case studies on the application of air quality assessment and management. It is not the intention of the book to address global air quality issues, therefore these are only detailed at a cursory level.

The book is intended as a practical guide for practitioners and undergraduate students studying air pollution assessment and management, or disciplines of environmental sciences, geography, environmental engineering and the broader social sciences.

It is hoped that this book will serve as a helpful guide. The topic of air pollution control is large and diversified. Therefore this book only provides specific elements of a myriad of strands that form a fascinating subject.

A car comes up, with lamps full-glare,
That flash upon a tree:
It has nothing to do with me,
And whangs along in a world of its own,
Leaving a blacker air

Thomas Hardy 1840 – 1928, Nobody Comes

This book would never have been completed without the patience of Sue. She has never complained and always given me her support. To her I owe everything.

Acknowledgements

I am grateful to many people who have helped in various ways in the preparation of this book and those who have given permission to cite their work. Whilst information in the book is from public documents the author has obtained, where possible, permission to use materials from the individuals who produced the original document. The opinions relating to the case studies cited are, however, those of the author and do not necessarily reflect the findings of the original studies.

Special acknowledgement should go to Annie Danskin and Matthew Shutt for giving up their time to critically review this book. I am grateful for their efforts.

I would also like to thank Haydn, Pat and Noel. They know very little about air pollution, however they are great people with whom to discuss the problems of writing a book, especially over a pint!

Every effort has been made to contact copyright holders and I apologise for any inadvertent omission. If any acknowledgement is missing it would be appreciated if contact could be made care of the publishers so that this can be rectified in any future edition.

Wild air, world mothering air,
Nestling me everywhere.

Gerald Manley Hopkins 1844 – 1889, The Blessed Virgin Compared to the Air We Breathe

CHAPTER ONE

Introduction

1.1 INTRODUCTION

Air pollution can be defined as the presence in the external atmosphere of one or more contaminants (pollutants), or combinations thereof, in such quantities and of such duration as may be or may cause injury to human health, plant or animal life, or property (materials), or which unreasonably interfere with the comfortable enjoyment of life, or property, or the conduct of business (Canter, 1996).

Breathing is not optional, it is essential even for a short time, and air has to be used as it is found. The Times newspaper (1881) quoted:

'The air we receive at our birth and resign only when we die is the first necessity of our existence'.

Despite its essential ingredient to life, air quality has been historically variable and frequently to the detriment of human health. Nevertheless, our quality of life dramatically improved during the twentieth century. Now, however, a growing body of research has found that certain pollutants may affect human health at lower concentrations than had previously been thought. This concern has heightened public anxiety to the importance of improving and managing air quality. It is paramount that a resource as important as air quality is protected and managed for future generations (Department of the Environment (DoE), 1993).

1.2 AIR POLLUTION - A CONCERN

Concerns about air quality have probably been around as long as mankind. From the moment fire was invented air pollution became a problem (Brimblecombe, 1987) and it has been a problem ever since. Concerns have occurred periodically throughout history, and are well documented (e.g. Ashby and Anderson, 1981; Brimblecombe, 1987; National Society for Clean Air and Environmental Protection (NSCA), 2000a).

Perhaps the historical air quality problems of the UK are the best documented. In 1257 in England, Queen Eleanor, wife of Henry III, was obliged to leave Nottingham on account of smoke nuisance. In 1273 a Royal proclamation was issued by Edward I to prohibit the use of sea coal in open furnaces because of the prejudicial effects to health from smoke emissions. Three hundred years later in 1578 it was written that Queen Elizabeth I:

'findeth hersealfe greatly greved and annoyed with the taste and smoke of sea-cooles'.

In subsequent centuries unfortunately, little positive attention was paid to air pollution control, despite the efforts of many individuals including the pamphleteer John Evelyn, who in 1661 published one of the first written accounts on air pollution entitled 'Fumifugium of the Aer and Smoake of London Dissipated'. Despite this early recognition of the problem, little attempt was made to control air pollution until the late 19th and early 20th centuries, even though many countries were gripped by the Industrial Revolution. The Revolution was based on coal as the energy source, and therefore brought worsening levels of air pollution (Fig. 1.1).

Figure 1.1 Smog episode (c1950) on the River Clyde, Glasgow, Scotland (courtesy of the Environmental Health Department, Glasgow City Council and the Scottish Environment Protection Agency (SEPA))

One of the first countries to attempt to tackle the problem of air quality was the UK. In 1863 the first Alkali etc. Works Act was passed. This, unfortunately, made no attempt to control smoke, but required 95% of offensive emissions to be arrested. The improvement in pollution levels was dramatic. A second Alkali Act in 1874 required the application of the Best Practicable Means (BPM) method to prevent the escape of noxious or offensive gases. The Act was subsequently extended to all the major industries that polluted air. Both Acts were eventually consolidated by the Alkali etc. Works Regulation Act 1906, which linked together a schedule of defined and chosen processes/works with a list of 'noxious and offensive gases'. The Act was a forerunner to the present day UK Environmental Protection Act (EPA 1990) (Chapter 7, Section 7.10.1).

Despite the powers of the Alkali Act to control industrial emissions, there was little measure to control the far more widespread problem of smoke and sulphur dioxide (SO_2) from domestic chimneys. The UK Government therefore appointed a committee with a remit to:

'examine the nature, causes and effects of air pollution and the efficacy of present preventive measures; to consider what further measures are practicable; and to make recommendations'.

This culminated in the introduction of the Clean Air Act (CAA) in 1956, which was later amended and extended by the 1968 CAA. The Acts constituted the operative legislation against pollution by smoke, grit and dust from domestic fires and other commercial and industrial processes not covered by the Alkali Acts and other subsequent pollution legislation. Again the improvement in air quality was discernible. Annual average SO_2 and smoke levels diminished considerably; the most dramatic reductions were achieved in urban areas (Fig. 1.2).

Figure 1.2 Historical smoke and SO_2 concentrations ($\mu g/m^3$) for Glasgow, Scotland (Data from the Department of the Environmental, Transport and the Regions (DETR), 2000b)

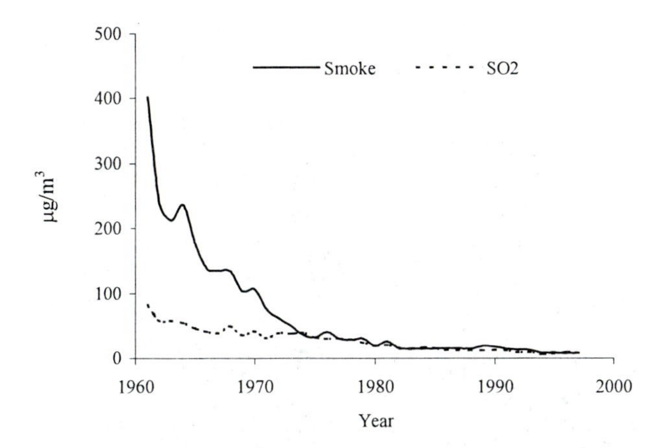

The reduction in the concentration of SO_2 and smoke was brought about by the burning of cleaner fuels, especially the use of gas, the use of tall stacks on power stations, their relocation outside cities, and the decline of heavy industry (DoE, 1993). The 1956 and 1968 Acts have subsequently been consolidated and their provisions re-enacted in the 1993 CAA. Occurrences of poor air quality have also been found in other countries (Chapter 2, Section 2.3 and Chapter 3, Section 3.2.1).

For most developed and developing countries motor traffic emissions now pose a principal threat to air quality, particularly in urban areas. Petrol and diesel engines emit a wide variety of pollutants, principally carbon monoxide (CO), oxides of nitrogen (NO_x), volatile organic compounds (VOCs) and particulates, which have an increasing impact on air quality. Whilst improvements in motor exhaust emission controls and fuel technology (Chapters 7 and 8) have resulted in an improvement in air quality (e.g. the elimination of lead (Pb) in fuel), concerns still persist about the elevated levels of pollution including the occurrence of photochemical smogs (or hazes) (Chapters 2 and 3). Photochemical reactions resulting from the action of sunlight on nitrogen dioxide (NO_2) and VOCs from vehicle emissions leads to the formation of ozone (O_3), a secondary long-range pollutant, which impacts in rural areas often far from the original emission site. These emissions of VOCs and NO_2 together with ammonia (NH_3) emissions also contribute to the formation of acid deposition, while emissions of oxides of carbon (CO_x) (e.g. carbon dioxde (CO_2)) from motor vehicles and industrial processes contribute to global warming concerns (Chapter 3). In all except worst case situations, emissions from industrial and domestic pollutant sources, together with their impact on air quality, tend to be steady state or have improved over time. In contrast, however, traffic pollution problems are worsening. The problem can be particularly acute in urban areas with dramatically increasing vehicle fleets and infrastructure limitations.

1.3 BOOK FORMAT

The principal objective of the book is to provide a guide to Air Quality Assessment (AQA) and Air Quality Management (AQM) at a practical level. The book is divided into two parts. Part 1 begins with details of the principal types of air pollution, their sources, and emissions trends (annual, diurnal, etc.) (Chapter 2) and outlines indoor air quality concerns. The effects of air pollution on human health, flora, fauna, materials, etc are detailed in Chapter 3, together with an overview of the formation and effects of acid rain, O_3 depletion and global warming. Emission sources and inventory compilation are given in Chapter 4 and air quality monitoring techniques, methods and protocols are detailed in Chapter 5. Chapter 6 reviews the statistical, mathematical and physical modelling techniques available to assess air quality impacts. An overview of air pollution control legislation is given in Chapter 7, including international protocols and agreements together with examples of air pollution control regimes in selected countries. Impact assessment criteria (i.e. air quality standards (AQS) and guidelines) are also given in Chapter 7. Chapter 8 provides a resume of air pollution mitigation and control techniques as well as air quality management and planning strategies. Chapter 9 outlines auditing and inspection techniques and project management issues. In Part 2, Chapter 10 provides a series of case studies to compliment the subject areas detailed in the book. A glossary of units and acronyms is provided in an appendix and references are detailed at the end of the book. The units expressed in the text are those given by the original data source, to aid the reader conversion factors are given in the appendix.

CHAPTER TWO

Air Pollution Sources and Types

2.1 INTRODUCTION

An analysis of air quality shows the presence of numerous substances in trace amounts. Some of these substances can be explained in terms of either natural or anthropogenic (man-made) source activities. Others are formed indirectly from chemical processes in the atmosphere. The contribution of a source to air quality varies according to its emission characteristics and the emitted pollutant. The two types of air pollutants that can therefore be distinguished are either primary or secondary pollutants.

Primary pollutants are those directly emitted to atmosphere from an emission source (e.g. SO_2, etc.). The term secondary air pollutant is generally applied to pollutants (gaseous and particulate) which are not emitted directly to atmosphere but are formed, in the atmosphere, by chemical reactions between other pollutants and atmospheric gases (e.g. ground level O_3) (Stedman, 2000).

A few air pollutants are called criteria pollutants and are generally those pollutants that are injurious to health, harm the environment and cause property damage. The current pollutants that are most commonly referred to are CO, Pb, NO_2, O_3, particulate matter less than 10 microns in diameter (PM_{10}) (or total suspended particulates (TSP)) and SO_2. Criteria pollutants are generally those that are given, at present, the greatest resources to reduce emissions and monitor them.

2.2 COMPOSITION OF THE ATMOSPHERE

'Pure' air comprises oxygen (O_2) (21%) and nitrogen (N_2) (78%) and a number of rarer gases, of which argon (Ar) is the most plentiful. CO_2 is present at a lower percentage concentration (0.03%) than argon (0.93%). Water vapour, up to 4% by volume, is also present. Plants as a by-product of photosynthesis produce oxygen. The atmosphere also contains a number of gases which, at higher than usual concentrations, are harmful to humans and animals and damaging to plants. These include O_3, SO_2, NO_2, CO and a wide range of VOCs. Some of the latter are carcinogenic, for example benzene and 1,3-butadiene. All these potentially toxic gases are generally referred to as 'air pollutants' (Chapter 1, Section 1.1).

Table 2.1 shows the concentrations of gases that comprise the atmosphere. The table also shows measured levels of polluted air for comparative purposes as well as the major elements in the atmosphere (O'Neill, 1985). Pollution levels may be several orders of magnitude above background (existing) concentrations.

Table 2.1 Concentration (ppm) of gases comprising the atmosphere (Stern, 1976; O'Neill, 1985)

Molecular species	*Background concentration*	*Polluted air*	*Percentage weight*
N_2	780840		75.5
O_2	209480		23.2
Ar	9340		1.3
CO_2	314-318[a]		
Ne	18.20		1.3×10^{-3}
He	5.24		
CH_4	1.0-2.0	1-10	72×10^{-6}
Kr	1.1		0.45×10^{-3}
H_2	0.5		23×10^{-6}
N_2O	0.25-0.5		
CO	0.1	5-10	
O_3	0.01-0.07	0.1-0.5	
NO_2	0.001-0.02	0.2-1.0	
NO	0.002-0.002		
SO_2	0.0002-1.0	0.02-2.0	
H_2S	0.0002		
C			9.3×10^{-3}
Xe			40×10^{-6}
S			70×10^{-9}

a) *Section 2.4.3*

2.3 AIR POLLUTION SOURCES

Air pollution sources may be either anthropogenic or natural. However, as human activity disturbs natural systems, the distinction may become blurred. Natural sources include dust storms, volcanic action, forest fires, etc. For some pollutants, e.g. SO_2, natural sources exceed anthropogenic sources on a global scale. Incursions from the stratosphere increase ground level (tropospheric) concentrations of O_3. However, when considering the effects of air pollutants on health, especially in urban areas where population densities are high, anthropogenic sources are very important and are those to which attention is usually directed with a view to control.

Many pollutants are emitted into the atmosphere from naturally occurring sources. An example of a natural pollution problem is dust storms. In north China (China Daily, 2000) in the Capital Circle, Beijing and Tianjin municipalities and Zhangjiakou and Chengde in Hebei Province, large areas are affected by strong winds and elevated dust levels from northern Inner Mongolia. The areas are also exposed to soil erosion, dry climates and water shortages. The number of dust storms has risen in recent years (Fig. 2.1). Efforts are currently being made to control desertification in the Capital Circle area by 2010 through tree and fauna planting schemes.

Figure 2.1 Frequency of dust storms in North China in the Capital Circle area (Data from the China Daily, 2000)

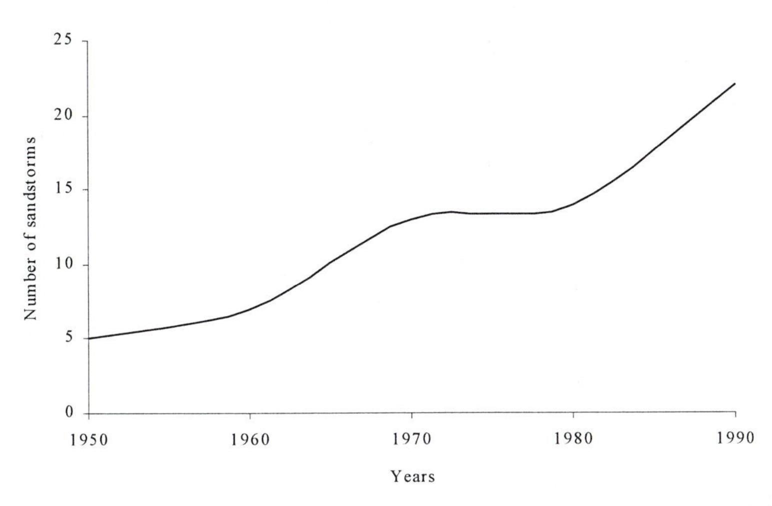

Pollutants can be emitted from point or stationary, area or mobile (linear) anthropogenic sources (Case Study 3, Chapter 10.4). Each source has its own distinct emission characteristics. Point sources include stacks, flues, etc. Area sources include groupings of usually small sources spread over a delineated area (e.g. industrial complexes) and mobile sources include motor vehicle, aircraft, etc. The form of the emission from these sources may be either controlled, uncontrolled, accidental, intentional or fugitive (e.g. uncontrolled minute leaks) (Harrop, 1999) (Chapter 4, Section 4.4). Case Study 12 (Chapter 10, Section 10.13) details the quantification of fugitive emissions from an oil terminal.

In the past, accidental releases of emissions to air have caused severe impacts on human health and air quality (Fig. 2.2). Some accidents have had extremely adverse effects on human health. Such well documented episodes include Bhopal (India) and Seveso (Italy). In December 1984, in Bhopal, approximately 2,500 people died when approximately 40t of methyl isocyanate was accidentally released from the Union Carbide chemical plant. Nearly 200,000 people were also injured, mainly from respiratory and eye injuries. In Seveso, in July 1976, an explosion occurred at a chemical plant making trichlorophenol as a herbicide, which released various chemicals, including dioxin (2,3,7,8-tetrachlorodibenzoparadioxine (TCDD). A cloud of dioxin, trichlorophenol, ethylene glycol and caustic soda dispersed into the surrounding environs. Within several weeks, fauna and flora died and residents surrounding the plant were admitted to hospital. No deaths were linked to the accident, but because of the possible link between dioxin and genetic mutations, 90 women decided to have abortions (Elsom, 1987). More than 700 people living close to the plant were evacuated and another 5000 people in a less contaminated area were permitted to stay at home but had restrictions placed on them with regard to raising animals, gardening and children playing (Fuller, 1977). As a result of the Seveso accident the European Union (EU) introduced the 'Seveso Directive' in 1984 (Chapter 7, Section 7.9.4).

Figure 2.2 Air pollution incident at a refinery (courtesy of Falkirk Council, Environmental Health Department)

Fortunately, large scale acts of deliberate pollution are rare. Nevertheless, when they occur they can be extremely detrimental to air quality. One such act was the deliberate burning of the Kuwaiti oilfields by Iraqi soldiers during the 1991 Gulf War. The Iraqi army ignited over 600 oil wells, storage tanks and refineries as they fled Kuwait. The fires produced large smoke plumes, which caused significant environmental effects on the Gulf region (Kuwaiti Data Archive (KuDA), 2000). Fig. 2.3 shows the dispersion of smoke for the Gulf region, where smoke was visible on satellite imagery. 'Heavy' smoke is defined as areas of smoke that are dark enough to obscure the surface, while 'moderate' smoke corresponds to the visible boundaries of the smoke plume.

Another example of deliberate air pollution is the large scale forest fires of Indonesia which caused elevated particulate levels (Schinder, 1998). The fires were caused by huge conversion burns in order to prepare land for pulp wood and oil palm plantations. The use of fire is forbidden, however many companies use fire because it is the only viable and economic method of reducing the huge amounts of biomass. The forest fires in Indonesia affected the whole ASEAN region (e.g. Indonesia, Malaysia, Thailand, Philippines and Vietnam) with Malaysia bearing the brunt of the impact. The dry season for the most fire prone islands of Indonesia (Sumatra and Kalimantan) began around mid-May 1997. In East Kalimantan an early warning was raised and all timber use licenses were revoked. This resulted in East Kalimantan having only moderate fire activity in 1997 compared to Kalimantan and Sumatra. By September the fires in Sumatra and Kalimantan had escalated. After months of extreme fire activity the rains began to fall again in mid-November and gave a respite to the situation. Unfortunately, dry weather conditions prevailed again in East Kalimantan in early 1998 and escalated to such an extent that extinguishing the fires was difficult with available resources. A thick

Figure 2.3 Dispersion of particulate emissions from the intentional burning of oil wells in the Gulf War, 1991 (KuDA, 2000)

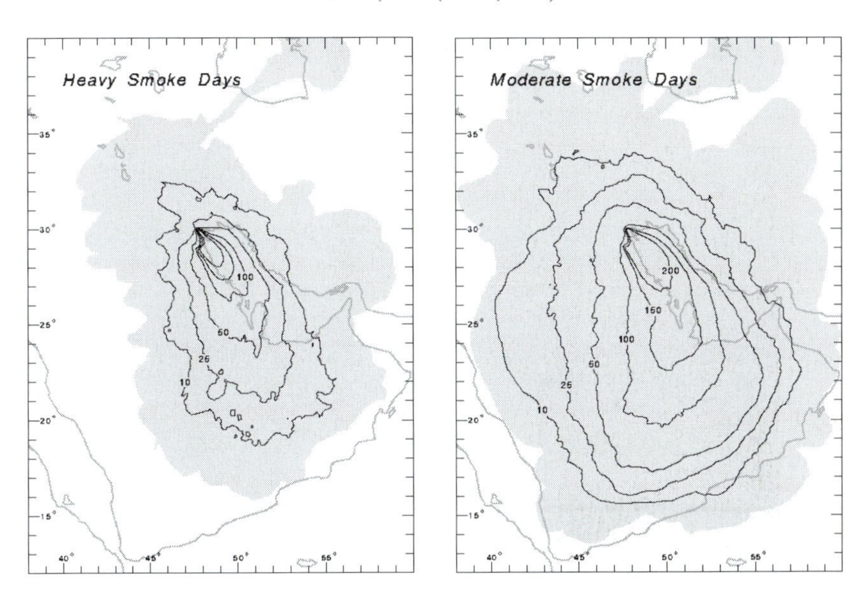

Contours indicate the number of days with smoke overhead (based on afternoon satellite image from the NOAA-11 polar orbiting meteorological satellite). The grey background areas indicate the areas for which smoke was detected overhead on at least one day. The analysis includes the period during the Gulf War, as well as for the rest of the year until the oil fires were extinguished.
Courtesy of Kuwaiti Data Archive (KuDA) Project 2000 at the National Centre for Atmospheric Research (NCAR) in Boulder, Colorado, USA. NCAR is sponsored by the National Science Foundation.

haze blanketed vast areas for weeks with reported extreme levels of pollution and visibility <200m in the interior of East Kalimantan. In Sarawak, where an emergency was declared, the air pollution index (API) exceeded 800 (Chapter 7, Section 7.3).

The deliberate burning of vegetation to rejuvenate vegetation growth is another contributor to air quality. The burning of grass on the Savannah (Fig. 2.4) is a widespread practice, however it causes extensive air quality and smoke/particulate nuisance problems to communities.

2.4 TYPES OF POLLUTANTS AND THEIR SOURCES

Generally air pollution may be either gaseous or particulate in nature. However, other forms of pollution of potentially equal concern may be physical (e.g. radiation) or heat related.

Fine and coarse particles generally have distinct sources and formation mechanisms, although there may be some overlap (Table 2.2). Primary fine particles are formed from condensation of high temperature vapours during

Figure 2.4 Particulate nuisance from vegetation burning on the Savannah, Venezuela

combustion. Secondary fine particles are usually formed in three ways (European Environment Agency, 2000a; World Health Organisation (WHO), 2000a):

- Nucleation mode (<0.2 μm in diameter) are particles recently emitted from a process or freshly formed within the atmosphere;
- Accumulation mode (>0.2-2 μm diameter) are particles which have grown from a nucleation mode by coagulation or condensation of vapours;
- By reaction of absorbed gases in liquid droplets.

Particles also form as a result of chemical reaction of gases in the atmosphere that lead to products that either have a low enough vapour pressure to form a particle, or react further to form a low vapour pressure substance. Some examples include (WHO, 2000a):

- The conversion of SO_2 to H_2SO_4.
- Reactions of H_2SO_4 with NH_3 to form NH_4HSO_4 and $(NH_4)_2SO_4$.
- The conversion of NO_2 to HNO_3, which reacts further with NH_3 to form particulate NH_4NO_3.

By contrast, most of the coarse fraction particles are formed directly as particles, and result from mechanical disruption such as crushing, grinding, evaporation of sprays, or suspension of dust from construction and agricultural operations. Energy considerations normally limit coarse particle sizes to greater than 1 μm in diameter. Some combustion-generated mineral particles, such as fly ash, are also found in the coarse fraction. Biological material such as bacteria, pollen, and spores may also be found in this fraction. Larger particles generally do not stay airborne for long periods, and are deposited close to their source; they are unlikely to be inhaled. Smaller particles can be carried long distances and inhaled. PM_{10} are associated with health effects and particles >2 μm diameter are associated with soiling. Studies on the mass concentrations of PM_{10}, $PM_{2.1}$ and $PM_{1.1}$ in urban air showed that the fine fraction dominated the PM_{10} mass concentration with $PM_{2.1}$ accounting for 58% of the total, and $PM_{1.1}$ accounting for 48% of the total on average (Rickard and Ashmore, 1996). These levels compare well with other studies that have found the percentage of $PM_{2.5}$ at an urban site was 60% (Quality of Urban Air Review Group, 1996).

Table 2.2 Comparisons of ambient fine and coarse mode particles (USEPA 1995a; 1995b)

	Fine mode	*Coarse mode*
	Gases	*Large solids/droplets*
Formed	Chemical reaction; nucleation; condensation; coagulation; evaporation of fog and cloud droplets in which gases have dissolved and reacted.	Mechanical disruption (e.g. crushing, grinding, abrasion of surfaces); evaporation of sprays; suspension of dusts.
Composed	Sulphate SO_4^-; nitrate NO_3^-; ammonium, NH_4^+; hydrogen ion, H^+; elemental carbon; organic compounds (e.g., PAHs); metals (e.g. Pb, Cd, V, Ni, Cu, Zn, Mn, Fe); particle-bound water.	Resuspended dusts (e.g., soil dusts, street dust); coal and oil fly ash, metal oxides of crustal elements (Si, Al, Ti, Fe); $CaCO_3$, NaCl, sea salt; pollen, mould spores; plant/animal fragments; tyre wear debris.
Solubility	Largely soluble, hygroscopic and deliquescent.	Largely insoluble and non-hygroscopic.
Sources	Combustion of coal, oil, gasoline, diesel, wood; atmospheric transformation products of NO_x, SO_2 and organic compounds including biogenic species (e.g. terpenes) high temperature processes, smelters, steel mills, etc.	Resuspension of industrial dust and soil tracked onto roads; suspension from disturbed soil (e.g. farming, mining, unpaved roads); biological sources; construction and demolition; coal and oil combustion; ocean spray.
Lifetimes	Days to weeks.	Minutes to hours.
Travel distance	100s to 1000s km	< 1 to 10s km

Other recognised classifications for particulates are fumes, smoke, dust and grit. Grit is defined in the UK Clean Air (Emissions of Grit and Dust from Furnaces) Regulations 1971 as particles >76 μm in diameter. The UK 1993 CAA defines fumes as any airborne solid matter smaller than dust. British Standard (BS) 3405 defines dust as small solid particles between 1–75 μm in diameter. Smoke is defined as the product of incomplete combustion consisting of minute carbonaceous particles less than 1 μm in diameter. Fine particulates are also host carriers for metals and for other pollutants, such as PAH.

Most gases are compounds, many of which are invisible at elevated concentrations. The principal pollutants (criteria pollutants) include NO_x (e.g. NO_2, NO), SO_2, CO, etc. The following sections provide a description of the sources of selected pollutants.

2.4.1 Carcinogenic pollutants

A considerable number of substances, carcinogenic or supposedly carcinogenic, occur as outdoor and indoor air pollutants (Section 2.6). Of these carcinogenic substances benzene, 1,3-butadiene and the PAHs are perhaps the best known. Some of them are genotoxic carcinogens and as such any level of exposure (no matter how small) may in theory be associated with an increased risk of cancer (Chapter 7).

The VOC protocol under the Convention on Long-range Transboundary Air Pollution (CLRTAP) (Chapter 7, Section 7.9.3) defines VOCs as all organic compounds of an anthropogenic nature other than methane, that are capable of producing photochemical oxidants by reactions with NO_x in the presence of sunlight.

2.4.1.1 Benzene

Benzene (C_6H_6) is a VOC and a minor constituent of petrol (typically between 2% and 5% by volume). It has a half-life in the environment of less than one day and consequently does not disperse significant distances from its source. It is readily washed from the atmosphere by precipitation, but can easily re-evaporate (Scottish Office, 1998). Table 2.3 shows typical levels of benzene in air levels for selected countries.

Table 2.3 Summary of benzene levels in urban areas (μg/m^3) (European Environment Agency, 1997; DETR, 2000b; USEPA, 2000)

Country	*Year of measurement*	*Concentration*
Belgium	1995	15.0
Denmark	1996	15.8
Germany	1994	10-12
Netherlands	1995/96	3.5-8.7
Sweden	1996	6.7-10.3
UK	1995	5.8[a]
US	1989	4.1[b]

a) Range of annual averages for UK sites for 1999 0.9 – 10. 6 μg/m^3
b) From USEPA Aerometric Information System (AIRS) (USEPA, 2000)

Motor vehicles, particularly in urban areas, are the principal source of benzene emissions. For example, in the UK motor vehicles account for 63.7% of total emissions (Fig. 2.5), although in the UK it is estimated that benzene levels will decline by almost 40% by 2010 on a 1995 base (DoE, 1997a). In Europe benzene emissions from existing petrol storage and handling facilities will come under the control of the EC Directive (94/63/EC) on controlling VOC emissions resulting from storage and distribution of petrol to service stations. Other EC Directives will further reduce motor vehicle emissions for cars, light vehicles and heavy goods vehicles (HGVs) sold from 2001 and 2006 as part of the EC Auto-Oil programme. The programme will reduce the amount of benzene and aromatics in petrol from the year 2000 as well as reduce the sulphur content of fuels from 2000 and again from 2005. The reduction of fuel sulphur content will help to reduce the deterioration in catalyst performance and therefore help to abate benzene emissions (DETR, 1998).

Figure 2.5 UK emission sources for benzene (Data from Goodwin *et al.*, 1999; SEPA, 2000)

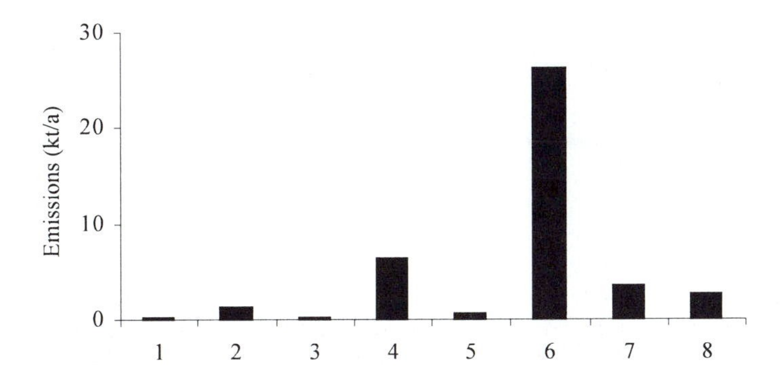

1. *Public power, co-generation and district heating*
2. *Commercial, institutional and residential combustion*
3. *Industrial combustion*
4. *Production process*
5. *Extraction and distribution of fossil fuels*
6. *Road transport*
7. *Other mobile sources and machinery*
8. *Waste treatment and disposal*

2.4.1.2 1,3-Butadiene

1,3-butadiene is a VOC arising from the combustion process of petroleum products. It disperses relatively rapidly within the atmosphere. Motor vehicles are the principal source of emissions. For example in the UK they account for 68% of total emissions whilst industrial chemical processes account for 18.3% (Fig. 2.6). In the UK emissions from petrol engine motor vehicles are anticipated to have declined by about 55% by 2000 on 1992 values and by 73% by 2010 (DoE, 1997a). Table 2.4 details typical levels of 1,3-butadiene in air for selected sites and countries. In the EU emissions of 1,3-butadiene will be reduced by the introduction of EC Directive (94/63/EC) and the Auto-Oil programme (DETR, 1998).

Table 2.4 Summary of 1,3-butadiene measurements (μg/m^3) (DETR, 2000b; USEPA, 2000)

Country	*Concentration*
UK	0.1-1.7[a, b]
Scotland (Edinburgh)	0.2
US	0.22[c]

a) Range of annual averages for UK sites for 1999
b) Includes Scotland
c) From USEPA Aerometric Information System (AIRS), USEPA (2000)

Figure 2.6 UK emission sources for 1,3 butadiene (Data from Goodwin *et al.*, 1999; SEPA, 2000)

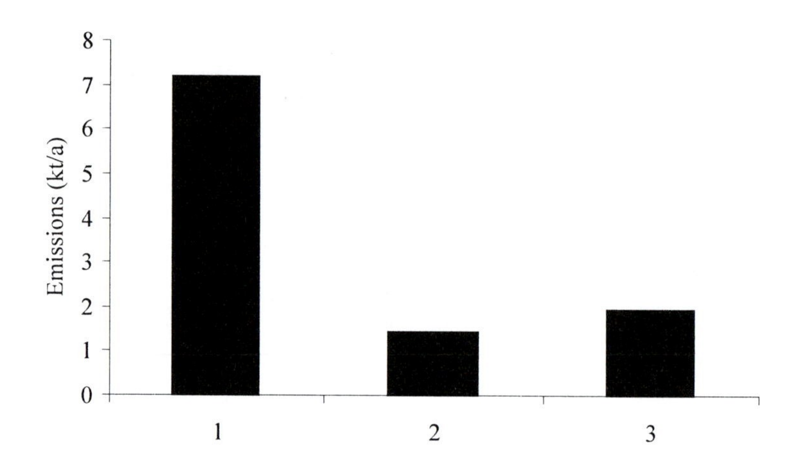

1. *Road transport*
2. *Other mobile sources and machinery*
3. *Chemical industry*

2.4.2 Carbon monoxide

Carbon monoxide (CO) is a colourless, odourless gas formed during the incomplete combustion of carbon containing materials. CO is a relatively stable compound which converts to CO_2 within the atmosphere as a result of a reaction with hydroxyl radicals (Scottish Office, 1998). Typically, concentrations fall relatively rapidly with distance from source. For example, CO levels resulting from motor vehicle emissions typically return to background levels within 200 m of a road (Hickman and Colwill, 1982). Consequently air quality impacts are generally localised.

Natural ambient concentrations of CO range between 0.01-0.23 mg/m^3 (WHO, 1987). In urban environments, mean concentrations over 8 hours are usually less than 20 mg/m^3 and 8 hour maximum levels are usually less than 60 mg/m^3. Concentrations of CO can be high in vehicles, underground car-parks, road tunnels and in other indoor environments where combustion engines operate with inadequate ventilation. Case Study 2 details the concentrations found in a poorly ventilated indoor area and the mitigation measures used to reduce concentrations to an acceptable level (Chapter 10, Section 10.3). In these circumstances, mean concentrations of CO can reach up to 115 mg/m^3 for several hours (WHO, 2000b). Table 2.5 shows typical levels of CO in air for selected countries.

Table 2.5 Summary of CO measurements

Country	*Concentration*	*Source*
Canada[a,b]	0.1-1.2	Environment Canada, 2000
Costa Rica[c]	9.3-10.3	Ministerio de Planificacion, 2000
Japan[a,d]	0.6	Japanese Government, 2000b
New Zealand[e]	5-10 urban <0.1 rural	New Zealand Government, 2000
Singapore[a] (1998)	0.6 (2.0)	Singapore Government, 2000
UK[a,f]	0.2-2.1	DETR, 2000b
US[g]	0.08-0.2	USEPA, 2000

a) Average, mg/m^3
b) Range of annual average levels for all NAP stations and cities for 1998
c) 1995 – 1997 San Jose (sampling duration not known) μg/m^3
d) Annual average levels from 145 general environmental monitoring stations for 1998
e) mg/m^3, 8 hour
f) Range of annual average values for UK automated sites for 1999
g) New York for 12 sites units are mg/m^3 (maximum 1 hr)

Figure 2.7 UK emission sources for CO (Data from Goodwin *et al.*, 1999; SEPA, 2000)

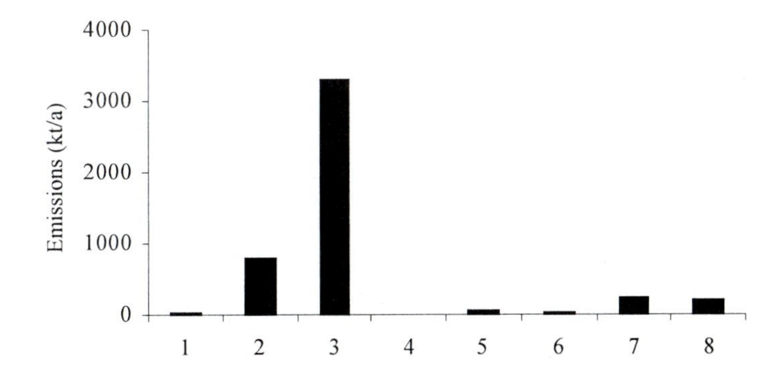

1. *Waste treatment and disposal*
2. *Other mobile sources and machinery*
3. *Road transport*
4. *Extraction and distribution of fossil fuels*
5. *Production process*
6. *Industrial combustion*
7. *Commercial/institutional/residential combustion*
8. *Public power/co-generation/district heating*

The main source of CO is road traffic. For example, in the UK it accounts for 71% of emissions (Fig. 2.7). In the EU reduced CO emission levels for motor vehicles as part of the Auto-Oil programme will result in a further reduction of emissions (DETR, 1998). Although road traffic contributes to the majority of CO emissions, large industrial combustion plants may also contribute significantly at a local level. It is anticipated that UK emissions of CO will reduce by 32% in 2000, 48% in 2005 and by 54% in 2010 compared with 1995 levels (DoE, 1997a).

2.4.3 Carbon dioxide

Carbon dioxide (CO_2) is a colourless and odourless gas which results anthropogenically from the combustion of fossil fuels. It is also an important natural greenhouse gas (Royal Commission on Environmental Protection (RCEP), 2000) (Chapter 3, Section 3.8). CO_2 occurs naturally from rocks, organic material and the oceans. They continuously release carbon into the atmosphere in the form of CO_2 and continuously reabsorb it. The present concentration of CO_2 in ambient air is approximately 370 ppm (Keeling and Whorf, 1998).

2.4.4 Lead

Lead is a heavy metal, which exists naturally in the earth's crust. It is found throughout the environment. Naturally occurring ambient concentrations of lead are low but its exploitation as a useful metal has increased the level of exposure. Lead inhaled as fine particles is deposited in the lungs. Since lead uptake by blood is dependent on deposition pattern and solubility (which is influenced by chemical form and particle size), total lead content is only a surrogate for the biologically effective dose. Furthermore, airborne lead can also reach humans indirectly via deposition on soil and vegetation, and through the food chain.

Since the introduction of unleaded petrol in the UK in the late 1980s, emissions have decreased markedly (Fig. 8.9). In the EU the agreement reached between the European Parliament and the Environment Council on the Directive on the Quality of Petrol and Diesel Fuels (part of the Auto-Oil programme) banned the sale of leaded petrol from 1 January 2000 in Europe (DETR, 1998). The increasing numbers of catalyst-equipped cars will ensure that emissions from petrol-driven vehicles decrease. For example, it has been estimated in the UK that lead emission reductions will be 80% by 2005 compared with 1995 levels and emissions will be very small by 2015 (DoE, 1997a). In those countries that still use leaded petrol, most of the lead in the air in cities comes from petrol-fuelled vehicles due to tetraethyl lead being used as a petrol additive to enhance the octane rating. For example, road transport in the UK accounts for 65.9% of emissions (Fig. 2.8).

Concentrations of lead found in air vary widely throughout the world and depend on the degree of industrial development, urbanisation and lifestyle factors. Ambient air levels over 10 $\mu g/m^3$ have been reported in urban areas near smelters, whereas lead levels below 0.1 $\mu g/m^3$ have been found in cities where leaded petrol is no longer used. In cities of developing countries traffic-related lead levels range between 0.3 and 1 $\mu g/m^3$ with extreme annual mean values between 1.5-2 $\mu g/m^3$

(WHO, 2000b). Table 2.6 shows typical levels of lead in air levels for selected sites and countries.

Table 2.6 Summary of lead measurements (μg/m³)

Country	*Annual concentration*	*Source*
Canada[a]	0-0.63	Environment Canada, 2000
Costa Rica[b]	0.2-0.5	Ministerio de Planificacion, 2000
Singapore (1998)	0.1-0.2	Singapore Government, 2000
UK[c]	0.004-1.43	DETR, 2000b
US[d]	0.02-0.2	USEPA, 2000
Venezuela[e]	0.4-0.9	MARNR[f], 1997

a) *Range of annual average levels for all NAP stations and cities for 1998*
b) *1995 – 1997 San Jose (sample duration not known)*
c) *Annual average levels for 1999 for UK monitoring sites*
d) *New York, average for 6 sites*
e) *San Cristobal, Valencia and Puerto Ordaz (1997)*
f) *Ministerio del Ambiente y de Los Recursos Naturales Renovables*

Figure 2.8 UK emission sources for lead in air (Data from Goodwin *et al.*, 1999; SEPA, 2000)

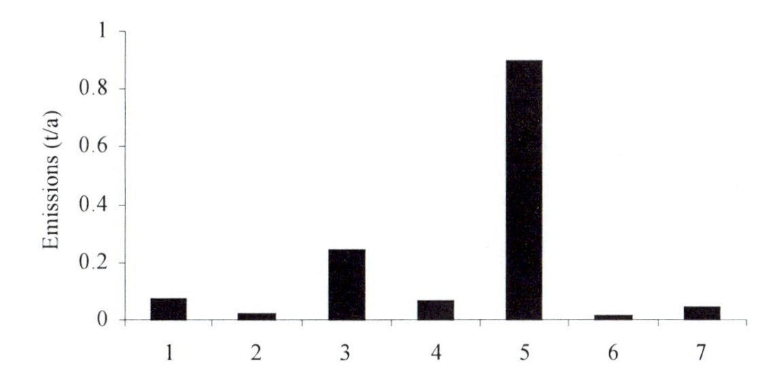

1. *Public power, co-generation and district heating*
2. *Commercial, institutional and residential combustion*
3. *Industrial combustion*
4. *Production process*
5. *Road transport*
6. *Other mobile sources and machinery*
7. *Waste treatment and disposal*

2.4.5 Nitrogen dioxide

Nitrogen is a constituent of both the natural atmosphere and of the biosphere. When industrial processes release nitrogen to atmosphere it is considered a 'pollutant' because of its chemical form (NO, NO_2, and N_2O). These NO_x can be toxic to humans, to biota, and they also affect the chemistry of the global atmosphere. Very little NO_2 is thought to be directly emitted into the atmosphere as a result of high temperature combustion processes; the bulk of NO_x emissions are in the form of NO, with subsequent chemical reactions to form NO_2. The dominant mechanism is the oxidation by O_3 (equation 2.1) (Stedman, 2000):

$$NO + O_3 = NO_2 + O_2 \quad (2.1)$$

Episodes of elevated NO_2 levels in the summer occur when photochemical O_3 episodes happen at the same time as relatively poor dispersion of NO_x emissions (Stedman, 2000). The mechanism for its formation is shown by the above reaction (2.1). For the winter formation of NO_2 an additional oxidation mechanism contributes to its formation at very high NO concentrations because the rate of this reaction depends on the square of the NO concentrations (equation 2.2) (Stedman, 2000):

$$NO + NO_x + O_2 = NO_2 + NO_2 \quad (2.2)$$

The speed of conversion of NO to NO_2 depends on the amount of oxidising pollutant available. Typically, conversion is greater in urban locations, owing to the availability of oxidising agents such as hydrocarbons. In many AQA a common assumption employed is that 100% of NO_x is NO_2. This is because the percentage of different oxides of nitrogen in NO_x is unknown, but NO_2 is the NO_x with the greatest likely health impact (Chapter 3, Section 3.2.4.2), so this assumption is worst case. It is likely, however, that 90% or more, of the emissions to atmosphere of the NO_x from a combustion process will comprise NO, which from a health perspective is relatively innocuous. Janssen (1988) have investigated the percentage of oxidation of NO_2 in power station plumes and derived an empirical relationship based on downwind distance, O_3 concentration, wind speed and season of the year. A worst case scenario of 50% conversion has previously been assumed. In reality the percentage oxidation to NO_2 will be significantly less than this value at the point of maximum impact under typical climatic conditions.

The principal sources of NO_x vary nationally. For example, in the UK the main emission sources of NO_2 are motor vehicles (49.6%); power generation (27.4%); large scale industrial combustion processes (8.2%); domestic and commercial sources (5.5%); and other mobile sources (11.8%) (Fig. 2.9). The EU's Acidification Strategy aims to reduce far beyond existing commitments emissions of NO_x in the EU by 2010. The main measures proposed are a Directive setting National Emission Ceilings for 2010 and a Large Combustion Plant (LCP) Directive. In addition, further reductions in emission levels for motor vehicles as part of the Auto-Oil programme have been agreed. In the UK it is estimated that current policies should deliver NO_x reductions of about 38% by 2005 based on 1995 values (DETR, 1998; DoE, 1997a).

Table 2.7 shows typical levels of NO_2 in air for selected countries. Ambient concentrations of NO_2 in air are variable. Natural background concentrations can be less that 1 $\mu g/m^3$ to more than 9 $\mu g/m^3$. In cities, annual mean concentrations can range from 20-90 $\mu g/m^3$ with hourly maximum concentrations from 75-1000 $\mu g/m^3$ (WHO, 1994). Fig. 2.10 shows the annual general trend in NO_2 levels for Scottish sites.

Figure 2.9 UK emission sources for NO_2 (Data from Goodwin *et al.*, 1999; SEPA, 2000b)

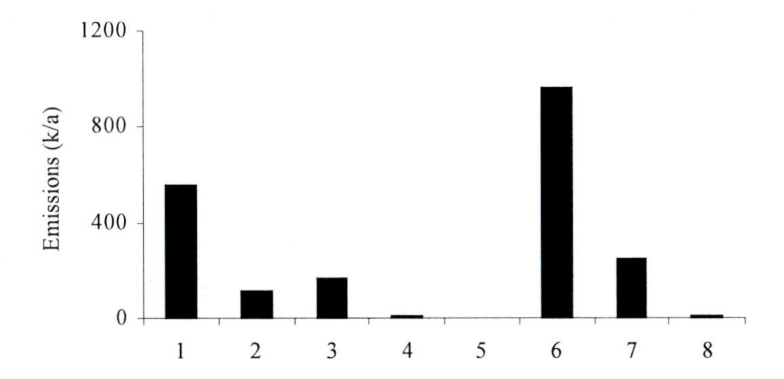

1. *Public power, co-generation and district heating*
2. *Commercial, institutional and residential combustion*
3. *Industrial combustion*
4. *Production process*
5. *Extraction and distribution of fossil fuels*
6. *Road transport*
7. *Other mobile sources and machinery*
8. *Waste treatment and disposal*

Table 2.7 Summary of NO_2 measurements (μg/m^3)

Country	*Annual concentration*	*Source*
Canada[a]	9.6-59.2	Environment Canada, 2000
Costa Rica[b]	40.3-46.4	Ministerio de Planificacion, 2000
Japan[c]	32.5	Japanese Government, 2000a
New Zealand	5-30 (24 hr) urban 0-1 (24 hr) rural	New Zealand Government, 2000
Singapore (1998)	34	Singapore Government, 2000
UK[d]	5.1-92.0	DETR, 2000b

a) Range of annual average levels for all NAP stations and cities for 1998
b) 1995 – 1997 San Jose (NO_x) (sampling duration not known)
c) Average levels from over 1400 general environmental monitoring stations for 1998
d) Range of annual average levels for UK automated sites for 1999

2.4.6 Ozone

There are two zones of O_3. O_3 in the stratosphere (15-50 km above the earth's surface) forms what is known as the 'ozone layer' and is essential in limiting the level of UV irradiation reaching the earth's surface (Chapter 3, Section 3.8.2). O_3 in the troposphere, the level that contains human life, is the other zone of interest.

Photochemical smog is frequently the most visible form of air pollution (Fig. 2.11) and is of particular concern in urban areas during spring and summer. O_3 is an acidic colourless gas, which acts as a very strong oxidising agent. There are no

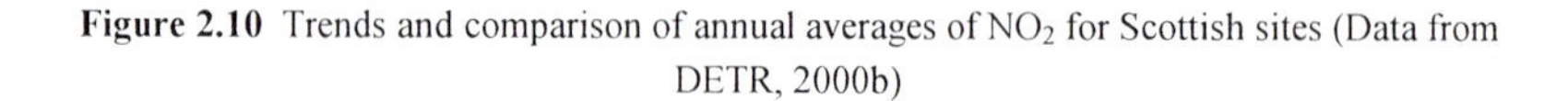

Figure 2.10 Trends and comparison of annual averages of NO_2 for Scottish sites (Data from DETR, 2000b)

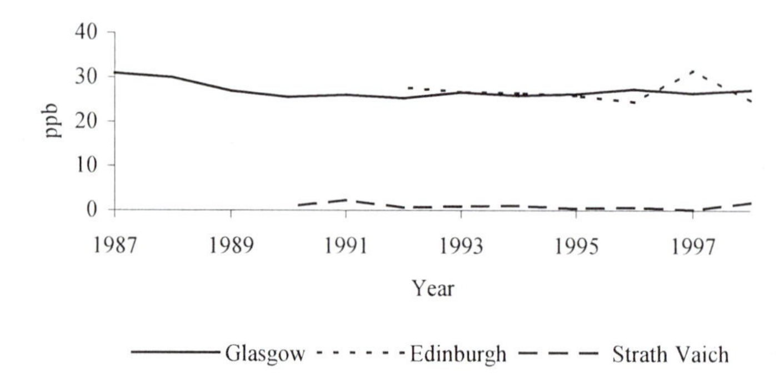

Figure 2.11 Smog in Lima, Peru (courtesy of Cordah Limited)

significant anthropogenic sources as it is a secondary pollutant formed from the reaction between NO_x (the sum of NO_2 and NO) and VOCs in sunlight to form photochemical smog (Photochemical Oxidants Research Group (PORG), 1997). The only significant reaction producing O_3 in the atmosphere is:

$$O + O_2 + M = O_3 + M \tag{2.3}$$

Where M is any molecule which can dissipate the energy released in the reaction, such as O_2 or N_2 (Stedman, 2000). Oxygen atoms are formed in the lower atmosphere by the photodissociation of NO_2:

$$NO_2 + \text{radiation} = NO + O \tag{2.4}$$

The NO formed in this reaction reacts rapidly with O_3 to produce NO_2 again:

$$NO + O_3 = NO + O_2 \tag{2.5}$$

resulting in no net O_3 production (Stedman, 2000). The photochemical oxidation and degradation of VOC species, which takes place in polluted atmospheres, however, provides an alternative route for the oxidation of NO to NO_2, removing the NO which would have reconverted the O_3 back to O_2, resulting in overall O_3 production (Stedman, 2000).

Background concentrations of O_3 in remote and relatively unpolluted parts of the world are often in the range of 40-70 $\mu g/m^3$ as a 1 hour average. In cities and areas downwind of cities, maximum mean hourly concentrations can be as high as 300-400 $\mu g/m^3$. High O_3 concentrations can persist for 8 to 12 hours per day for several days, when atmospheric conditions favour O_3 formation and poor dispersion conditions exist. O_3 is normally at higher concentrations in ambient air outdoors than in indoor air (WHO, 2000a). Table 2.8 shows typical levels of O_3 in air levels for selected sites and countries.

Table 2.8 Summary of O_3 measurements ($\mu g/m^3$)

Country	*Level*	*Source*
Canada[a]	14-72	Environment Canada, 2000
Costa Rica[b]	36.8-52	Ministerio de Planificacion, 2000
UK[c]	13-74	DETR, 2000b
US[d]	20-310	USEPA, 2000

a) Range of average levels for all NAP stations and cities for 1998
b) 1995 – 1997 San Jose (sampling duration not known)
c) Range of annual average levels for UK automated sites for 1999
d) New York 25 sites (1 hr average) (year)

In areas with high motor vehicle flows O_3 concentrations tend to be low, as any O_3 generated is utilised up in the conversion of NO to NO_2. In comparison, in rural areas O_3 concentrations are frequently higher as there are less vehicle emissions to ‘soak-up’ the excess O_3. The formation of O_3 may take several hours or possibly days to complete depending on the amount of VOCs available, and may remain present in the atmosphere for several days after formation. O_3 can be formed at considerable distances from its individual constituent component parts and consequently control of the formation of O_3 requires provision in the main at national or even international administrative levels.

High concentrations of ground level O_3 are often a particular problem in hot sunny climates such as in southern Europe, with cities like Athens and Rome having a particular problem. Other areas of the world are also prone to photochemical smog (e.g. Los Angeles). For example, in Ontario, Canada (Environment Canada, 1999) much of the smog is generated locally. However, air pollution from the US contributes to about 50% of the ground level O_3. Whereas the southern Atlantic region, parts of Nova Scotia and New Brunswick receive air pollution from the eastern US, where cross border pollution contributes to between 50% and 80% of the region's smog, in British Columbia, 80% of ground level O_3 originates from local sources.

In Europe reductions in the emissions of the O_3 precursors (NO_x and VOCs) have been agreed in emission levels for motor vehicles as part of the Auto-Oil programme. In addition, O_3 formation will be reduced by the introduction of the EC Directive (94/63/EC) on controlling VOC emissions resulting from the storage and distribution of petrol.

2.4.7 Particulates

Particulates can be emitted from a considerable range of sources, both natural (e.g. pollen, fungal spores, sea salt, etc.) and anthropogenic (e.g. agriculture, combustion, construction activity, quarrying, etc.) and can comprise a significant variety of materials (Scottish Office, 1998).

Resuspension of particles may also occur in very windy and dry conditions. Secondary sources of particles (e.g. aerosols of sulphates and nitrates) can also make up a significant contribution to overall concentrations of particulates. Secondary particles are formed from gaseous species in the atmosphere by chemical and physical processes (Stedman, 2000). The main secondary particle contributions to PM_{10} and $PM_{2.5}$ are ammonium sulphate, ammonium nitrate, sodium nitrate and secondary organic carbon, since these pollutants are formed relatively slowly in the atmosphere and are transboundary in nature. Fig. 2.12 shows the annual variation in sulphate levels for Eskdalemuir, Scotland.

Figure 2.12 Variations in annual average sulphate particle concentration at Eskdalemuir, Scotland, 1973-1997 (Data from DETR, 2000b)

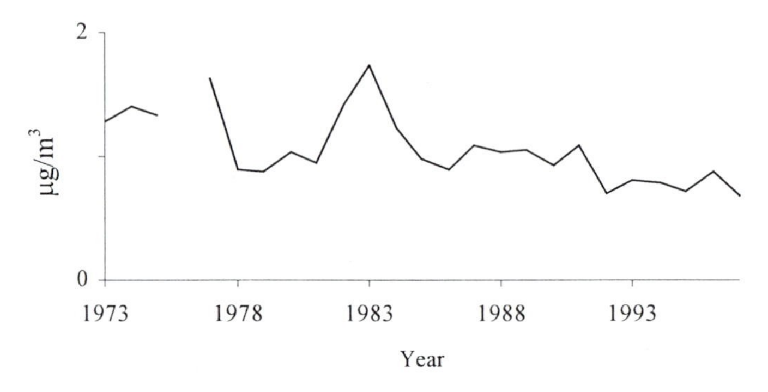

Concentrations of particulate matter in air are highly variable. In some areas very high levels occur naturally due to wind-blown dust from arid soils. Human activities, such as fires, overgrazing, agricultural practices and mining, can increase concentrations. In Western Europe and North America efforts to control emissions of particulate matter have generally resulted in lower levels of particles in ambient air. In many cities the annual average concentrations of PM_{10} are in the range 20-50 $\mu g/m^3$ (WHO, 1994). However, annual average concentrations in some cities in Eastern Europe and in some developing countries can be above 100 $\mu g/m^3$. Concentrations of $PM_{2.5}$ are usually about 45 to 65% of the concentrations of PM_{10} (WHO, 2000b). Table 2.9 shows typical levels of PM_{10} in air levels for selected countries.

The sources of particulate emissions are variable. For example in the UK, the emission sources include motor vehicles (23.7%); general industrial combustion plants (7.8%) including large power plants (17.4%); uncontrolled domestic coal burning and major mining, quarrying and construction activities (production processes) (27.4%) (Fig. 2.13). EU Directives will further reduce the contribution of motor vehicle emissions as part of the Auto-Oil programme. This programme has led to an agreement to reduce the sulphur content of fuels, which will also lead to some reductions in emissions of PM_{10}. Policies in place in the UK are expected to deliver particulate reductions of about 40% by 2005 compared with 1995 values (DoE, 1997a). The EU's Acidification Strategy aims to reduce emissions of SO_2, NO_x and NH_3 in the EU by 2010. The main measures will be a Directive setting National Emission Ceilings for 2010; Directive on the Sulphur Content of Liquid Fuels and a LCP Directive. These instruments will each have an impact on the levels of secondary particles.

Table 2.9 Summary of PM_{10} measurements ($\mu g/m^3$)

Country	*Annual concentration*	*Source*
Canada[a]	9-32	Environment Canada, 2000
Costa Rica[b]	38-54	Ministerio de Planificacion, 2000
Japan[c]	32	Japanese Government, 2000b
New Zealand	25-35 (24 hr) urban 2-10 (24 hr) rural	New Zealand Government, 2000
Singapore (1998)	35	Singapore Government, 2000
UK[d]	12-46	DETR, 2000b
US[e]	10.0-46.9	USEPA, 2000

a) Range of annual average levels for all NAP stations and cities for 1998
b) 1995 – 1997 San Jose (sampling duration not known)
c) Average levels from over 1500 general environmental monitoring stations for 1998
d) Range of annual average levels for UK automated sites for 1999
e) New York range of 11 sites

2.4.8 Sulphur dioxide

Sulphur dioxide (SO_2) is a colourless soluble gas with a characteristic pungent smell. It is produced by the combustion of fossil fuels containing sulphur. Approximately 80% of world-wide emissions come from coal and lignite burning and 20% from oil. Coal typically contains about 2% sulphur and heavy fuel oil about 3% by weight.

Figure 2.13 UK emission sources for particulates (Data from Goodwin *et al.*, 1999; SEPA, 2000)

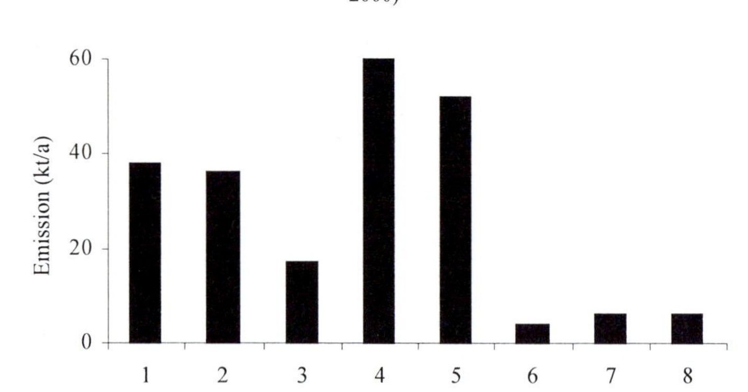

1. *Public power, co-generation and district heating*
2. *Commercial, institutional and residential combustion*
3. *Industrial combustion*
4. *Production process*
5. *Road transport*
6. *Other mobile sources and machinery*
7. *Waste treatment and disposal*
8. *Agriculture*

A review of national sulphur emissions from 1850 to 1990 (Lefohn *et al.*, 1999) showed that the former USSR, China, and US were the main sulphur emitters (i.e. approximately 50% of total emissions) in the world in 1990. The USSR and the US have stabilised their sulphur emissions over the past 20 years, and the recent increases in global sulphur emissions are linked to the rapid increases in emissions from China (Lefohn *et al.*, 1999). US sulphur emissions have shown moderate fluctuation since the beginning of the 20th century. The causes of the long-term fluctuations include recessions, major wars, fuel switching and environmental concerns. Over the years there has also been a shift from manufacturing to power plants as the main emitters of sulphur. Fig. 2.14 shows changes that have occurred in US emission trends.

In the UK the main source of SO_2 is from the combustion of sulphur-containing fossil fuels, namely coal and oil. Large combustion plants, in particular power stations, are the main industrial source of SO_2 (71.5%) (Fig. 2.15). Unauthorised industrial boilers, those of insufficient size to require authorisation (i.e. permit), burning coal and fuel oil can also make a significant contribution to SO_2 concentrations particularly in urban areas. Motor vehicles (i.e. diesel) also contribute to SO_2 concentrations, but in comparison to other sources their contribution is relatively minor. Total UK emissions of SO_2 are the subject of the Second Sulphur Protocol under the United Nations Economic Commission for Europe Convention on Long-range Transboundary Air Pollution (UNECE CLRTAP) (Chapter 7). The UK committed itself to reducing emissions by 50% by 2000, 70% by 2005 and 80% by 2010 on a 1980 base (DoE, 1997a). The EU's Acidification Strategy aims to reduce emissions of SO_2 by 2010. The main measures will be a Directive setting National Emission Ceilings for 2010; a Directive on the Sulphur Content of Liquid Fuels setting a maximum permissible sulphur content of heavy fuel oil of 1% from 2003 and gas oil (0.1% from 2008) and a LCP Directive. In addition, further reductions have been agreed in emission levels for motor vehicles as part of the Auto-Oil programme. This programme has also led to an agreement to reduce the fuel sulphur content from 2000 and 2005.

Figure 2.14 Trend of total sulphur emissions (thousand short tons) for the US (Data from USEPA, 2000)

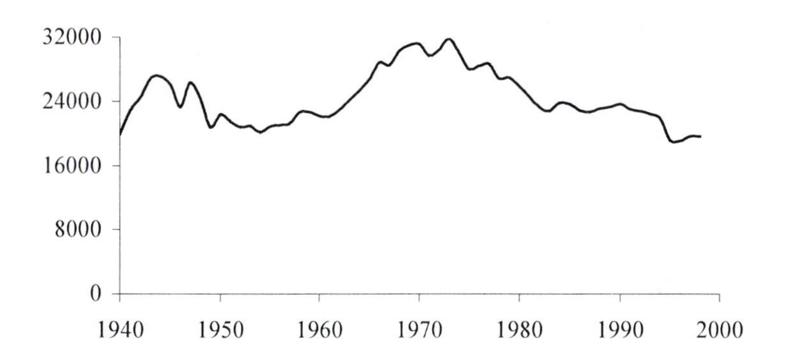

Figure 2.15 UK emission sources for SO_2 (Data from Goodwin *et al.*, 1999; SEPA, 2000)

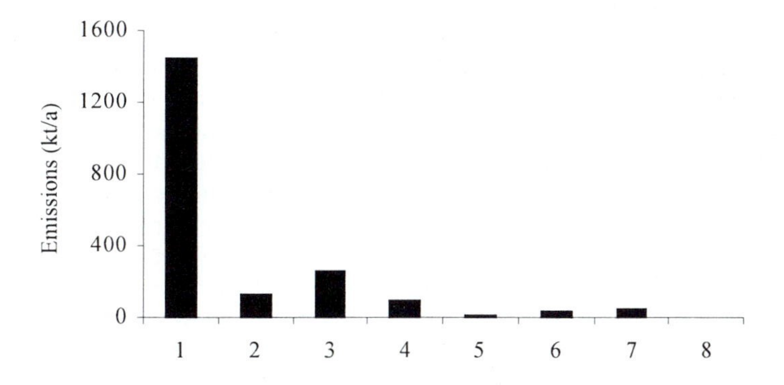

1. *Public power, co-generation and district heating*
2. *Commercial, institutional and residential combustion*
3. *Industrial combustion*
4. *Production process*
5. *Extraction and distribution of fossil fuels*
6. *Road transport*
7. *Other mobile sources and machinery*
8. *Waste treatment and disposal*

Table 2.10 Summary of SO_2 measurements ($\mu g/m^3$)

Country	*Annual concentration*	*Source*
Canada[a]	2.7-58.5	Environment Canada, 2000
Japan[b]	10.6	Japanese Government, 2000b
Singapore (1998)	20	Singapore Government, 2000
UK[c]	1.6-21	DETR, 2000b
US [d]	2.7-37.2	USEPA, 2000

a) Range of annual average levels for all NAP stations and cities for 1998
b) Annual average levels from over 1500 general environmental monitoring stations for 1998
c) Range of annual average levels for UK automated sites for 1999
d) New York range of 25 sites (ppm)

Table 2.10 shows typical levels of SO_2 in air levels for selected sites and countries. Concentrations in ambient air in cities of developed countries have mostly decreased in the last two to four decades due to tighter emission control, increased use of low sulphur fuels and industrial restructuring. Consequently, previously high ambient concentrations observed in earlier decades have been replaced by annual mean concentrations of about 20-40 $\mu g/m^3$ in most cities in developed countries and daily means rarely exceed 125 $\mu g/m^3$ (WHO, 2000b). However, the situation is more complex in developing countries. In cities, the annual mean concentrations of SO_2 in ambient air may range from very low levels up to 300 $\mu g/m^3$ (WHO, 1998). Peak concentrations measured as 10 minute averages may exceed 2000 $\mu g/m^3$ under conditions of poor atmospheric dispersion such as inversions, or when emissions from a major source are brought to ground levels by certain atmospheric conditions.

2.4.9 Dioxins and furans

Dioxins and furans are part of a group of pollutants often called toxic organic micro-pollutants (TOMPS). Of principal concern are polychlorinated dibenzo-para-dioxins (PCDDs) and dibenzofurans (PCDFs) (e.g. PCDD/Fs). These are often called dioxins for the sake of simplicity, although they are actually two separate groups of substances with similar effects. They comprise altogether 210 different chemical compounds, of which 12 are especially toxic. There are 75 PCDDs and 135 PCDFs. The molecules are flat, tricyclic aromatic hydrocarbons, all with similar chemical characteristics. They vary greatly, however, in toxicity and biological properties. Individual PCDD/Fs are referred to as congeners, while PCDD or PCDF congeners with identical chemical formulae are called isomers - a homologue group is a set of PCDD/F isomers. The best known and probably the most studied of these substances is 2,3,7,8 tetrachlorodibenzo-*p*-dioxin, or TCDD. This is probably also the most toxic of them all. Next in toxicity are those compounds that are chlorinated in all the lateral positions (2,3,7 and 8), and contain altogether 4,5 or 6 chlorine atoms (National Swedish Environmental Protection Board, 1988).

Dioxins are highly resistant to acids, but also to oxidising and reducing agents. TCDD is stable in the presence of bases while the hepta- and octachloro compounds react with bases even at room temperature. Dioxins are highly heat resistant, solubility is low in water, medium in solvents. UV radiation from the sun can break them down (National Swedish Environmental Protection Board, 1988).

Since the effects of most toxic dioxins and furans are similar but of greatly varying strength, the attempt has been made to rank them by comparing their toxicity with that of TCDD. The factor thus obtained (TCDD equivalent factor - TEF) for each substance is multiplied by its amount or concentration and the results added together. The total is 'TCDD equivalents' (TEQs). By definition the factor for TCDD is 1.0, while the other dioxins have lower factors. There are several systems of equivalents. Their relevancy for low dose exposure is, however, still uncertain (National Swedish Environmental Protection Board, 1988).

A review has been provided of natural and anthropogenic sources of dioxins by Environmental Resources Management Ltd (2000a). Natural sources have been identified as releasing PCDD/Fs. An analysis of pre-1900 soil samples before large scale manufacture and use of chlorinated chemicals, confirmed the presence of these compounds (Alcock *et al.*, 1998). Samples from 2800 year old Chilean mummies (Ligon and Dorn, 1989) and from Japanese sediment samples dating from at least 6100 years BC have also confirmed the presence of PCDD/Fs in historical times. In addition, it has been postulated that biological formation of PCDDs and PCDFs is possible in sediments and soils, especially forest soils and sediments (Rappe *et al.,* 1997; Hoekstra *et al.*, 1999). There is however, general agreement that anthropogenic sources and activities are far greater emitters of PCDD/Fs than natural sources. Environmental levels of PCDD/Fs have increased since the 1930s, this has coincided with the large scale production and use of chlorinated chemicals (Fortin and Caldbick, 1997; Alcock *et al.*, 1998). Anthropogenic sources of PCDD/Fs can be divided into three main categories (Fiedler, 1993).

- Chemical processes where PCDD/Fs can be formed during the manufacture of organic and inorganic chemicals and chemical products where chlorine is present, either as a reactant or as a constituent of a reactant.
- Thermal combustion processes include:
  - Stationary sources: incineration of various fuels (wood, coal, oil, etc) and wastes (chemical, municipal, clinical, sewage sludge, straw, etc), as well as foundries, sinter and secondary metal recovery facilities, etc.
  - Diffuse sources: automobile exhausts, private home heating, smoking of cigarettes, etc.
  - Accidental releases: warehouse fires, fires involving PCBs, fires involving contaminated wood, etc.

  Most of the thermal processes listed in Table 2.11 (AEA Technology, 1995) have the potential to emit these pollutants.
- Secondary sources occur where PCDD/F releases include emissions from contaminated sites and landfills. In addition, soil and sediment in which PCDD/Fs can accumulate can also act as secondary sources of these chemicals by redistributing them via wind blown dust and resuspension of sediments. The cycling of PCDD/Fs is thought to account for the majority of the measured environmental burden in the atmosphere. Approximately 90% of the measured deposition is believed to originate from PCDD/Fs that have been redistributed

between air, water and land, rather than arising as new emissions from the sources listed above (Rappe, 1992).

Table 2.11 Sources of dioxins and furans (AEA Technology, 1995)

Power stations (coal combustion)	Sinter plants
Industrial and commercial coal combustion	Domestic coal combustion
Industrial wood combustion	Domestic wood combustion
Coke production plants	Other iron and steel processes
Chemical industry	Asphalt production
Municipal solid waste incineration	Chemical waste incineration
Clinical waste incineration	Sewage sludge incineration
Landfill gas	Straw burning furnaces
Combustion of waste oil	Combustion of scrap tyres
Natural fires	Crematoria
Cement manufacture	Lime manufacture
Brick and breezeblock manufacture	Total vehicle emissions
Ceramics and glass manufacture	Regeneration of activated carbon
Primary and secondary non-ferrous metal production	

Dioxin and furans are present in the atmosphere in trace quantities (e.g. pg/m^3 or fg/m^3). However, dioxins and furans are extremely persistent and toxic compared to most other pollutants and can therefore still cause a large environmental impact in trace quantities. Concentrations in air vary according to location of emissions sources to the sampling location. Duarte-Davidson *et al.* (1994) have monitored dioxin levels in air for four urban UK cities of non-detectable to 62 pg/m^3. Deposition levels ranged from non-detectable to 31 $ng/m^2/day$. The concentrations were a sum of the 2,3,7,8-substituted PCDD/Fs only. Measured levels of dioxins and furans for selected UK sites (data from Duarte-Davidson *et al.*, 1994 and Davis, 1993 given in AEA Technology, 1995) show median Σ2,3,7,8- PCDD/F concentrations of 3.2, 4.0, 3.5 and 2.7 pg/m^3 for London, Cardiff, Manchester and Hazelrigg respectively. The median total TEQ values (ΣTEQ) for Manchester, Cardiff and London were essentially the same (0.1, 0.1 and 0.07 pg/m^3 respectively). The values for Bowland Fell were found to be as low as 0.008 pg/m^3.

A study measuring a full range of 2,3,7,8 substituted congeners and homologue groups in air at a rural site in Derbyshire, England, observed a median Σ2,3,7,8-PCDD/F concentration of 4.3 pg/m^3 with a range of 1.7-17 pg/m^3 over a 1 year period (data from Jones, 1995 given in AEA Technology, 1995). The ΣTEQ value for the rural site in Derbyshire was 0.18 pg/m^3. The furans made a greater contribution to the ΣTEQ than the dioxins. Individual compounds (2,3,7,8-T_4CDF, 2,3,4,7,8-P_5CDF and 1,2,3,4,7,8- and 1,2,3,6,7,8-H_xCDF) generally made a substantial contribution to the ΣTEQ. Harrad and Jones (1992) observed a mean urban concentration for PCDD/Fs in the UK of 3.4 pg/m^3 (2,3,7,8-substituted PCDD/Fs only). A study undertaken in Northern Ireland (Aspinwall & Company, 1994) observed a PCDD/F concentration of 0.12 $pg\text{-}TEQ/m^3$.

The annual average ΣTEQ values derived by various authors for Europe range 0.003-1.6 pg/m^3 (Table 2.12). Heister *et al.* (1995) presented the annual mean air PCDD/F concentrations for four cities in North Rhine Westfalia (Koln, Duisburg, Essen and Dortmund), Germany, in 1987/1988 and again in 1993/1994. During this period annual averages ΣTEQs decreased by between 46 to 69% for the four locations as follows: Koln 0.13-0.04 pg/m^3, Duisburg 0.33-0.12 pg/m^3, Essen 0.2-0.076 pg/m^3 and Dortmund 0.22-0.14 pg/m^3 (Alcock and Jones, 1996). The annual average ΣTEQ values derived by various authors for Asia and Australia range from 0.02 to 2.9 pg/m^3 (Table 2.13).

Table 2.12 Summary data for PCDD/Fs for various locations in Europe (pg/m^3) (AEA Technology, 1995)

Country	*Levels*[a]	*ΣTEQs*	*Reference*
Belgium			
Ambient air		0.02-0.59	Wevers *et al.* (1993)
Germany			
Rural		<0.07	Rappe (1993)
Urban		0.07-0.35	Rappe (1993)
Close to major sources		0.35-1.60	Rappe (1993)
Rural	0.66	0.05	Konig *et al.* (1993)
Industrial or rural	1.05-1.99	0.08-0.15	Konig *et al.* (1993)
Sweden			
Urban	0.45-0.57	0.013-0.024	Broman *et al.* (1991)
Remote/coastal	0.08-0.15	0.003-0.004	Broman *et al.* (1991)
Rural	0.3-5.2		Tysklind *et al.* (1993)
Urban	1.0-1.8		Tysklind *et al.* (1993)

a) *Concentrations expressed as the sum of the 2,3,7,8-substituted PCDD/Fs.*

Table 2.13 Summary data for PCDD/Fs in air for Asia and Australia (pg/m^3) (AEA Technology, 1995)

Country	*Concentrations*[a]	*ΣTEQs*	*Reference*
Japan			
Urban (summer)	22.8 (11-36)[b]	0.79 (0.4-1.3)	Sugita *et al.* (1993)
Urban (winter)	36.0 (6.2-80)[b]	1.46 (0.3-2.9)	Sugita *et al.* (1993)
Australia			
Sydney - 4 sites		0.02-0.06	Taucher *et al.* (1992)

a) *Concentrations expressed as the sum of the 2,3,7,8-substituted PCDD/Fs only, unless otherwise specified.*

b) *Values calculated from the sum of the minimum and maximum values reported for each individual PCDD/F.*
mean (range)

Other studies (Japan Environmental Management Association for Industry, 1997) have stated that the present position of ambient air quality for PCDD/Fs in Japan is:

- Residential area near industrial area: average 1.0 pg-TEQ/m^3;
- City area: average 1.02 pg-TEQ/m^3;
- Suburban city area: average 0.82 pg-TEQ/m^3;
- Background area: average 0.07 pg-TEQ/m^3;
- Ambient air quality levels of PCDD/Fs in Japan are 0.8 pg-TEQ/m^3 or under as an annual average.

The annual average ΣTEQ values derived by various authors for the US are 0.1 pg/m^3 (Table 2.14).

Table 2.14 Summary data for PCDD/Fs for various locations in the US (pg/m^3) (AEA Technology, 1995)

Site	*Level*[a]	*ΣTEQs*	*Reference*
Coastal environment (winter)	3.61	0.10	Hunt and Maisel (1990)
Bloomington (mean of 4 sites)	1.04[b]		Eitzer and Hites (1989)
Bloomington (range)	0.74-12[c]		

a) Concentrations expressed as the sum of the 2,3,7,8-substituted PCDD/Fs only, unless otherwise specified.
b) Mean value
c) Concentrations expressed for the sum of the total tetra-through octachlorinated PCDD/F.

2.4.10 Other pollutants

Table 2.15 compares VOC levels around a refinery to levels found in urban air quality, levels are comparable. Table 2.16 details concentrations of NH_3, HF, HCl, H_2S and F. Table 2.17 details typical heavy metal concentrations in air for UK sites. An extensive listing of air pollutants and their respective measured concentrations is given elsewhere (DoE, 1995c).

Table 2.15 Annual concentrations (ppb) of VOCs around a refinery compared to urban concentrations (1994/95) (BP Research, 1995 from Cordah Limited, 1997)

Pollutant	*Concentration*		*Pollutant*	*Concentration*	
	Refinery	*Urban*		*Refinery*	*Urban*
Benzene	<1-2	0.7-0.8	Propane	<1	1.0-4.2[a]
Toluene	<1-3	0.7-1.8	Acrylonitrile	<1	
Xylene	<1-2	0.3-1.8	Ethylacrylate	<1	
Styrene	<1		Methylmetacrylate	<1	
THC	13-47		1,3-Butadiene	<1	0.08-0.3[a]
Ethylene	<1-2	0.3-3.0[a]	Iso-butane	<1-2	
Ethane	<1-8	0.9-5.4[a]	n-butane	<1-3	
Propylene	<1	0.6-1.3[a]	n-Hexane	<1-7	

Urban concentrations for Edinburgh and Glasgow
a) range
THC – total hydrocarbons

Table 2.16 Measured pollutant concentrations for selected pollutants (ppb)

Pollutant	*Concentration*
NH_3	2.8-6.2 (range of means)[a*]
H_2S	<3[b]
Particulate fluoride (μg/m^3)	<0.2-7.4[c]
F as gas (μg/m^3)	<0.2-10.8[c]
HCl (μg/m^3)	2.5[a], 0.3-1.8[d],1.6(range 0.96-2.23)[e]
HF (μg/m^3)	0.50[e] urban areas not adjacent to known sources, 0.29 to 2.00[f], 0.02-2.43[e]

* *Edinburgh, Glasgow, Grangemouth and Motherwell*

a) *DoE (1995c)*
b) *BP Research (1992)*
c) *Clyde Analytical (1993)*
d) *Sturges and Harrison (1989)*
e) *Aspinwall & Company (1992)*
f) *Robins and Clark (1987)*

Table 2.17 Metals in air concentrations (ng/m^3) (WHO, 1987; DoE NI, 1990; Hester, 1987; DETR, 2000b)

	Site	*Concentration*[a]	*Concentration*[b,c]
Cr	Range in ambient air	5-100	1.9-4.3
Mn	Urban areas and rural areas without significant pollution	10-70	5-16
Ni	Rural areas	1-10	<9.4
	Urban areas	20	
	Industrial areas	110-180	
As	Rural areas	1-10	
	Cities	up to few hundred	
Cd	Rural areas	<1-5	0.87-1.0
	Urban areas	5-50	
Hg	Remote areas	2-4	
	Urban areas	10	
	Industrial areas	0.5-20	
Sb	Urban area	8.5 (mean)	
Cu	Urban area	28 (mean)	8.1-52
V	Urban area	13-23	1.7-4.7
Zn	Urban area	93-240	19-58
Co			0.89-1.2[d]
Fe			291-1212

a) *WHO (1987); DoE NI (1990) and Hester (1987)*
b) *DETR (2000b)*
c) *Range of annual average levels for UK sites for 1999 (rural and urban)*
d) *1995*

2.5 TRENDS IN AIR QUALITY

The combination of variable emissions patterns coupled with prevailing meteorological conditions can give rise to air pollution trends in air quality. The trends may be annual, seasonal, diurnal, hourly, etc. Examples are given for CO, Pb, NO_x, SO_2, O_3 and CO_2 (Figs 2.16–2.22 and 8.9).

2.5.1 Carbon monoxide

Since the main source of CO is motor traffic, concentrations are highest near to heavily trafficked roads. Concentrations fall rapidly with distance away from roads so that it is a local, rather than a transboundary pollutant. A typical example of the effect of road traffic on CO levels is seen for Union Street in Aberdeen, Scotland. CO and NO_x concentrations were seen to increase throughout the daytime period with increased traffic levels and concentrations increased to over 2300 $\mu g/m^3$ (2000 ppb) and 303 $\mu g/m^3$ (263 ppb) respectively, against significantly reduced night-time levels (Aberdeen City Council, 1999) (Fig. 2.16). A comparison of weekday and weekend CO levels in Glasgow shows that average concentrations may be up to 2 to 3 fold greater for weekdays than weekends due to increased traffic levels (City of Glasgow Council, 1999) (Fig. 2.17).

Figure 2.16 CO and NO_x concentrations and traffic flow for Union Street Aberdeen, Scotland (28/11/97) (Data from SEPA, 2000; Aberdeen City Council, 1999)

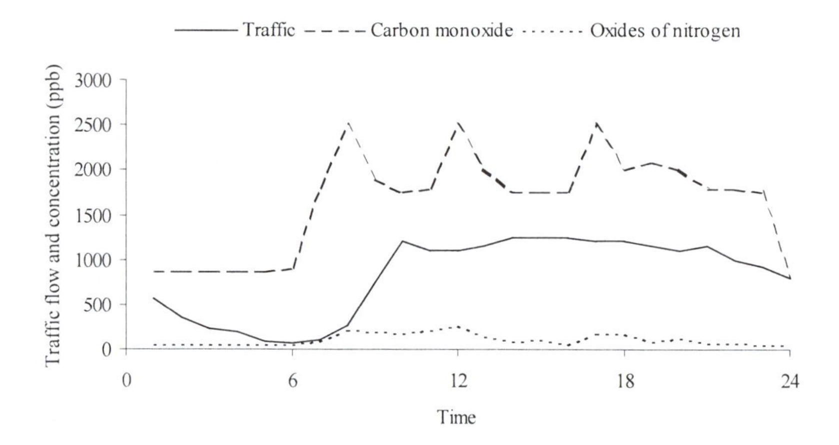

2.5.2 Lead

In many countries during the 1970s and early 1980s the lead content of petrol gradually reduced, maintaining total emissions from vehicles broadly constant. For example, in the UK at the end of 1985 the maximum permitted content of petrol was significantly reduced from 0.4 to 0.15 g/l, and in 1987 unleaded petrol was

Figure 2.17 CO concentrations at Glasgow, Scotland kerbside for weekdays and weekends (Data from City of Glasgow Council, 1999)

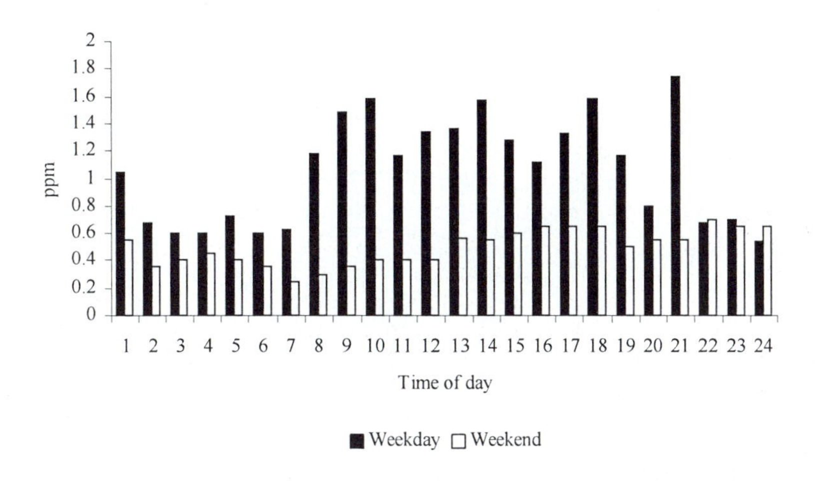

introduced. The reduction in the lead content of petrol showed a very pronounced reduction in concentrations. For example, average annual lead concentrations have reduced from 1170 ng/m^3 in 1976 to 40 ng/m^3 in 1997 for Glasgow, Scotland (City of Glasgow Council, 1999) (Fig. 8.9).

2.5.3 Nitrogen dioxide

In many countries motor vehicles are the principal source of NO_x (NO and NO_2) emissions. In the UK they account for nearly half of all NO_x emissions. Within urban areas the percentage contribution increases significantly (e.g. 76% for Glasgow and 96% for Edinburgh (Table 4.11, Chapter 4) compared to other sources. The greater percentage of low lying emission sources further exasperate elevated NO_2 levels in city areas and by busy roads. A contour plot of recent NO_2 average levels (1997–1998) for sites in Glasgow illustrates the extent of the area that may exceed the annual UK AQS of 21 ppb (City of Glasgow Council, 1999) (Fig. 2.18) primarily due to motor vehicle emissions. Similar patterns have been found in other cities.

2.5.4 Sulphur dioxide

The principal anthropogenic source of SO_2 emissions is the combustion of fossil fuels. The use and control of fossil fuels has been reflected in monitored air quality levels. SO_2 levels exhibit discernible annual and seasonal trends. Fig. 2.19 shows that annual SO_2 levels in Japan have decreased since the 1970s. A similar pattern is seen in Europe (Scotland (Fig. 2.20)) and North America.

Figure 2.18 Spatial variation of annual NO_2 concentrations (ppb) in the City of Glasgow, Scotland (SEPA, 2000). Concentrations greater than 21 ppb indicate possible exceedence of the UK annual NO_2 AQS (courtesy of the Scottish Environment Protection Agency). Data from information supplied courtesy of the Environmental Health Department, The City of Glasgow Council, Scotland)

Figure 2.19 Annual average SO_2 levels (ppm) for 14 Sites in Japan (Data from Japanese Government, 2000a)

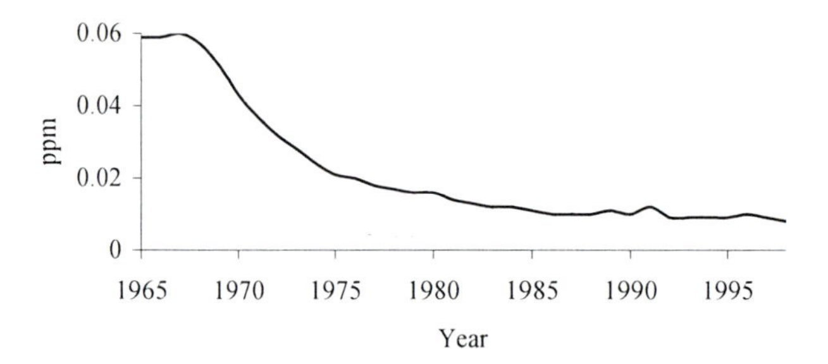

2.5.5 Carbon dioxide

Recent research (RCEP, 2000) has revealed that the CO_2 levels in air have risen significantly since the beginning of the industrial revolution in the mid 18th century, when CO_2 concentrations were 270–280 ppm. Over the past 250 years levels of CO_2 have risen to about 370 ppm (Keeling and Whorf, 1998) (Fig. 2.21). Concentrations are now increasing by 0.4% a year on average (RCEP, 2000). Nearly four fifths of the extra CO_2 entering the atmosphere since 1750 is estimated to have come from fossil fuel burning (Rotty and Marland, 1986).

Figure 2.20 Annual average SO_2 levels (ppb) for selected sites in Scotland 1961-1997 (Data from DETR, 2000b)

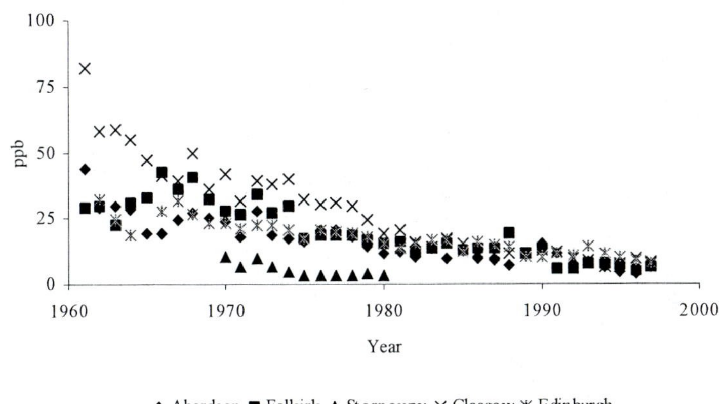

Figure 2.21 CO_2 levels (ppm) for Mauna Loa Observatory (Data from Keeling and Whorf, 1998)

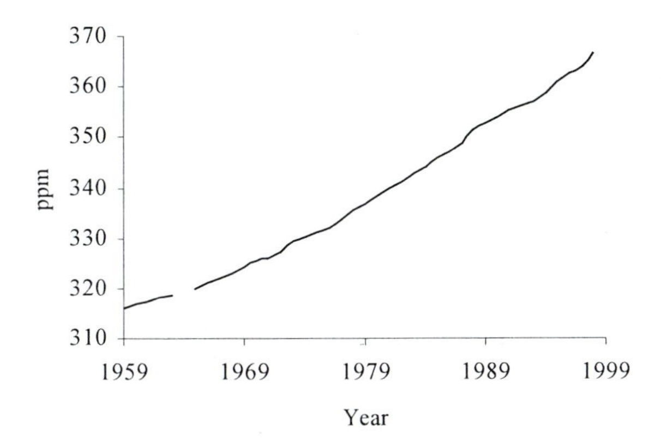

2.5.6 Ozone

Because O_3 is formed by chemical reaction in the atmosphere it is less dependent on emission patterns and tends to be more strongly influenced by meteorology. UV radiation drives these reactions and, as a result, its production is highest in hot, sunny weather. The process can take from hours to days to complete. As a result, O_3 can be formed considerable distances downwind of the primary emission source. Concentrations in urban areas are suppressed as a result of O_3 reaction with local emissions (e.g. motor vehicle emissions (i.e. NO)). The net result is that the highest O_3 concentrations may be seen in rural areas (Stedman, 2000) (Fig. 2.22).

Figure 2.22 Comparison of O_3 levels (ppb) monitored in rural and urban areas in Scotland simultaneously over a diurnal cycle (Data from DETR, 2000b)

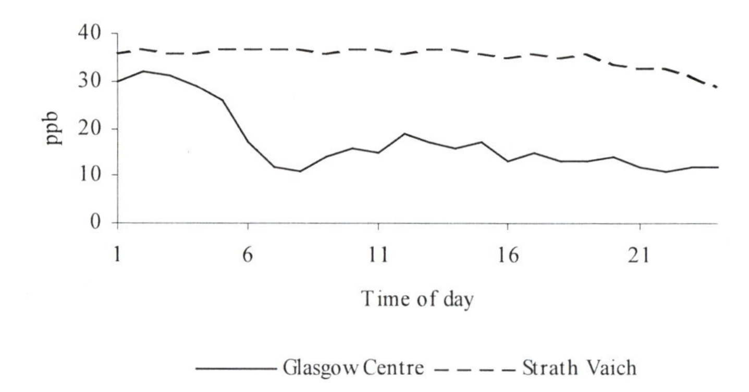

2.6 INDOOR AIR POLLUTION

As most people spend approximately 90% of their time indoors (NSCA, 2000a; Australian Government, The National Health and Medical Research Council, 2000) the issue of indoor air quality is a principal concern as air pollution levels indoors can be elevated (European Environment Agency, 2000a). Most of a person's daily exposure to many air pollutants comes through inhalation of indoor air pollution, both because of the amount of time spent indoors and because of the higher pollution levels found indoors (WHO, 2000a). The occupier largely controls the presence of air pollutants in the home environment. For most countries there are no regulations governing indoor air pollution, although there are regulations and/or guidance governing ventilation in homes (NSCA, 2000a). Available indoor AQSs are detailed in Chapter 7 (Section 7.7). The problems of indoor air pollution have been long recognised (WHO, 1982b; USEPA, 1994).

Indoor air can be defined as air within a building occupied for at least one hour by people of varying states of health. This can include the office, classroom, transport facility, shopping centre, hospital and home. Indoor air quality can also be defined as the totality of attributes of indoor air that affect a person's health and well being (Australian Government, The National Health and Medical Research Council, 2000). There are a number of factors, such as social and economic status, eating and drinking habits, whether people smoke or live with a smoker and the presence of other compounds in the air, that complicate any assessment of the health effects of indoor air pollutants. Such confounding factors have to be considered when looking at the health risks of indoor air quality and attempting to link substances and their effects (Institute of Environment and Health, 1997).

Indoor air pollutants usually differ in type and concentration from that found outdoors, although some are found outdoors. Indoor pollutants include environmental tobacco smoke, biological particles (such as pollen, mites, moulds, insects, micro-organisms, pet allergens, etc), non-biological particles (such as smoke), VOC, NO_x, Pb, radon, CO, asbestos, various synthetic chemicals and others (Table 2.18).

Table 2.18 Principal pollutants and source of indoor air pollution grouped by origin (WHO, 2000a)

Principal pollutants	*Sources, predominantly outdoor*
SO_2, SPM, RSP	Fuel combustion, smelters
O_3	Photochemical reactions
Pollens	Trees, grass, weeds, plants
Pb, Mn	Automobiles
Pb, Cd	Industrial emissions
VOCs, PAH	Petrochemical solvents, vaporisation of unburnt fuel
Principal pollutants	*Sources, both indoor and outdoor*
NO, CO	Fuel burning
CO_2	Fuel burning, metabolic activity
SPM, RSP	Environmental tobacco smoke, resuspension, condensation of vapours and combustion products
Water vapour	Biological activity, combustion, evaporation
VOC	Volatilisation, fuel burning, paint, metabolic action, pesticides, insecticides, fungicides
Spore	Fungi moulds
Principal pollutants	*Sources, predominantly indoor*
Radon	Soil, building construction materials, water
HCHO	Insulation, furnishing, environmental tobacco smoke
Asbestos	Fire-retardant, insulation
NH_3	Cleaning products, metabolic activity
PAH, As, nicotine, acrolein	Environmental tobacco smoke
VOCs	Adhesives, solvents, cooking, cosmetics
Hg	Fungicides, paints, spills or breakage of Hg-containing products
Aerosols	Consumer products, house dust
Allergens	House dust, animal dander
Viable organisms	Infections

Degradation of indoor air quality has been associated with a range of health effects, including discomfort, irritation, chronic pathologies and various cancers. Clearly, indoor air pollution can affect people both in their homes and at the workplace. In the latter case, workers cannot choose the air they breathe and there is a need for legislative or regulatory action. For the protection of employees occupational exposure standards (OES) (Health & Safety Executive (HSE), 2000; Occupational Safety & Health Administration (OSHA), 2000) (Chapter 7) have been promulgated.

Indoor pollution can arise from heating, ventilation or cooking appliances or from the structure or furnishings of the building. Indoor air pollutants can be classified in different ways. One approach is to divide them into chemical, physical and biological agents. Another approach is to classify them according to their origin (Table 2.18) (WHO, 2000a). The other sources and effects of indoor pollutants include the following:

- Formaldehyde is emitted mainly from particle board, carpets and insulation foams. Beside eye, nose and throat irritation, it may in some allergic people cause wheezing, coughing, skin rash and other severe allergic reactions. High concentrations may trigger attacks in people with asthma. Average formaldehyde concentrations of 0.025 mg/m^3 in dwellings have been found (Berry *et al.*, 1996).
- VOCs originate mainly from solvents and chemicals used at home or in offices. The main indoor sources include perfumes, hair-spray, furniture polish, glues, paints, stains and varnishes, wood preservatives, pesticides, air fresheners, dry cleaning, moth repellents, etc. The main health effects are eye, nose and throat irritation. In more severe cases there may be headaches, loss of co-ordination and nausea. In the long-term, some VOCs are suspected to damage the liver, the kidneys and the central nervous system as well as being potential carcinogens.
- O_3 can damage crops and vegetation (Chapter 3), although its activity is such that concentrations indoors are rapidly reduced by reaction with plastics. Exposure to O_3 is frequently an outdoor problem although elevated concentrations have been recorded in the vicinity of photocopiers and office printers.
- Biological pollutants include pollen from plants, mites, pet hair, fungi, parasites and some bacteria that are found indoors. Most of them are allergens and can cause asthma, hay fever and other allergic diseases.

  House dust mites are microscopic spiders that inhabit beds, carpets and soft furnishings. They feed on human skin scales and on fungi and bacteria. Exposure to house dust mites represents a potential hazard to health because the mites produce a protein (allergen) in their droppings to which some people may develop an allergy. Symptoms may include asthma, rhinitis (runny nose) and eczema (Institute for Environment and Health, 1997).

  Micro-organisms, including fungi and bacteria, are important factors in influencing indoor air quality. A wide range of fungal and bacterial species can be isolated in indoor air. The variety of species found alters with nutrient source, water availability and temperature (Institute for Environment and Health, 1997). Studies have found *Penicillium* to be the most frequently isolated fungus in indoor air; the dominant bacteria were *Bacillus, Staphylococcus* and *Micrococcus* (Berry *et al.*, 1996). Health effects associated with fungal and bacterial species, other than infections, include rhinitis, upper respiratory symptoms, asthma, allergic skin reactions, tiredness and headaches (Institute for Environment and Health, 1997).

- NO_2 is emitted from gas cookers and people using them have been shown to have an increased risk of respiratory infections compared to those living in homes with electric cookers. Indeed it has been observed that 70% of a person's exposure to NO_2 occurs in the home (Institute for Environment and Health, 1997). Studies in the UK have found higher average NO_2 concentrations in homes with gas cookers (ranging from 25 to 70 $\mu g/m^3$) than in homes without gas cookers (13–40 $\mu g/m^3$). A maximum 1 hour average concentration of 1115 $\mu g/m^3$ has been recorded in a gas cooker-equipped kitchen (Institute for Environment and Health, 1997).
- CO is produced from gas and paraffin heaters, stoves and cookers. If ventilation is inadequate or appliances poorly maintained, CO levels may accumulate in dangerous concentrations.
- Radon is found in specific geographical areas, and in buildings, which have not been designed to be radon proof, concentrations of radon gas may reach levels that cause lung cancer. Guidance on the protective measures for dwellings is provided elsewhere (Building Research Establishment, 1991).
- Asbestos-based insulating materials are still widely found in dwellings. They are found in roof and wall cladding, some types of insulation, flue pipes, storage heaters, etc. Asbestos can cause scarring of lung tissue (fibrosis); there is also a risk of lung, chest and abdominal cancer. Mineral fibres usually cause eye and skin irritation and there is the potential for lung damage (NSCA, 1995).

Indoor concentrations of air pollutants are influenced by outdoor air quality, indoor sources, the rate of exchange between indoor and outdoor air, and the characteristics and furnishings of buildings. Indoor air pollution concentrations are subject to geographical, seasonal and diurnal variations (WHO, 2000a). Indoor air quality is also influenced by human work and cleaning activities. For example, furniture polishing activities will cause resuspension of particulate materials.

Perhaps the most important factor that causes different exposures to air pollutants across different countries is that of indoor heating and cooking by solid fuel burning. On a global scale, biomass fuels (wood, crop residues, dung, grass, etc) are used in about half the world's households for cooking and/or heating. In China, for example, it has been estimated that coal burning results in indoor particulate concentrations up to 5000 $\mu g/m^3$, whereas residences in Nepal and Papua New Guinea have peak levels of $>10{,}000$ $\mu g/m^3$ (WHO, 2000a cites Smith, 1996). Biomass smoke contains significant amounts of pollutants including CO, particles, hydrocarbons and to a lesser extent NO_x. The smoke also contains many organic compounds, including PAH, that are thought to be toxic, carcinogenic, mutagenic, etc. Coal smoke contains all of these as well as additional pollutants such as SO_x and heavy metals. Table 2.19 details the levels of concentrations of air pollutants found indoors. Table 2.20 shows the global particulate levels and exposures in urban and rural outdoor and indoor environments.

The design of modern buildings has created another type of indoor air quality concern. People working in virtually airtight buildings with the supply of air controlled by air conditioning and heating systems has promulgated 'sick building syndrome' (SBS) and 'building-related illness' (BRI). The symptoms of SBS include lethargy, stuffy or runny nose, dry throat, headache, eye irritation, chest tightness and dry skin (NSCA, 1995). What actually causes the syndrome is still a matter of some debate. However, it seems that a combination of pollutants that occur in the indoor environment may be a contributing factor. Nevertheless, where people feel unable to control air quality, or where their complaints may be ignored, a human factor may also play a contributing part in causing the syndrome (NSCA, 1995). A room's noise, odour, temperature and humidity may also be a contributing factor. BRI is an illness related to indoor exposures to biological and chemical substances and is experienced by some people working or living in a particular building and it does not appear after leaving it (WHO, 2000a). Illnesses include respiratory tract infections and diseases, legionnaire's disease, cardiovascular diseases and lung cancer.

Table 2.19 Air pollution levels ($\mu g/m^3$) found indoors (from WHO, 2000a which cites Smith, 1996)

Country	*Pollutant*	*Level*
Brazil	Particle	90-1100
China	Particle	80-10,900
Europe	NO_2	average 20-40 (living room) [a] average 40–70 (kitchens)[a] 10–20 (dwellings)[b]
Europe	HCHO	9-70
Gambia	Particle	2000-2100
Guatemala	Particle	90-1200
India	Particle	630-21,000
Kenya	Particle	800-4000
Mexico	Particle	335-439
Nepal	Particle	4400
Netherlands	NO_2	peak levels in kitchen of up to 3800 for 1 minute
Papua New Guinea	Particle	1300-5200
South Africa	Particle	1020-1720
Zimbabwe	Particle	1300
	Radon[c]	average 20–70 Bq/m^3 although levels 10 times higher have been found in certain areas
	CO[c]	average 10–20 mg/m^3 peak levels up to 30 mg/m^3

a) Gas appliance
b) Without gas appliances
c) Concentrations not related to specific country
Particles are a combination of TSP, inhalable particles (cut-off 10 mm) and respirable particles (cut-off 5 mm). Samples were taken over variable sampling periods, times of the day, differing living accommodation and using different biomass fuels.

Case Study 2 (Chapter 10, Section 10.3) details the consequences of the build up of air pollution levels (i.e. CO) in a garage and the associated human health effects. Mitigation measures to reduce accumulated air pollution levels are detailed.

2.6.1 Smoking

Smoking generates a wide range of harmful pollutants including nicotine, tar, HCHO, NO_x and CO. It is now accepted that smoking causes cancer. It is also established that passive smoking causes a wide range of problems to the passive smoker ranging from being disturbed by the odour, and eye, nose and throat irritation in recipients, to cancer, increased risk of bronchitis and pneumonia, severe asthma crises and decreases in lung function.

Exposure to environmental tobacco smoke (ETS) is an important factor of indoor air quality. The particle and vapour phases of ETS are complex mixtures of many chemicals, including known carcinogens, such as nitrosamines and benzene (WHO, 2000a). One of the commonly used indicators of environmental pollution by tobacco smoke is PM_{10}; this has been found to be 2 to 3 times higher in houses with smokers than in other houses (Schwartz and Zeger, 1990). Nicotine concentrations of up to 10 $\mu g/m^3$ have previously been found in houses with smokers (WHO, 2000a).

Table 2.20 Particle concentrations ($\mu g/m^3$) and exposures in eight major global environments (WHO, 2000a which cites Smith, 1996)

Region	*Concentration*		*Exposure (%)*		*Total*
	Indoor	*Outdoor*	*Indoor*	*Outdoor*	
Developed					
Urban	100	70	7	1	7
Rural	80	40	2	0	2
Developing					
Urban	250	280	25	9	34
Rural	400	70	52	5	57
Total			86	14[a]	100

a) Variance in summation of individual values to total value was observed in cited reference

Note: population exposures expressed as a percentage of the world total. Here exposure is defined to equal to the number of people exposed multiplied by the duration of exposure and the concentration breathed during that time.

CHAPTER THREE

Effects of Air Pollution

3.1 INTRODUCTION

Poor air quality can affect human health, cause damage to crops and materials and degrade sensitive ecosystems. The severity of impact is dependent upon the concentration and mixture of the pollutants, the duration of exposure and the susceptibility of the sensitive receptor (i.e. human, flora, etc.) (SEPA, 2000). Inevitably the focus of attention in many AQAs is for the protection of human health. Nevertheless, equal consideration should be given to all receptors whether they are ecological, the built environment or human health. This chapter seeks to detail some of the most tangible effects of air pollution.

3.2 HUMAN HEALTH

One of the major purposes of an air quality strategy is to ensure a high degree of protection against risks to public health. An understanding of air pollution on public health relies on scientific evidence obtained from two types of studies (i.e. epidemiological and toxicological). Epidemiological studies are concerned with the effects occurring in human communities exposed under natural conditions. Toxicological studies are investigations on humans or animals in which the level, duration and conditions of exposure are laboratory controlled (Elsom, 1987).

The effects of air pollution on human health are variable, although the relationship between human disease and exposure is neither simple nor fully understood. It is often fairly easy to identify a measure of ill health, e.g. the number of admissions to hospital per day associated with an air pollution event. To show real cause and effect such as that a causal relationship exists, a number of guidelines or tests have to be undertaken. These include an assessment of the consistency of the results of a number of studies, the way in which the results of different studies relate to each other (coherence), whether there is a 'dose response relationship' and whether the sequence of events is logical, i.e. the cause always precedes the effect. Death and disease represent only the extreme of a spectrum of responses (Fig. 3.1).

As well as causing immediate (acute) symptoms with severe effects, air pollutants can result in long-term (chronic) effects or even premature death. The most severe effects such as death and chronic illness will be manifested in a relatively small proportion of the population (Elsom, 1987). The effect of air quality on an individual is subject to their age and general health, nature of the pollutant(s), duration of exposure and level of activity. It is evident from the latter that the exposure pathway is a key factor. Potential exposure pathways from air pollution may be direct (i.e. inhalation) or indirect (i.e. ingestion) (Fig. 3.2).

Figure 3.1 Health effects of air pollution (Reprinted with permission from Shy and Finklea, 1973, copyright 1973 American Chemical Society)

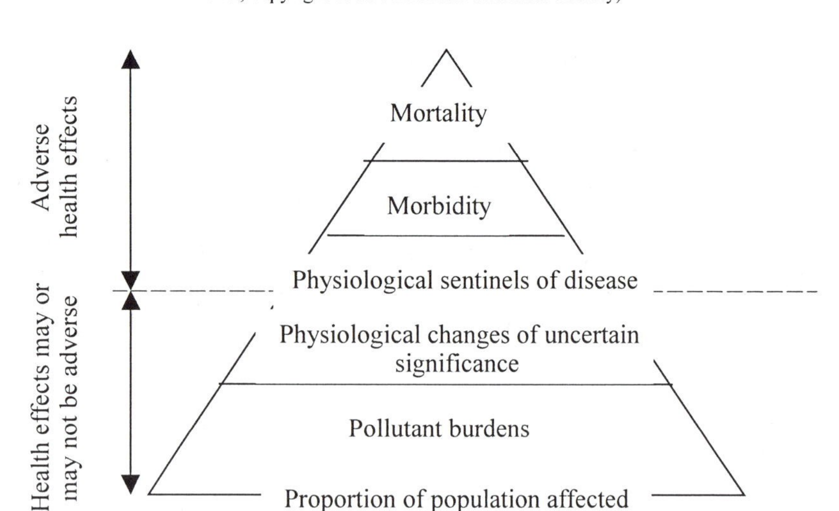

3.2.1 Smogs and air pollution episodes

An air pollution episode is the general term applied to a period of poor air quality, sometimes lasting for several days, often extending over a large geographical area. During an air pollution episode, concentrations of all the measured pollutant substances may be increased, or only one may be affected (Environment Agency, 2000). Air pollution episodes can vary widely in terms of spatial coverage, duration, pollutants affected and incidence throughout the year (Department of Health, 2000). During the summer months photochemical episodes (e.g. O_3) can occur during hot and sunny weather, from the precursor pollutants NO_x and VOCs. During the winter months, cold stable weather conditions can contain pollutants close to the ground near to sources and inhibit their dispersion (Environment Agency, 2000). Studies have identified three common types of air pollution episodes relating to primary emissions (Table 3.1) (Environment Agency, 2000). The result is frequently elevated concentrations of a number of primary pollutants building up over several days (DoE, 1997a).

Perhaps the most widely documented type of air pollution episode is smog. The Industrial Revolution introduced large amounts of particulate emissions from coal and wood burning for energy. The emissions resulted in high concentrations of SO_2 and smoke, which usually occurred at the same time. Dr Des Voeux first used the term 'smog' to describe the conditions of sooty or smoky fogs in 1905. The term is derived from the words **sm**oke and f**og** (Brimblecombe, 1987). Today the term is also applied to the photochemical haze produced by the action of sunlight or UV radiation on motor vehicle and industrial emissions, principally hydrocarbons and NO_2.

Figure 3.2 Exposure pathways from atmospheric concentrations of organic pollutants to human exposure (from AEA Technology, 1995)

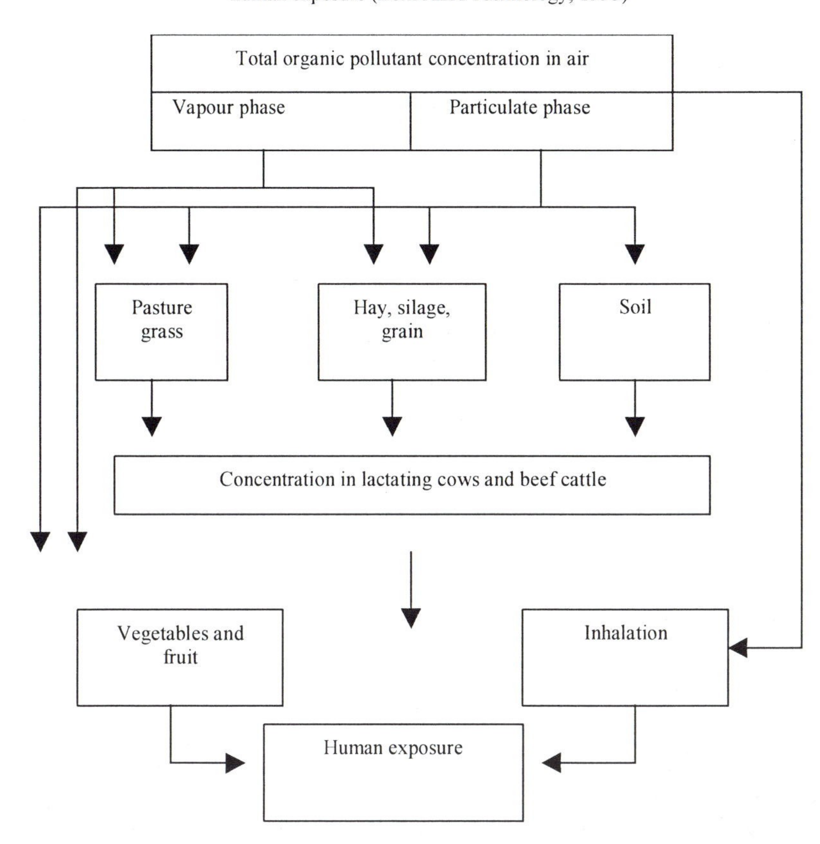

Smog occurrences and their effect on human health have been widely chronicled, for example in Scotland by such bodies as the Scottish Glasgow Corporation Medical Officers of Health (Chalmers, 1930) and the Glasgow smog of 1909 received particular attention. A series of exceptionally bad smogs in that year contributed to a rise in the death rate from respiratory ailments from 35 in October to 233 in November. This accounted for 49% of the total death rate. Well documented episodes also occurred in 1912 and 1929 (Hampton *et al.*, 1993). Smog episodes were also reported for other Scottish cities. Despite these smog occurrences, cities like Glasgow began to show an improvement in air quality during the 20th century particularly with respect to particulates (Fig. 1.2).

In London, England, three air pollution episodes, during which heavy fogs and air pollution were associated, resulted in the death of nearly 5000 people in 1948, 1952 and 1956. The episode in December of 1952 alone accounted for 3000 to 4000 deaths more than expected for that time of year. Although the increase was present in every age group, the greatest increase was in the age group of 45 years and over with more than 80% of deaths occurring among individuals with known heart and respiratory disease. A similar event in 1962 contributed to the excess deaths of a further 340-700 people (Brimblecombe, 1987). During each of these incidents, comparable conditions were present and included limited air mixing as a result of low lying temperature inversions and calm winds, and a continuing emission of air pollution from multiple sources.

Table 3.1 Types of air pollution episodes (Environment Agency, 2000)

Type	*Description*
A1	A plume from a point source reaching the ground during periods of unstable atmospheric conditions. A daytime occurrence associated with light moderate wind speeds, and most frequent in summer.
A2	A plume from a point source reaching the ground during periods of neutral or slightly unstable atmospheric conditions. Associated with moderate strong winds.
A3	A plume from a point source being trapped above the boundary layer by a low level temperature inversion. As the inversion breaks up, the plume is mixed down to ground level.
A4	Plumes from a number of point sources dispersing over a wide area during a period of light wind speed and restricted dispersion.
B1	Build up of pollution beneath an overnight radiation inversion, associated with low wind speeds.
B2	Build up of pollution over several days beneath a persistent inversion – an extended version of B1. Often associated with poor visibility.
C	Episode related to limited dilution/dispersion from a particular source (e.g. emissions from traffic during periods of light wind speed).

Other well documented events have occurred elsewhere in Europe. Among these include the Meuse Valley in Belgium which experienced elevated levels of SO_2 of 25,000 $\mu g/m^3$. During a five-day fog in December 1930 63 people died. Elderly people with previously known heart and lung diseases accounted for the majority of deaths. The signs and symptoms were primarily those caused by a respiratory irritant, including chest pains, cough, shortness of breath and irritation of the eyes. Air pollution episodes have also been documented in the US at Donora and Pittsburgh, Pennsylvania; Ducktown, Tennessee and St Louis, Missouri. In 1948 in Donora 1948 20 people died and approximately 7000, or 50% of the population, experienced acute illness (Faith and Atkisson, 1972).

The occurrence of smogs for pollutants other than smoke and SO_2 is also a concern (CO (Laxen, 1989) and NO_2 (Laxen, 1985)). However, the occurrence of high pollution episodes is now almost synonymous with any conurbation area in which the motor vehicle is now becoming a principal pollutant emitter. The motor vehicle has significantly contributed to air quality concerns in cities such as Los Angles, the Nephos (cloud) in Athens (Greece) and Mexico City (Mexico), where photochemical pollution (e.g. O_3, peroxyacetylnitrate (PAN), etc.) pose a serious threat to human health. The now growing positive associations between particulate concentrations (PM_{10} and $PM_{2.5}$) and mortality and morbidity (Pope *et al.*, 1995), and the increasing levels of recorded cases of asthma, have further heightened the public's anxiety about air pollution and high pollution episodes.

A dramatic example of the effect that weather can have on the dispersal of pollution occurred in London in December 1991 (European Environment Agency, 2000a). From the morning of the 12th December until the evening of the 15th December, London was covered in smog. During the period of the smog NO_2 levels for most of the time were over 190 μg/m^3, peaking at 800 μg/m^3 in the south west of London on the 13th December. Up to 160 additional deaths were thought to have occurred during this period. Deaths from respiratory disease (including asthma) were 22% higher than expected when compared with the same period for the previous year.

A recent high pollution episode (Environment Agency, 2000) occurred on the 2nd September 1998 in the Midlands and South Yorkshire in the UK. SO_2 concentrations (Fig. 3.3) exceeded the UK AQS (Chapter 7, Section 7.2.3) 6.5 times. The high pollution episode coincided with adverse meteorological conditions (i.e. low wind speed) which allowed pollutants from a range of industrial sources to accumulate in the atmosphere before dispersal. Maximum levels were recorded in Nottingham (1737 μg/m^3 (653 ppb)) during an episode lasting from 14:00 to 21:00 hours and peaking at 17:30 hours. The elevated SO_2 levels resulted in the UK 15 minute 266 μg/m^3 (100 ppb) AQS being exceeded 19 times, the World Health Organisation (WHO) 1 hour and 24 hour guidelines being exceeded 4 times and once respectively (Chapter 7, Section 7.2.2). Table 3.2 shows the peak SO_2 concentrations for monitoring stations in the UK at the time of the episode. Case Study 13 (Chapter 10, Section 10.14) shows elevated air pollution levels (i.e. SO_2) in Shenyang, People's Republic of China.

Figure 3.3 SO_2 high pollution episode on 2nd September 1998 in Nottingham, UK (ppb) (Data from Environment Agency, 2000)

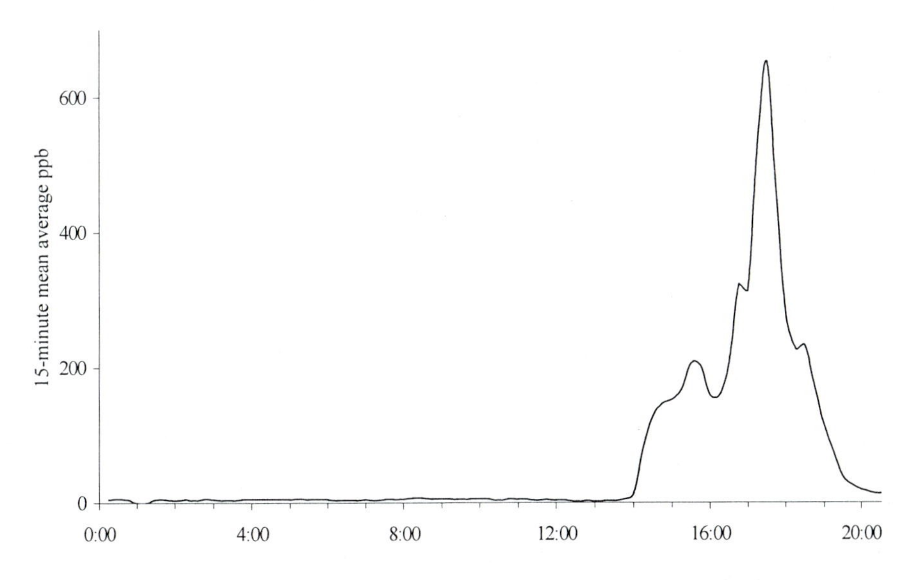

Another example of elevated SO_2 levels occurred in the Grangemouth area of central Scotland between 2nd and 8th August 1999 (SEPA, 2000). Peak SO_2 levels reached 1100 μg/m^3 (380 ppb) due to the close proximity of a coal-fired power station and petrol chemical refinery. An example of a large geographical-scaled episode occurred in April 1988. An odour was reported in various places in the UK Home Counties. Back trajectory analysis of meteorological data over the 13th and 14th April traced the source of pollution back to muck spreading in the Netherlands (File, 1988).

Table 3.2 Comparison of peak SO_2 concentrations (ppb) for UK sites during a high pollution episode on September 2nd 1998 (Environment Agency, 2000)

Site	*Level*	*Site*	*Level*
Barnsley 12	278	Norwich Centre	11
Barnsley Gawber	237	Nottingham Centre	653
Birmingham Centre	211	Oxford Centre	8
Birmingham East	179	Narbeth	3
Bolton	21	Plymouth Centre	13
Bradford Centre	15	Port Talbot	21
Bristol Centre	7	Reading	9
Bury Roadside	129	Redcar	97
Cardiff Centre	14	Rotherham Centre	10
Coventry Centre	19	Salford Eccles	72
Exeter Roadside	7	Sandwell Oldbury	206
Hull Centre	91	Scunthorpe	14
Ladybower	105	Sheffield Centre	90
Leamington Spa	10	Stockport	79
Leeds Centre	9	Stoke-on-Trent Centre	74
Leicester Centre	74	Sunderland	4
Liverpool	46	Swansea	10
Manchester South	54	Thurrock	121
Mansfield DC	20	Wolverhampton	88
Middlesbrough	76		

3.2.2 Assessing health effects

One way of classifying the health effects of pollutants is to make a distinction between acute (short-term) and chronic (long-term) effects. For each of these categories, the effects can range in severity from death to minor illness or discomfort. For example, particles in air can have an acute effect, such as immediate irritations to eyes and throat, or hospitalisation and even death from respiratory failure or heart attacks, caused by severe episodes of air pollution. For some pollutants there may be a threshold level of exposure, below which no health effect is evident. For others, there may be no safe threshold, and some effect may occur whatever the level of exposure might be. Studies (WHO European Centre for Environment and Health, 1995) have estimated the impact of ambient air pollution in Europe (Table 3.3) and have identified that it does have a discernible effect on human health.

In a given population, however, not all individuals are affected equally by the same hazard. Variations in sensitivity to an exposure may occur due to age, nutritional status, genetic predisposition and state of general health. Risk assessments therefore need to be made for particular high-risk groups such as infants and young children, the elderly, pregnant women and their foetuses, the nutritionally deprived, and individuals suffering from some diseases. It is especially important to identify these 'high-risk' groups because they may be the first to experience adverse health effects as the level of air pollution increases. Often, however, only a minority of the population may experience high levels of exposure, and the doses received by the majority of the population are so low that only vulnerable high-risk groups are likely to be affected. Therefore any excess mortality due to a pollutant is likely to be restricted to a small group of the population, so mortality rates for entire populations can often be weak and insensitive indicators of health effects. Estimates of health effects due to air pollution are therefore difficult to calculate. Studies by the World Bank (2000) estimate global mortality rates of 4 to 8% due to air pollution. Fig 3.4 shows the percentage of air pollution-related diseases (World Bank, 2000). Some epidemiological evidence on the health effects of air pollution in children is given in Table 3.4.

Table 3.3 Estimated health impact of ambient air pollution in Europe (WHO European Centre for Environment and Health, 1995)

Indicator of health deficiency	*Proportion of the health deficiency attributed to the pollution (%)*	*Estimated number of cases (annual)*
Cough and eye irritation in children	0.4-0.6	2.6-4 million
Lower respiratory illness in children	7-10	4-6 million
Lower respiratory illness in children causing a medical visit	0.3-0.5	17-29 thousand
Ambulatory visits due to respiratory disease	0.2-0.4	90-200 thousand
Decrease of pulmonary function by more than 5%	19	14 million
Incidence of chronic obstructive pulmonary disease	3-7	18-42 thousand
Hospital admissions due to respiratory disease	0.2-0.4	4-8 thousand

Healthy individuals are not thought to be at significant risk from current levels of air pollution in most ambient situations, but studies have indicated associations which persist at relatively low levels, between daily variations in levels of some pollutants and daily variations in mortality and hospital admissions for acute respiratory conditions. In some cases the mechanisms are not yet known, but the UK Department of Health's Committee on the Medical Effects of Air Pollutants

(COMEAP) (Department of Health, 1998a; 1998b) has advised that it would be imprudent not to regard the associations as indicating causal links. An assessment of the health benefits that are likely to result from the reductions in air pollutant concentrations as a result of the implementation of pollution control measures is therefore an important component of any air quality strategy (Department of Health, 1998a; 1998b).

Table 3.4 Some epidemiological evidence on the health effects of air pollution in children (Briggs *et al.*, 1996)

Country or area exposure	*Association found with increased exposure*
Alpine Region, 1993	SO_2, NO_2, O_3. Reduced lung function. Higher asthma
Czech Republic, 1992	PM_{10}, SO_2, NO_x. Infant and respiratory mortality increased with PM_{10} exposure
Finland, 1991	SO_2, NO_x, H_2S, TSP. Increased respiratory infections
France, 1989	Mixed industrial pollution. Increased rhinitis and school absenteeism
Germany, 1991	TSP, NO_2. Increased croup
Italy, 1992	Outdoor air pollution and passive smoking. Increased asthma
Netherlands, 1993	SO_2, NO_2, PM_{10} and black smoke. Increased wheezing and use of bronchodilator
Switzerland, 1989	NO_2. Increased respiratory symptoms

Figure 3.4 Percentage of air pollution-related diseases (Data from World Bank, 2000)

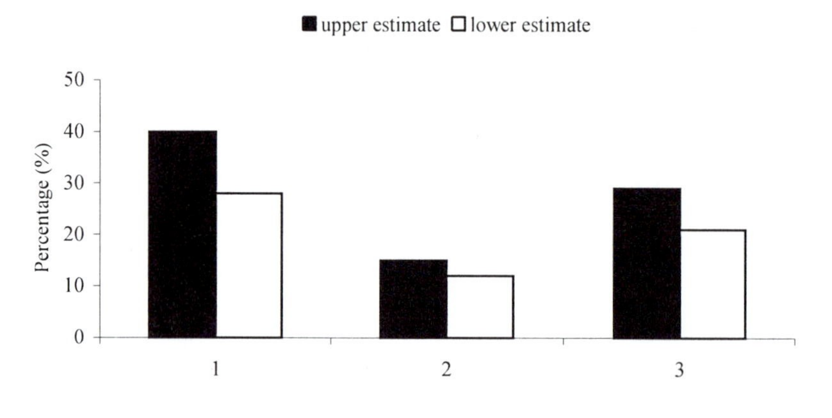

Studies undertaken by COMEAP have shown that numbers of deaths and respiratory hospital emissions brought forward by current levels of PM_{10}, O_3 and SO_2 are significant. It should be noted however, that they comprise only a small percentage (<2%) of the total deaths and hospital emissions occurring each year in the UK (Table 3.5). A sensitivity analysis suggests that 1996 levels of NO_2 might have contributed to around 8700 respiratory hospital emissions (brought forward or additional) (DETR, 1999b). COMEAP (Department of Health, 1998b) have suggested that short-term air pollution events may cause the premature deaths of 12,000 to 24,000 vulnerable people (i.e. old, young, and sick).

Table 3.5 Number of deaths and hospital admissions for respiratory diseases affected per year (DETR, 1999b; SEPA, 2000)

Pollutants	*Health outcomes (brought forward)*	
	Deaths (all cause)	*Hospital admissions (respiratory) and additional*
PM_{10}	8100	10500
SO_2	3500	3500
O_3	700-12500	500-9900
NO_2		8700

3.2.3 Asthma

A common ailment associated with air pollution is asthma. Asthma is a disorder of the airways in which inflammation obstructs airflow, causing breathlessness and wheezing. It is an inflammatory disease, which causes the bronchi to over-react to a stimulus. To counteract the irritant, the airways become inflamed and swollen and produce excess mucus causing narrowing of the airways in vulnerable individuals.

In the UK asthma is thought to affect between 4-6% of children and 4% of adults, where in Finland, over 2.5% of the population suffer from asthma. A common feature throughout all Western countries is a strong increase in the number of asthma cases (European Environment Agency, 2000a). Table 3.6 shows the observed effects an asthmatic suffers from different air pollutants (NSCA, 1996).

3.2.4 Health effects of specific air pollutants

The human health effects are particularly well defined for the criteria air pollutants. It is not coincidental that these pollutants most often have national or international AQSs (Chapter 7, Section 7.2). Detailed below is a summary of health effects.

3.2.4.1 Sulphur dioxide

It has long been recognised that SO_2 is a potent respiratory irritant (Environment Agency, 2000). Exposure to SO_2 stimulates nerves in the lining of the nose, throat and airways of the lungs. Studies have shown that people suffering from asthma may be especially susceptible to the adverse effects of SO_2. SO_2 can also be converted in the atmosphere to sulphuric acid aerosols and particulate sulphate compounds, which are corrosive and potentially carcinogenic. Historically the health effects of SO_2 have been associated with high levels of particulates or other

Table 3.6 The observed effects on asthma suffers for selected pollutants (NSCA, 1996)

Pollutant	*Observed effect on asthma sufferer*
Allergens (tree and grass pollen)	Pollens can trigger an allergic response in the airways of susceptible asthmatics, causing inflammation and consequent narrowing. At high concentrations air pollutants can enhance the effect of some pollens. Air pollution may also affect the concentration of pollens
SO_2	Causes narrowing of the airways of asthma suffers, through the immune response of the airway to the irritant
NO_x	No significant effects have been found. In some cases increased response to allergens in asthma sufferers
O_3	Causes inflammation and consequent narrowing of the airways after short exposure and can increase response to irritants, exacerbating asthma symptoms
Particles	It is suspected that PM_{10} can impair lung function, and that increased levels of particulate can affect asthma sufferers

Table 3.7 Summary of SO_2 human health effects experienced at various exposure levels (ppm) (Utah State Department for the Environment, 2000)

Level	*Exposure*	*Health symptoms*
400	-	Lung edema; bronchial inflammation
20	-	Eye irritation; coughing in healthy adults
15	1 hr	Decreased mucoduary activity
10	10 min	Bronchospasm
8	-	Throat irritation in healthy adults
5	10 min	Increased air way resistance in healthy adults at rest
1	10 min	Increased airway resistance in asthmatics at rest and in healthy adults at exercise
0.5	10 min	Increased airway resistance in asthmatics at exercise
0.5	-	Odour threshold
0.19	24 hr	Aggravation of chronic respiratory disease in adults
0.07	Annual	Aggravation of chronic respiratory disease in children

pollutants. The world's major air pollution disasters have been associated with high levels of SO_2 and particulates (Section 3.2.1). The deaths attributed to these pollutants were due to respiratory failure and occurred predominantly, but not exclusively, in the elderly and infirm. Table 3.7 details a summary of SO_2 health effects experienced at various exposure levels (Utah State Department for the Environment, 2000).

The Air Pollution and Health: A European Approach (APHEA) showed the relative risks associated with a rise in 50 $\mu g/m^3$ in the daily average of SO_2 concentrations (Table 3.8). For all cause mortality the estimate was based on seven western European cities and corresponded to a 3% rise in total deaths. Although this is very unlikely to be a chance effect (the confidence limits show that the true effect is probably between 2.3% and 3.5% (Table 3.8)), there was a significant degree of residual heterogeneity in the modes (i.e. the relationship may not be the same in all seven cities).

Table 3.8 Changes consequent upon a rise in SO_2 of 50 μg/m^3 (24 hr) (summary of APHEA estimates for SO_2 in western European cites) (Environment Agency, 2000)

Outcome	*Age*	*Number of cities*	*95% CL*
All cause mortality		7	1.035 1.023
Cardiovascular mortality		5	1.060 1.100
Respiratory mortality		5	1.030 1.070
Respiratory admissions	15-64	5	0.992 1.025
	65+	5	1.005 1.046

CL = confidence limits

3.2.4.2 Nitrogen dioxide

NO_2 exposure can bring about reversible effects on lung function and airway responsiveness and increase reactivity to natural allergens. NO_2 exposure may also put children at an increased risk of respiratory infection and may lead to poorer lung function in later life. At relatively high concentrations NO_2 causes acute inflammation of the airways. Short-term exposure can affect the immune cells of the airways in a manner that might predispose people to an increased risk of respiratory infections. Table 3.9 contains a summary of NO_2 health effects experienced at various exposure levels (Utah State Department for the Environment, 2000).

Table 3.9 Summary of NO_2 health effects experienced at various exposure levels (ppm) (Utah State Department for the Environment, 2000)

Level	*Exposure*	*Human symptoms*
300	-	Rapid death
150	-	Death after 2 or 3 weeks by bronchiolitis fibrosa obliterans
50	-	Reversible, non-fatal bronchiolitis
10	-	Impairment of ability to detect odour of NO_2
5	15 min	Impairment of normal transport of gases between the blood and lungs in healthy adults
2.5	2 hr	Increased airway resistance in healthy adults
1.0	15 min	Increased airway resistance in bronchitics
0.12	-	Odour perception threshold of NO_2

Essentially NO_2 is an irritant gas and exposure to high concentrations produces narrowing of the airways in both asthmatic and non-asthmatic individuals. Asthmatics are more sensitive to NO_2 than non-asthmatics with exposure to concentrations of about 560 μg/m^3 for 30 minutes producing a small change in standard indices of lung function; in non-asthmatics exposure to about 1800 μg/m^3

would be necessary to produce a similar response. The exposure response relationship for NO_2 is erratic. Exposure to concentrations of 560 μg/m^3 may produce a response whilst exposure to double that may not; the response does seem to reappear and remain as concentrations approach and exceed 1800 μg/m^3, the explanation for this is not clear (European Environment Agency, 2000a).

3.2.4.3 Ozone

O_3 reacts and produces toxic effects on small airway surfaces. The dose delivery is greatest in terminal and respiratory bronchioles. Unlike NO_2 and SO_2, there is very little difference in lung function responsiveness between asthmatics and healthy subjects. There is, however, a great variability in individual responsiveness that is not yet understood. The importance of duration of exposure, and the fact that this may extend over 8 hours on a sunny day, has led to guidelines and AQS for O_3 usually being defined in terms of an 8 hour average concentration (Chapter 7). According to the WHO hourly concentrations of 200 μg/m^3 can cause eye, nose and throat irritation, chest discomfort, cough and headache; exposure for about six hours to concentrations of 160 μg/m^3 have been shown to produce inflammation of the airways and changes in standard indices of lung function (WHO, 1987). Short-term exposure (300 to 500 μg/m^3) to O_3 may impair mechanical functions of the lung and may induce respiratory and related symptoms in sensitive individuals (those with asthma, emphysema, or reduced lung function) (Utah State Department for the Environment, 2000). Symptoms and effects of O_3 exposure are more readily induced in exercising subjects. Table 3.10 shows a summary of O_3 health effects for various exposure levels (Utah State Department for the Environment, 2000).

Table 3.10 Summary of O_3 health effects expected at various exposure levels (ppm) (Utah State Department for the Environment, 2000)

Level	*Human health symptoms*
10.0	Severe pulmonary edema; possible acute bronchiolitis; decreased blood pressure; rapid weak pulse
1.0	Coughing; extreme fatigue; lack of co-ordination; increased airway resistance; decreased forced expiratory volume
0.5	Chest constriction; impaired CO diffusions capacity; decrease in lung function without exercise
0.3	Headache; chest discomfort sufficient to prevent completion of exercise; decrease in lung function in exercising subjects
0.25	Increase in incidence and severity of asthma attacks; moderate eye irritation
0.15	For sensitive individuals, reduction in pulmonary lung function; chest discomfort; irritation of the respiratory tract, coughing and wheezing

3.2.4.4 Particulates

Although many of the obvious effects of particulate air pollution disappeared with historical smog events (Section 3.2.1), recent research has suggested that, even at the much lower levels now found in the UK, parts of Europe and the US, particulate air pollution appears to be associated with a range of measures of ill health including effects on the respiratory and cardiovascular systems, asthma and mortality. Indeed there is now substantial evidence linking ambient PM_{10} levels with respiratory symptoms, decreased lung function, hospital visits, school absenteeism and other health outcomes (Department of Health, 1999; Pope and Dockery, 1992; Pope *et al.*, 1995; Department of Health, 1998b). Particles can cause eye, nose and throat irritation.

Particles inhaled by humans are segregated by size during deposition within the respiratory system. Larger particles deposit in the upper respiratory tract, while smaller inhalable particulates travel deeper into the lungs and are retained for longer periods of time. This is why PM_{10}, and particularly $PM_{2.5}$, are of primary concern with regard to health effects. Not only do they penetrate deeper and remain longer in the lungs than larger particles, but the particulates also contain large quantities of organic materials that may have significant long-term health effects.

Studies in the US have shown that death rates from respiratory and cardiovascular disease increase with increased concentrations of PM_{10} and monitoring shows significant levels in many countries (Pope and Dockery, 1992; Pope *et al.*, 1995). Estimated health effects for PM_{10} exposures are listed in Table 3.11 (Utah State Department for the Environment, 2000). The toxicity of particles retained in the lungs varies with chemical composition. Some chemicals such as sulphuric acid may react directly with the human respiratory system, while others may act to retard clearance of other particles from the lungs. Particulates may also act as carriers for gaseous pollutants and can cause synergistic effects, such as when SO_2 and particulate exposures occur simultaneously. Carbon particles are the most common carrier for gaseous and semi-gaseous pollutants.

Table 3.11 Estimated health effects for PM_{10} ($\mu g/m^3$) exposure (Utah State Department for the Environment, 2000)

Effects	*Effects possible*	*Effects likely*
Reduced lung function in children	140	350
Aggravation of bronchitis	350	600

Associations between ambient concentrations of particulates and other indices of ill health have been reported (DoE, 1995b). Evidence shows that consistent rises in PM_{10} levels may be associated with an increase in the numbers of admissions to hospital, increases in reported symptoms, and decreases in lung function. This effect has been observed in studies by the WHO in estimating the approximate effects of different concentrations of PM_{10} on some indices of human health (Table 3.12). Case Study 9 (Chapter 10, Section 10.10) details the effects of particulate emissions on human health from an open cast coal site (OCCS) and the mitigation measures to reduce particulate levels.

3.2.4.5 Carbon monoxide

Carbon monoxide (CO) is a colourless, odourless, tasteless and non-irritant gas that is slightly less dense than air and is sparingly soluble in water. It is a product of the incomplete combustion of carbonaceous fuels and may be used as a fuel, burning in air to yield CO_2 (the product of complete combustion) (Health & Safety Executive (HSE), 1984). CO is a problem both as an indoor and as an outdoor air pollutant. In terms of accidental deaths, indoor exposure presents the major problem (Chapter 2). Case Study 2 (Chapter 10, Section 10.3) details the build up of CO in a garage and the associated human health effects.

Table 3.12 Summary of short-term exposure response relationships of PM_{10} with different health indicators ($\mu g/m^3$) (DoE, 1995b)

Health effect indicator		*Estimated change in daily average PM_{10} concentration needed for a given effect*
Daily mortality	5% change	50
	10% change	100
	20% change	200
Hospital admissions for respiratory conditions	5% change	25
	10% change	50
	20% change	100
Numbers of asthmatic patients using extra bronchodilators	5% change	7
	10% change	14
	20% change	29
Numbers of asthmatic patients noting exacerbation of symptoms	5% change	10
	10% change	20
	20% change	40

CO enters the body through the lungs, where it is absorbed by the bloodstream and combines with haemoglobin, the substance that carries oxygen to the cells. Haemoglobin that is bound with CO is called carboxyhaemoglobin (COHb). Haemoglobin binds approximately 240 times more readily with CO than with oxygen. Thus, the amount of oxygen being distributed throughout the body by the bloodstream is reduced in the presence of CO. Blood laden with CO can weaken heart contractions, lowering the volume of blood distributed to various parts of the body. It can also significantly reduce a healthy person's ability to perform manual tasks, such as working, jogging and walking (Utah State Department for the Environment, 2000). CO is exhausted from the body at varying rates depending on physiological as well as external factors. The general guideline is that 20-40% of the CO is lost from the system after 2 to 3 hours following exposure. Because it takes time for CO to be removed from the blood stream, the severity of health effects depend both on the concentration being breathed and the length of time the person is exposed (Utah State Department for the Environment, 2000). Cigarette smokers (Chapter 2, Section 2.6.1) seem to adapt to high levels of COHb. Often as much as 10% of total haemoglobin is combined with CO as COHb in the blood of heavy smokers. Indeed, smokers are on average net contributors to ambient CO levels. In healthy, non smoking, individuals the effects of CO exposure appear at a COHb concentration of about 5% (WHO, 1987). Tables 3.13 and 3.14 detail the relationship between health effects and exposure to COHb (Utah State

Department for the Environment, 2000; HSE, 1984). Table 3.15 shows the health effects of CO by concentration on human health (HSE, 1984).

Several empirical equations have been derived for estimating COHb levels from ambient CO exposure levels (WHO, 1979a). The simplest equations describe COHb levels as a linear function of CO concentration in the inspired air (CO), and of exposure time (t). Such a relationship is given by equation 3.1.

Table 3.13 Relationship between health effects and exposure to COHb (%) (Utah Department for the Environment, 2000)

COHb	*Health symptoms associated with COHb level*
80	Death
60	Loss of consciousness; death if exposure is continued
40	Collapse on exercise; confusion
30	Headache; fatigue; judgement disturbed
20	Cardiovascular damage; electrocardiographic abnormalities
5	Decline (linear with increasing COHb level) in maximal oxygen uptake of healthy young men undergoing strenuous exercise; decrements in visual perception, manual dexterity, and performance of complex sensorimotor tasks
4	Decrements in vigilance (i.e. ability to detect small changes in one's environment that occur at unpredictable times); decreased exercise performance in both healthy persons and those with chronic obstructive pulmonary disease
2.5	Aggravation of cardiovascular disease (i.e. decreased exercise capacity in patients with angina pectoris, intermittent claudication, or peripheral arteriosclerosis)

$$[COHb](\%) = k.CO.T \qquad (3.1)$$

Where k is a constant equal to 0.0003 for an individual at rest with a ventilation rate (V_A) of 6 l/min and a pulse rate of 70, 0.0005 for light activity (V_A of 9.5 l/min and a pulse rate of 80), 0.0008 for light work (50 watts, V_A of 18 l/min and a pulse rate of 110) and 0.0011 for heavy work (about 100 watts, V_A of 30 l/min and a pulse rate of 135). Another expression to estimate the COHb level is given by equation 3.2:

$$\Delta[COHb](\%) = ([CO]\ (ppm).V_A\ (l/min).T\ (min))/(4650.V_B\ (l)) \qquad (3.2)$$

Where V_B is the blood volume. Another expression for the prediction of COHb levels is given below (equation 3.3).

$$[COHb](\%) = 0.147[CO]\ (1\text{-}e^{-0.00289T}) \qquad (3.3)$$

Table 3.14 Human health effects of different blood COHb levels (%) (HSE, 1984; DoE, 1994b)

Level	*Symptoms*[a]	*Level*	*Symptoms*[b]
0-10	None	2.5-4	Decreased short-term maximal exercise duration in young healthy men
10-20	Tightness across forehead	2.7-5.1	Decreased exercise duration due to increased chest pain (angina) in patients with ischaemic heart disease
20-30	Headache	2.0-20	Equivocal effects on visual perception, audition, motor and sensorimotor performance, vigilance, and other measures of neurobehavioural performance
30-40	Severe headache, weakness, dizziness, nausea, vomiting	4.0-33	Decreased maximal oxygen consumption with short-term strenuous exercise in young healthy men
40-50	Collapse, increased pulse rate and respiratory rate	20-30	Throbbing headache
50-60	Coma, intermittent convulsions	30-50	Dizziness, nausea, weakness, collapse
60-70	Depressed heart action, death possible	>50	Unconsciousness and death
70-80	Weak pulse, slowed respiration, death likely		

a) *HSE (1984)*
b) *DoE (1994b)*

Table 3.15 Human health effects of CO concentrations (ppm) in air (HSE, 1984)

Level	*Health effect*
50	Recommended occupational exposure limit (OEL). 8 hour time-weighted average
200	Headache after about 7 hours if resting or after 2 hours exertion
400	Headache with discomfort with possibility of collapse after 2 hours at rest and 45 minutes exertion
1200	Palpitation after 30 minutes at rest or 10 minutes exertion
2000	Unconscious after 30 minutes at rest 10 minutes exertion

Where CO is in ppm and t is in minutes. The equation is valid for subjects ventilating about 6 l/min. WHO (1979a) provides a greater explanation of the expressions and their respective data requirements. An application of the methodology to calculate the COHb level is given in Case Study 2 (Chapter 10, Section 10.3) relating to accumulated levels of CO in a garage. The high 8 hour mean concentrations were calculated to be equivalent to levels of approximately 8.3 COHb% above normal background levels (0.1–1% in a non-smoker and 6–7% in a smoker).

3.2.4.6 Dioxins and furans

The effects of dioxins have primarily been studied through experiments with animals. Animals exposed in experiments to higher doses show a number of pathological changes. The effects from lower dose ranges are thought to be cancerous and damage the immune and reproductive systems. Human beings that have been exposed to large amounts of dioxins as a result of accidents (e.g. Seveso) show a great variety of symptoms, such as mental disturbances, various systematic disturbances, neurological problems, and skin damage. Due to the long latency period and indeterminate symptoms it is difficult to attribute directly dioxin poisoning to causes of human death although some individuals are, however, believed to have died of acute dioxin poisoning following industrial accidents.

3.2.4.7 Lead

Lead exhibits toxic effects in humans, which are seen in the synthesis of haemoglobin, acute or chronic damage to the nervous system, effects in the kidneys, gastrointestinal tract, joints and reproductive system. Lead is absorbed into the body both through the stomach and intestines after being taken in through the mouth and through the lungs when inhaled. Once absorbed it spreads around the body and accumulates particularly in bone, teeth, skin and muscle. In these tissues it is relatively stable and released only over months or years (DETR, 2000b).

The toxic effects of lead are a consequence of its ability to inhibit the actions of certain enzymes and to damage chemicals in the nuclei of cells. In workers with high exposure of lead poisoning there is acute brain damage, causing delirium and fits. Severe poisoning can also induce many other symptoms, and damage to organs such as the kidney can occur when concentrations in the blood exceed 100 μg/dl (DETR, 2000b). At somewhat lower concentrations, above about 80 μg/dl, colicky intestinal pains may occur. Above about 50 μg/dl anaemia can arise due to an inability to produce haemoglobin. Reversible effects on the kidneys and male reproductive organs have been described at blood concentrations greater than 40 μg/dl, as have effects on nerve functions in the limbs at concentrations above 30 μg/dl. Above a level of 10 μg/dl studies of large groups of children have shown subtle evidence of changes in brain development, and this is also the lowest concentration at which biochemical evidence of interference with blood pigment synthesis has been described (DETR, 2000b). Table 3.16 details the health effects of other non-criteria air pollutants (Harrop, 1999; WHO, 1987).

Table 3.16 Observed effects of non-criteria air pollutants on human health (Harrop, 1999; WHO, 1987)

Pollutant	*Effect*
Benzene	Exposure of greater than 3200 mg/m^3 causes neuro-toxic symptoms. Persistent exposure to toxic levels may cause injury to bone marrow, resulting in persistent pancytopenia. Early manifestations of toxicity are anaemia, leucocytopenia or thrombocytopenia. Benzene is a known human carcinogen. A large number of cases of nyeloblastic and erythroblastic leukaemia have been associated with benzene exposure
PAH	On the basis of experiment results, toxic effects other than carcinogenicity are not to be expected
Toluene	The health effect of primary concern is dysfunction of the central nervous system (CNS). Acute experimental and repeated occupational exposure at levels >375 mg/m^3 has elicited dose related CNS alterations (e.g. fatigue, confusion, etc.)
As	The clinical picture of chronic poisoning with As varies widely. It is usually dominated by changes in the skin and mucous membranes and by neurological vascular and haematological lesions. Involvement of the gastrointestinal tract, increased salivation, irregular dyspepsia, abdominal cramps and loss of weight may also occur. Other symptoms may also include vestibular functions, optic nerve degeneration. Inorganic compounds are established human carcinogens
Asbestos	Effects include asbestosis, mesothelioma, and lung cancer
Cd	Acute respiratory effects with concentrations >1 mg/m^3. Chronic effects with long-term exposure (1 year) to 20 ug/m^3
Cr	Chrome ulcers, corrosive reaction in the nasal septum acute irritative dermatitis and allergic eczematous dermatitis. Slight effects on respiratory tract due to Cr^{IV}. Cr^{IV} is carcinogenic
H_2S	Acute intoxication is mainly the result of action of the nervous system. Greater than 15 mg/m^3 causes conjunctival irritation. Affects sensory nerves in the conjunctivae. Serious eye damage at 70 mg/m^3, >225 mg/m^3 paralysing effect on olfactory perception. High concentration respiratory irritation is the predominant symptom with risk of pulmonary oedema at 400 mg/m^3. At >1400 mg/m^3 immediate collapse
Hg	Toxic effects of Hg and its compounds depend upon the chemical form of Hg. Hg vapour damage is mainly to the nervous system, but effects are seen, depending on dose, in the oral mucosa and the kidneys. Inorganic divalent Hg compounds are corrosive poisons that can cause death in acute doses. Methymercury compounds damage is almost exclusively limited to the nervous system
1,3-butadiene	Studies have shown that 1,3-butadiene causes a variety of cancers in rodents and damages the genetic structure of the cell. It is thus a genotoxic carcinogen and as such absolute safe levels cannot be defined
Ni	In human acute intoxication with nickel carbonyl, allergy dermatitis, asthma and mucosal are reported. Ni has carcinogenic properties

3.3 FLORA

The possible injury and damage to plant communities by air pollution is a combination of a range of physical, chemical and biological stresses which may affect a plant's physiology. The visible symptoms produced by these various stresses would need to be distinguished, as do the very different symptoms, which can be produced in different plant species by the same factor. Further difficulties are introduced by the fact that plants are commonly subjected to more than one stress, either simultaneously or successively, and that the sensitivity of plants to a particular stress will be altered by other environmental factors. Air pollution impacts involve a combination of factors including the concentration and the exposure period to the pollutant (or pollutants), the plant species, the plant's age, and other environmental conditions. It is therefore difficult to identify direct cause and effect relationships between a pollutant concentration and its effect on vegetation (Taylor *et al.*, 1988). Table 3.17 details the ranges of air pollution concentrations that affect vegetation.

Table 3.17 Air pollution concentrations (mg/m^3) that affect vegetation (Taylor *et al.*, 1988)

Pollutant	*Concentration*	*Time*
NO_x	1.9-1900	0.5 to 8 hr
SO_2	0.28-12.0	1 to 8 hr
CO	112	1 week

Taylor *et al.* (1988) have previously detailed the effects of air pollution on plants. The most common and widely investigated air pollutant and its effects on plants is SO_2. Leaf injury symptoms may appear within 1 hour of a severe exposure but it will often take 2 to 8 days to develop fully, especially in cool, dull weather conditions. SO_2 concentrations known to cause plant injury range from 0.28 to 12.0 mg/m^3 over exposure times of 1 to 8 hours (Taylor *et al.*, 1988). Other studies (Treshow, 1984) have noted that to prevent SO_2 injury to most species, concentrations should not exceed 1.93 mg/m^3 for 1 hour, 1.1 mg/m^3 for 2 hours, 0.7 mg/m^3 for 4 hours, 0.28 mg/m^3 for 5 hours. The WHO (1987) notes that it is generally accepted that the SO_2 concentrations prevailing in most agricultural regions of Europe are unlikely to reduce cereal yield levels. However, there is evidence that some other species are affected by concentrations less than 100 μg/m^3. Unfortunately, there are many discrepancies between various SO_2 fumigation experiments on agricultural crops, which are probably due to interactions with environmental conditions and factors relating to the plants themselves.

Harrop (1999) has previously summarised from the literature the effects of air pollution on flora. Although NO_x includes N_2O_3, N_2O_4 and N_2O_5 it is unlikely that any except NO and NO_2 are ever present in the atmosphere at high enough concentrations to cause plant injury. NO is fairly rapidly oxidised to NO_2 in the atmosphere. Relatively high concentrations of NO_2 are needed to produce acute symptoms in plants. Acute symptoms develop within 2 to 48 hours in most species although it may take a week for some deciduous trees. Concentrations known to cause plant injury range between 1.9-1900 mg/m^3 from 0.5 to 8 hours exposure (Taylor *et al.*, 1988). These levels are far in excess of existing air quality levels. For forests, symptoms would be expected at doses of approximately 3-5 mg/m^3 for periods of up to 48 hours. The threshold for leaf injury may require exposure to 38 mg/m^3 if the exposure is only for 1 hour, while a concentration of 1900 μg/m^3 might require up to 100 hours to produce symptoms. NO injury to trees in the field is unknown, while NO_2 injury would be expected in the vicinity of an excessive industrial source (Smith, 1981).

CO produces very similar effects to VOCs on plants (e.g. epinasty, chlorosis and abscission) (Taylor *et al.*, 1988). Plants are relatively insensitive to CO at the low concentrations that have been found to be toxic to animals (US Department of HEW, 1970). There are no reports of an effect of CO on plants below 112 mg/m^3 for 1 week (Taylor *et al.*, 1988) to 115 mg/m^3 (1 to 3 weeks) (US Department of Health, Education and Welfare (HEW), 1970). Case Study 10 (Chapter 10, Section 10.11) details the assessment process to evaluate the impact of industrial emissions on flora. Critical loads were calculated using air dispersion modelling techniques and predicted levels were compared to recommended assessment criteria.

For O_3 the critical levels are defined using the WHO AOT40 index - the accumulative exposure over a threshold of 40 ppb during the growing season. The Kuopio Workshop in Finland (1996) revised critical levels of O_3 to prevent damage to most sensitive crops, forests and semi-natural vegetation. This has provided the necessary patterns of exceedence of these critical levels in Europe. The area exceedence of the critical level represents 81% of UK land area (91% and 76% of arable crops and semi-natural vegetation area respectively) (Institute of Terrestrial Ecology, 1999). Exceedence of the critical level does not necessary mean that there will be damage to vegetation, but only that the risk of damage exists for sensitive species and conditions.

Troyanowsky (1985), Gruderian (1977), Treshow (1984), Georgii (1986), Thomas (1961) and Taylor *et al.* (1988) provide additional general texts on the effects of air pollution on flora.

3.4 FAUNA

Parker (1978) has identified that airborne pollutants may enter animals through two principal pathways, namely ingestion of contaminated material, and inhalation. Other less obvious pathways include ingestion of pollutants through licking or grooming the coat or feathers, and/or direct deposition causing irritation of the eyes or skin. Pollutants entering through inhalation may cause responses in various parts of the respiratory tract. SO_2 and O_3 produce changes in the pulmonary capillaries, which reduce absorption. Other pollutants may influence the bronchial blood vessels so that absorption through mucosa is reduced, or they may affect the circulatory blood system more generally so that the distribution of an absorbed compound is retarded. Interference with the function of macrophage cells, which are responsible for clearing the alveoli and other lung tissue of foreign matter, may be another important effect of inhaled pollutants. Toxic effects in animals may show chemical, physiological and morphological changes in tissues and organs, but there may also be sensory effects and overt behavioural signs, which are as obvious as visual symptoms of injury in plants. However, with animals it is more difficult to diagnose the cause of such symptoms, because there have been few definitive experiments involving exposure to specific pollutants.

There is general agreement that most animals die when COHb levels reach about 70% and that the rate of administration of the gas is important in determining the outcome. It is also agreed that COHb levels exceeding 50% are often associated with damage to organs, including the brain and the heart (WHO, 1979a). WHO (1979b) studies show that the lowest adverse effect SO_2 concentration on animals varies considerably from study to study. In general, however, it has been noted that sulphuric acid aerosols and some sulphate salts such as zinc ammonium sulphate are more irritating to respiratory organs than SO_2, and that some aerosols, particularly those in the sub micron range, enhance the effect of SO_2 when they are present simultaneously (WHO, 1979b). Fibrosis in many animal species, and bronchial carcinomas and pleural mesotheliomas in the rat, have been observed following inhalation of both chrysotile and amphibole asbestos (WHO, 1986). There are differences in susceptibility to injury both between and within species of animals. These may sometimes be due to differences in mobility and powers of selecting food (Parker, 1978). Catcott (1961) also provides a description of the effects of air quality on animals.

3.5 ECOSYSTEMS

The effects of air pollution on ecosystems ranging from forests to freshwaters are well known. However, to gain a broader appreciation of the effects of air pollution national governments, such as the UK Government, have set up advisory groups (e.g. Critical Loads Advisory Group (CLAG)) to develop national critical loads and levels programmes. Such groups review the impacts of air pollutants on specific parts of the environment including soils, freshwater, vegetation, buildings and materials and to provide information about input fluxes and estimates of pollutant exposure (SEPA, 2000). Their aim is to assess the sensitivity of each receptor across the regions and hence define the environmental capacities available to absorb pollutant loads or to withstand gaseous concentration levels for critical pollutants. Effectively the groups assess the level of pollutant that a receptor (e.g. ecosystem, plant etc.) can tolerate without suffering long-term adverse effects according to current knowledge (Doe, 1994a).

For many years, there has been considerable concern about the effects of acid deposition on areas of the EU (amongst others) and various measures aimed at reducing discharges of pollutants (e.g. NO_x, SO_2, etc.) thought to contribute to the problem have been established. They provide a framework for the reduction of industrial air pollution and propose specific emission limits for these pollutants (Chapter 7, Section 7.9).

Following the 5th Environmental Action Programme (5EAP) the EC confirmed the political long-term goal of no exceedence of the critical loads and levels. The objectives of the strategy were to reduce beyond existing commitments emissions of SO_2, NO_x and NH_3. The EC proposed to reduce the area of ecosystem in the EU at risk from acid rain from 6.5%, on the basis of existing commitments, to 3.3% by 2010 (SEPA, 2000). The EC's study showed that even with ambitious abatement programmes for reducing acidifying pollutants the ultimate target of never exceeding critical loads could not be achieved by 2010. They therefore followed the 'gap closure' approach which aimed to reduce the difference between the level of ecosystem protection in 1990 and the 100 percent ecosystem protection by the year 2000 (SEPA, 2000). The EC felt that the most appropriate interim target was a 50% gap closure (European Environment Agency, 1998a; 1998b). The European Monitoring and Evaluation Programme (EMEP) regularly updates these predictions using revised emission inventory values to evaluate progress.

3.6 MATERIALS

The effects of air pollution on materials are well documented (Harter, 1986; Lee and McMullen, 1994). The types of damage caused by air pollution include the corrosion of metals, decay of building stone (Fig. 3.5), erosion and discoloration of paints, etc. Table 3.18 summarises the potential damaging effects of acidic air pollutants and other environmental factors on various types of material at risk. All these deleterious effects occur as a result of natural environmental conditions as well as by interaction with pollutants. Factors which contribute to natural damage are water (in the form of atmospheric humidity and surface wetness), solar radiation, atmospheric gases such as CO_2, O_2, naturally occurring acidic species, temperature fluctuations and the activities of micro-organisms (Harter, 1986). These factors produce the same types of damage as do anthropogenic pollutants, which therefore make it difficult to determine how damage may be caused by either natural or anthropogenic air pollution sources. The effects of atmospheric pollutants on buildings provide some of the clearest examples of atmospheric pollutant damage. Impacts fall vary broadly into two main categories: material loss/erosion from acidic deposition and soiling from particle deposition. For most materials, the dry deposition of SO_2 exerts the strongest corrosive effect of atmospheric pollutants. Wet deposition of secondary pollutants (formed from SO_2 and NO_x emissions), expressed as rain acidity, has a corrosive effect on certain materials but is generally weaker. Particles can also act as a catalyst for stone

Figure 3.5 Decay of building stone owing to air pollution in northern Scotland (courtesy of the Scottish Environment Protection Agency (SEPA, 2000))

Table 3.18 Air pollution damage to materials (from Harter, 1986, based on Altshuller *et al.*, 1983)

Material	*Type of impact*	*Pollutant*	*Other environmental factors*	*Measures of measurement*
Metals	Corrosion, tarnishing	SO_x, other acidic gases	Cycles of relative humidity (RH), air salt, particulate matter	Weight loss after removal of corrosion products, reduced physical strength, change in surface characteristics
Building stone	Surface erosion, soiling, black crust formation	SO_x, other acidic gases	Mechanical erosion, particulate matter, cycles of RH, temperature fluctuations, salts, vibrations, CO_2, micro-organisms	Weight loss of sample, surface reflectivity, measurement of dimensional changes, chemical analysis
Ceramics and glass	Surface erosion, surface crust formation	Acid gases, especially fluoride-containing	Cycles of RH	Loss in surface reflectivity and light transmission, change in thickness, chemical analysis
Paints and organic coatings	Surface erosion, discoloration, soiling	SO_x, H_2S	Cycles of RH, UV and visible light, particulate matter, mechanical erosion, micro-organisms, O_3	Weight loss of exposed painted panels, surface reflectivity, loss of thickness
Paper	Embrittlement, discoloration	SO_x	Cycles of RH, physical wear, acidic materials introduced in manufacture	Decreased folding endurance, pH change, molecular weight measurement, tensile strength
Photographic materials	Micro-blemishes	SO_x	Particulate mater, cycles of RH	Visual and microscopic examination
Textiles	Reduced tensile strength, soiling	SO_x, NO_x	Particulate matter, cycles of RH, UV and visible light, physical wear, washing	Reduced tensile strength, chemical analysis (e.g. molecular weight), surface reflectivity
Textile dyes	Fading, colour change	NO_x	Temperature fluctuations, UV and visible light, O_3	Reflectance and colour value measurements
Leather	Weakening, powdered surface	SO_x	Physical wear, residual acids introduced in manufacture	Loss in tensile strength, chemical analysis
Rubber	Cracking		UV and visible light, O_3, physical wear	Loss in elasticity and strength, measurement of crack frequency and depth

erosion, and may have synergistic effects with NO_2 and SO_2. The UK National Material Exposure Programme (NMEP) (NMEP, 1999; SEPA, 2000) assesses the rates of corrosion (metal loss) for bare mild steel, galvanised steel, copper and aluminium. Results show variable rates of corrosion depending on the metal and location. Corrosion rates can be up to 46 μm over an 8 year period (Table 3.19).

Soiling of buildings by combustion particles is one of the most obvious signs of pollution in urban areas. Soiling of buildings results primarily from the deposition of particles on external surfaces and includes residential dwellings, commercial premises and historic buildings. Studies (Mansfield, 1987) have shown a decline in reflectance of white surfaces with duration of exposure (Fig. 3.6). Studies have suggested that the soiling of a white painted surface by smoke deposition can be described by the following relationship:

$$\text{Decrease in reflectance} = 0.41 \sqrt{\text{dose}} \qquad (3.4)$$

Where dose equals the product of the TSP level (μg/m^3) and the exposure time (months). Equation 3.4 shows that the reflectance of white painted wood would fall by 2%, 3%, 3.5% and 4% after 50, 100, 150 and 200 days exposure respectively in an environment where the TSP level was 15 μg/m^3 (Mansfield, 1987).

Table 3.19 Metal loss due to air pollution in Scotland (μm) (1987-95) (SEPA, 2000)

Site	*Mild steel*	*Galvanised steel*	*Copper*	*Aluminium*
Glasgow	32	9.2	4	1
Clatteringshaws	46	21.5	5	1
Eskdalemuir		11.0	4	1
Strath Vaich Dam	30	9.4	3	1

3.7 VISIBILITY (PARTICLE HAZE)

Haze is caused when sunlight interacts with small pollution particles in the air. Some sunlight is absorbed by these particles. Other light is scattered away before it reaches an observer. The greater the pollution then the greater the absorption and scattering of light affecting the clarity and colour of the view. The air pollutants causing the problem come from a variety of sources both anthropogenic and natural (Chapter 2, Section 2.3). The spatial and temporal distribution as well as the anthropogenic causes of atmospheric haze have received considerable attention (e.g. Miller *et al.*, 1972; Munn, 1973; Weiss *et al.*, 1977; Leaderer and Stolwijk, 1979; Ferman *et al.*, 1981; Robinson and Valente, 1982; Trijonis, 1982; Wolff *et al.*, 1982; Sloane, 1984). Much of this research deals with the physico chemical properties of haze, with the aim of understanding its sources and formation mechanisms.

The occurrence of haze (Fig 2.4) is widespread, for example it has occurred in Indonesia, Malaysia, Singapore, Brunei and the surrounding region as a result of forest fires from Kalimantan and Sumatra in Indonesia. It has also affected parts of the US as a result of forest fires from Mexico and Florida. Russia, Greece and Canada have also experienced forest fires, as well as Brazil amongst many other

Figure 3.6 Reduction in reflectance of surfaces[a] due to particulate deposition (Mansfield, 1987)

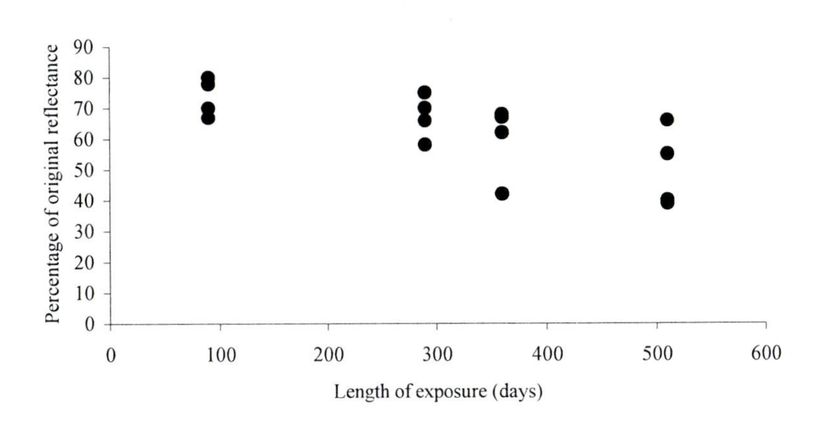

a) Reflectance surfaces were white painted wood, ceramic tiles, white formica and cotton

countries. Whilst forest fires are a common source of haze, haze from anthropogenic activities also contributes to the concern. For example, studies have observed that sunshine levels have decreased to 58% in Shenyang, China (Sustainable Shenyang Project Office, 1998). Previously (1988-1998) there was a mean of 33.5 misted (haze) days in each year, and 61% of these days occurred in winter (Sustainable Shenyang Project Office, 1998).

3.7.1 Visual range

Visual range to the human eye is a subjective concept, being the maximum distance at which an observer can discern the outline of an object. The limitations in actually making a judgement of visual range include the viewers' visual acuity, the number, configuration, and physical and optical properties of the visible targets. Viewers' subjectivity imposes a random component on the observed signal. The lower contrast of real targets compared to black objects imposes a systematic underestimate of visual range. In addition, visibility is reported in quantified units, depending on the availability of visible targets. The visual range, or visibility, is an understandable and for many purposes, an appropriate measure of the optical environment (Griffing, 1980). It is inversely proportional to aerosol concentration. Another measure of haziness is the extinction coefficient, Bext, defined as Bext = k/visual range. Where k is the constant. The value of k is determined by both the threshold sensitivity of the human eye as well as by the contrast of the visible objects against the horizon sky. The extinction coefficient is in units of km^{-1} and is proportional to the concentration of light scattering and absorbing aerosols and gases.

3.8 STRATEGIC AIR QUALITY ISSUES

Many air pollution problems are recognised as being global/strategic and requiring international air quality management solutions (e.g. depletion of the O_3 layer, global warming, acid rain, etc). The introduction of international air pollution protocols and conventions has greatly strengthened the control of air quality (Chapter 7, Section 7.9). However, despite international pressure to reduce these global effects, concern still persists about their long-term effects. The following text provides an overview of these concerns.

3.8.1 Acid rain

A Scotsman, Angus Smith, the first UK Alkali Inspector, first used the term 'Acid Rain' in the 19th century whilst studying the composition of rainwater across the UK. The term 'acid rain' was used as long ago as 1858 to describe rain made more acidic by acid gas pollution (SEPA, 2000). It has been argued that the natural acidity of unpolluted rainfall is pH 5.6 (Innes, 1987). A more accurate term, however, is acid deposition. Wet deposition occurs when pollutants are carried in rain, snow, mist and low cloud; pollutants may be wet deposited after being carried long distances. Dry deposition is the direct fallout of acid pollutants and mostly occurs close to the source of emission.

The acidity of rainwater is determined by the balance between cations (positively charged ions) and anions (negatively charged ions) in precipitation. In Europe the main cations are H^+ and NH_4^+ and to a lesser extent, Ca^{2+}, Na^+, K^+ and Mg^{2+} and the main anions are SO_4^{2-}, NO_3^- and Cl^-. The primary pollutants of principal concern in the formation of acid rain are SO_2, NO_x and NH_3 (Fig. 3.7).

Figure 3.7 Formation of acid rain, Swedish Ministry of Agriculture (1982) (courtesy of the Swedish Ministry of Agriculture)

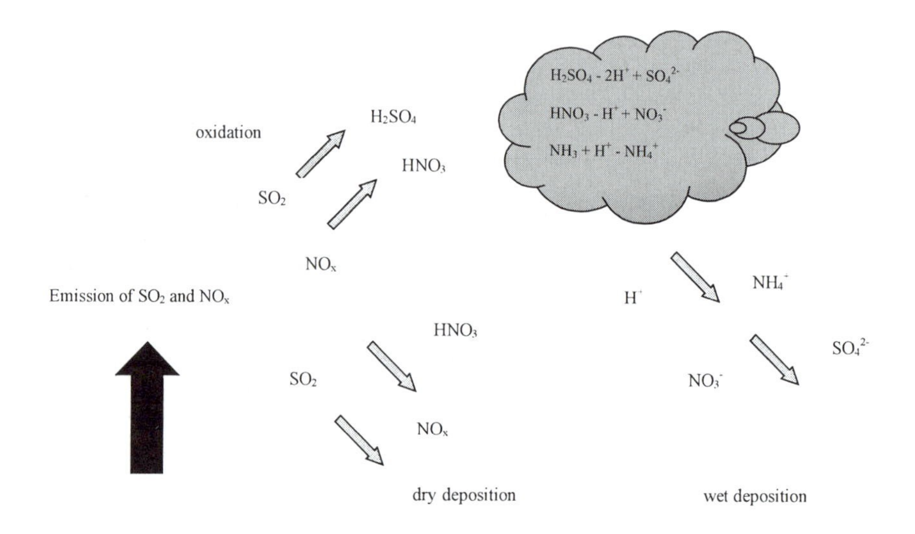

SO_2 and NO_x are oxidised to form H_2SO_4 and NHO_3, either in the atmosphere or after deposition. NH_3 may react with H_2SO_4 and NHO_3 to form ammonium sulphate and ammonium nitrate particles (Swedish Ministry of Agriculture, 1982; Smith *et al.*, 1996). Once released into the atmosphere, acidic pollutants can be transported long distances. For example, more than 50% of acid rain in eastern Canada comes from sources in the US (Environment Canada, 1999).

The largest deposition of sulphur occurs in areas with the largest emissions and is largely due to the dry deposition of SO_2. High rates of sulphur deposition also occur in areas of high precipitation. Similar patterns are found for the deposition of oxidised nitrogen, although relatively smaller amounts are deposited close to the emission sources. Oxidised nitrogen is transported over longer distances and contributes to the tropospheric O_3 problem because NO_x is a major precursor in its formation (SEPA, 2000). The deposition of reduced nitrogen compounds, which originate from NH_3 emissions, is to a larger extent than sulphur dominated by high deposition rates near the sources. There is, therefore, less long-range transport of NH_3 than for SO_x or NO_x (European Environment Agency, 1998a; 1998b). An example of this effect is the breakdown of acid precipitation falling on Scotland arising from sources outside the country. For example, 45% of sulphur deposition comes from elsewhere in the UK, and 45% from elsewhere in Europe, with only 10% coming from Scottish sources (Smith *et al.*, 1996). The majority of acid precipitation compounds emitted in Scotland are likewise deposited elsewhere. Dry deposition of SO_2 accounts for about one third and wet deposition of sulphate about two thirds of the total deposition of sulphur in Scotland. Wet deposition is the dominant pathway except in the central belt. Areas of high sulphur deposition in Scotland are in the south west Highlands (from wet deposition), the central belt (from dry deposition) and Galloway and the Borders (from a combination of wet and dry deposition) where deposition rates are over 15 kg S/ha/yr. Cloud droplet deposition increases total deposition on the eastern Border Hills, in the central Highlands and on the Cairngorm and Grampian Mountains, but contributes less than 4% of total deposition to Scotland as a whole. Deposition rates in these hill and mountain areas can reach about 12 kgS/ha/yr. Particulate aerosol deposition only effects some forests and contributes 1% of total deposition. Deposition rates for sulphur to many areas of Scotland are less than 6 kgS/ha/yr (Smith *et al.*, 1996) (Fig. 4.6).

Extensive damage to trees, in the form of defoliation and discoloration, has been reported due to acid deposition. However, the damage does not necessarily relate to acidification as other environmental stresses, weather or intrinsic features may also lead to defoliation and reduced vitality. A causal connection cannot, therefore, always be established between an input of acid deposition in excess critical load and observed foliage reduction, even in areas where the neutralising capacity of the soil is likely to be an important influence on the growth and ageing of forest stands. Despite pollutant emission reductions for NO_x and SO_2, studies show a general increase in defoliation. For example, in the UK 13.9% and 15% of conifer and deciduous trees had >25% to 100% needle or leaf loss (European Environment Agency, 1998a; 1998b). Fig. 3.8 shows the critical loads map for Scotland (Smith *et al.*, 1996) depicting levels of deposition at which ecosystems are vulnerable to acidity.

Figure 3.8 Critical loads maps for Scotland (Smith *et al.*, 1996) (Permission from The Centre for Ecology and Hydrology, Edinburgh)

Figure 3.9 Percentage (%) of lakes with exceedence of the critical load for sulphur (Data from European Environment Agency, 1998a)

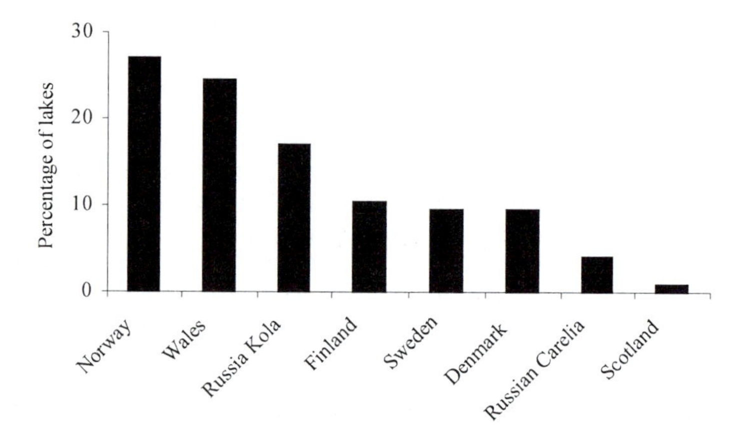

Many lakes in Europe have been affected by acid deposition. The effects can be direct, because of toxicity, or indirect, because of acid sensitive prey or plant food or because of complex changes in water chemistry caused by increased acidity. In the late 1960s and early 1970s, Scandinavian rivers and lakes and their aquatic life began to show signs of being adversely affected by pollution, and in the late 1970s trees in Central European forests showed signs of being similarly affected. The percentage of lakes with exceedence of the critical load for sulphur varies for different countries. For example in Scotland the figure is just over 1%, for Wales it is approximately 24% and Norway 27% (European Environment Agency, 1998) (Fig. 3.9).

3.8.2 Ozone depletion

Ozone (O_3) is a naturally occurring form of oxygen, which forms a layer in the upper levels of the earth's atmosphere (the stratosphere). The O_3 screens out a high proportion of the UV in sunlight and prevents it reaching the lower levels of the atmosphere (the troposphere). Without this protection, the surface of the earth would be exposed to elevated levels of UV solar radiation, which would be harmful to almost all forms of life. Increased levels of UV light have been linked to the recent growth in skin cancers (SEPA, 1999b). UV radiation can also depress the human immune system, and harm aquatic systems and crops. Scientists now believe that over the past ten years (i.e. since the late 1980s) average global O_3 concentrations have decreased by 3% (SEPA, 1999b).

During the 1960s industrialised countries manufactured increasing quantities of compounds containing chlorine and bromine used in a wide range of applications. CFCs and HCFCs are used as refrigerants, as solvents and as foam-blowing agents. Halons are used as fire suppressants. Methyl bromide is used as a fumigant to destroy pests found in soil, in structures and stored products. Once released to atmosphere some of these chemicals including CFCs, HCFCs, carbon tetrachloride, 1,1,1 trichloroethane, halons and methyl bromide reacted with the O_3 layer causing it to deplete.

Concern over the depletion of the stratospheric O_3 layer, particularly over Antarctica, was first raised in the 1970s due to the experience of a 'hole' over the area. The extent of the problem is not, however, only confined to the earth's polar regions. Satellite observations, for example, show that the total atmospheric O_3 levels over Scotland have declined by around 10% over the past 20 years (SEPA, 1999b).

The UK Stratospheric Ozone Review Group (SORG), summarised in DETR (2000b), has presented assessments of the scientific issues underlying the worldwide changes taking place in the O_3 layer, and the causes of these changes. The SORG reports showed that there had been a steady decrease in stratospheric O_3 over all latitudes outside the tropics since the late 1970s, with the largest reductions in total O_3 occurring over Antarctica each spring. Over the same period there has been a steady increase in the amounts of chlorine and bromine present in the stratosphere as a result of human activity. In 1993 it was reported that the rate of increase of chlorine had slowed in response to regulation under the Montreal Protocol. The most recent SORG assessment (1996) concluded that at mid latitudes an annually averaged decrease in total O_3 of 4 to 5% per decade had taken place since 1979. Smaller losses occurred during the 1970s. In the northern hemisphere the decreases were largest in winter and spring, up to 7% per decade. The decreases were primarily the result of O_3 destruction in the lower stratosphere, between altitudes of 15 and 25 km. The largest O_3 losses continue to be observed over Antarctica each spring, when total O_3 is 60% lower than in the 1960s, and there is almost complete removal of O_3 from much of the lower stratosphere. The decline in total O_3 in the Antarctic summer has continued, and amounts are now 20% less than in the 1960s. In 1996, total O_3 in high northern latitudes was about 20% below the average for 1979 to 1986. Losses of O_3 between January and March are estimated to have been up to 50% in the Arctic lower stratosphere in both 1995 and 1996, partly as a result of the record low temperatures during those winters.

3.8.3 Greenhouse effect

Like a window pane in a greenhouse, a number of gases in the earth's atmosphere let solar radiation (visible light) pass to the surface of the earth while trapping infrared radiation, also known as heat radiation, that is re-emitted by the surface of the earth, which would have otherwise escaped to space. It is the trapping of infrared radiation that is generally referred to as the 'greenhouse effect' (Krause, Bach and Koomey, 1990). The gases that do this naturally are mainly water vapour and CO_2. Without this natural greenhouse effect, the earth would be over 30°C cooler (DETR, 2000d). However, the principal concern is that greenhouse gas concentrations are rising well above their natural levels causing additional global warming.

The gases that influence the surface atmosphere radiation balance are also called radiatively active (e.g. CO_2, CH_4, etc.) (Krause, Bach and Koomey, 1990). As well as nitrogen and oxygen, the atmosphere also contains small amounts of water vapour, CO_2, CH_4, N_2O and O_3. These are sometimes referred to as 'greenhouse gases' (Fig. 3.10). CH_4 is thought to contribute about a fifth of the current enhancement to the greenhouse effect (Intergovernmental Panel on Climate Change (IPCC), 1996). Since pre-industrial times CO_2 concentrations in the atmosphere have increased by about 28% (Fig. 2.21 shows an example of atmospheric CO_2 increase for Hawaii), CH_4 by 145% and N_2O by 13% (IPCC, 1996). CO_2 is increasing due to the burning of coal, gas and oil (fossil fuels) and the destruction of forests. CH_4 is generated through modern agricultural practices ranging from rice growing (e.g. paddy fields) to livestock farming and termites, as well as emissions from coal mining, natural gas production and distribution, refuse and sewage disposal. Fluorinated compounds, including the CFCs, PFCs, HFCs and SF6 are also greenhouse gases. Although their atmospheric concentrations are small, they are strong greenhouse gases and have very long atmospheric lifetimes. Therefore, they play a potential role in climate change. Each greenhouse gas has a different capacity to cause global warming, depending on its radiative properties, its molecular weight and its lifetime in the atmosphere. Its so called global warming potential (GWP) encapsulates these criteria (Table 3.20).

Figure 3.10 Formation of the greenhouse effect (after RCEP, 2000) (Crown copyright is reproduced with the permission of the Controller of Her Majesty's Stationary Office)

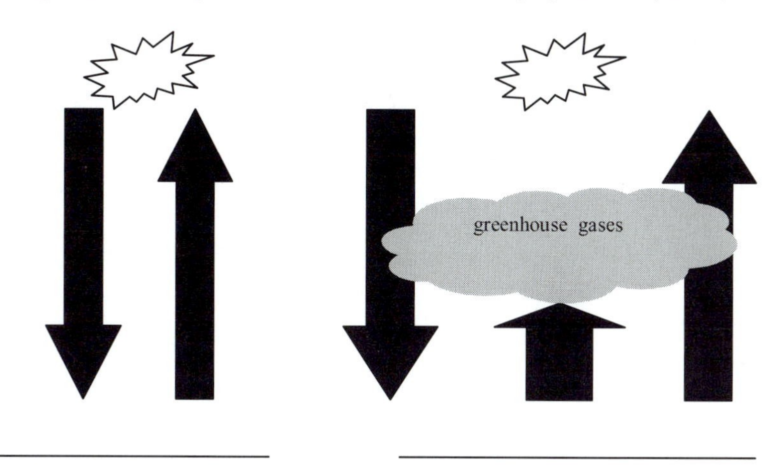

The GWP is defined as the warming influence over a set time period (e.g. 100 years) of a gas relative to that of CO_2 (Table 3.20). It means, for example, that 1 tonne of HFC-134 emitted to the atmosphere has 1000 times the GWP over 100 years of 1 tonne of CO_2. To compute the CO_2 equivalent of the emission of any gas, its emission is multiplied by the GWP. This is often expressed as the carbon equivalent. The value is then multiplied by 12/44, the ratio of the atomic weights of carbon and CO_2. Thus, for example, an emission of 1 tonne of HFC-134 is equivalent to 1 x 1000 x 12/44 = 273 tonnes of carbon. Table 3.21 shows the differences in the amount of CO_2 emitted in different EU countries as well as the different trends regarding CO_2 emissions.

There are still uncertainties about the effects on the climate of increasing concentrations of greenhouse gases and indeed the extent and rate at which the climate is changing. Climate change is likely to affect human health through the shifts in the distribution of diseases, such as malaria and respiratory disorders (IPCC, 1996). Despite these uncertainties governments worldwide have agreed that precautions need to be taken now (RCEP, 2000).

Despite these uncertainties there is now sufficient evidence that human activity is having an effect on the earth's climate. For example, the most recent forecasts (1998) by the UK Impacts Programme of Climate Change in Scotland are that by 2050 (SEPA, 1999c):

- Mean temperature will have increased by between 0.8 and 2.0°C;
- Annual precipitation will have increased by between 3% and 6%, although the winter increases may be up to 13%;
- The intensity of precipitation will have increased;

- Net sea level will have risen by somewhere between 5 and 65 cm;
- Storm damage, flooding and erosion in coastal zones will have increased.

Table 3.20 Global warming potential of greenhouse gases (DETR, 2000d)

Greenhouse gas	*Lifetime (years)*	*100 years GWP*[a]
CO_2	50-200	1
CH_4	12.23	21
N_2O	120	310
HFC-23	264	11,700
HFC-32	5.60	650
HFC-41	3.70	150
HFC-43-10mee	17.10	1300
HFC-125	32.60	2800
HFC-134	10.60	1000
HFC-134a	14.60	1300
HFC-152a	1.50	140
HFC-143	3.80	300
HFC-143a	48.30	3800
HFC-227ea	36.50	2900
HFC-236fa	209	6300
HFC-245ca	6.60	560
Chloroform	0.51	4
Methylene chloride	0.46	9
Sulphur hexafluoride	3200	23,900
Perfluoromethane	50,000	6500
Perfluorethane	10,000	9200
Perfluoropropane	2600	7000
Perfluorobutane	2600	7000
Perfluoropentane	4100	7500
Perfluorohexane	3200	7400
Perfluorocyclobutane	3200	8700
Trifluoroiodomethane	<0.005	<1

a) A 100 year time horizon has been chosen by the United Nations Framework Convention on Climate Change (UNFCCC) in view of the relatively long time scale for addressing climate change.

Other evidence of the effects of global warming show that globally, seven of the ten warmest years on record were in the 1990s and in the future, the earth's climate could warm by as much as 3°C over the next 100 years (DETR, 2000d). UK studies have shown that the predicted effects from global warming will include (DETR, 2000d):

- Sea level is expected to rise by over 40 cm by 2080 because of thermal expansion of the oceans as temperatures rise and because of melting of land ice. Of the additional 80 million people projected to be at risk of flooding, 60% are expected to be in southern Asia (Pakistan, India, Sri Lanka,

Bangladesh and Myanmar) and 20% in South East Asia (from Thailand to Vietnam, including Indonesia and the Philippines);
- Africa is expected to experience significant reductions in cereal yields, as are the Middle East and India and an additional 290 million people could be exposed to malaria by 2080, with China and central Asia likely to see the largest increase in risk;
- Water resources for drinking and irrigation will be affected by reduced rainfall or as ground water in coastal zones suffers from salination as sea levels rise. An additional three billion people could suffer increased water stress by 2080. Northern Africa, the Middle East and the Indian subcontinent will be the worst affected;
- By 2070, large parts of northern Brazil and central southern Africa could lose their tropical forests because of reduced rainfall and increased temperatures. Should this happen, global vegetation which currently absorbs CO_2 at the rate of some 2-3 gigatonnes of carbon (GtC) per year will become a carbon source generating about 2 GtC per year by 2070 and further adding to CO_2 build up in the atmosphere (current global man-made emissions are about 6-7 GtC per year).

Table 3.21 CO_2 emissions in the EU (European Environment Agency, 2000a)

Country	*Millions tons in 1990*	*Millions tons in 1993*	*Variation*
Belgium	110.7	109.1	- 1.4
Denmark	52.2	57.8	+ 10.7
France	367.4	356.0	- 3.1
Germany	993.3	916.1	- 7.7
Greece	73.3	72.7	- 0.8
Ireland	30.6	29.9	- 2.3
Italy	401.6	390.6	- 2.7
Luxembourg	12.2	12.7	+ 4.1
Netherlands	156.7	160.3	+ 2.3
Portugal	39.8	44.4	+ 11.6
Spain	209.4	223.5	+ 6.7
UK	578.8	556.2	- 3.9

Studies of future anthropogenic emission scenarios of greenhouse gases (IPCC, 1996; RCEP, 2000) have predicted a rise in ambient CO_2 levels of between 450 – 750 ppm by 2100. The RCEP (2000) report has extended projections to 2300. The study results show elevated future levels of CO_2 and significant increases in the earth's surface temperature and sea levels. To limit the damage of greenhouse gas emissions will require reductions in global emissions.

In response to increasing concerns about climate change, the United Nations Framework Convention on Climate Change (UNFCCC) was agreed at the Earth Summit in Rio de Janeiro in 1992 (Chapter 7, Section 7.9) and 184 countries have now signed it. Under the Convention, all developed countries agreed to aim to return their greenhouse gas emissions to 1990 levels by 2000. It was quickly recognised that the Convention commitments could only be a first step in the international response to climate change. Climate prediction models show that deeper cuts in emissions will be needed to prevent serious interference with the climate. The Kyoto Protocol, agreed in December 1997, was designed to address this issue. Developed countries agreed to targets that would reduce their overall emissions of a basket of six greenhouse gases (CO_2, CH_4, N_2O, PFCs, HFCs and SF6) by 5.2% below 1990 levels over the period 2008-2012. For the first time these targets will be legally binding, and differentiated between parties to the Convention. For example, the EU and its Member States agreed to -8%, US to -7%, Japan to -6%, Russia and the Ukraine to return to 1990 levels, and Australia was allowed an 8% increase. Unfortunately, following the Kyoto Protocol the US announced in 2001 that it would not be a signatory the Protocol as it considered, amongst a series of reasons, that the reductions in CO_2 levels were too harsh for American industry to comply with.

Under the Kyoto Protocol, the EU and its Member States can agree to meet their commitments jointly. This arrangement allows the EU's target to be redistributed between Member States to reflect their national circumstances, requirements for economic growth, and the scope for further emission reductions (Chapter 7). In June 1998 countries agreed how the target should be shared out. The UK agreed to reduce its emissions by 12.5%. Targets for other member states ranged from –21% for Germany and Denmark, to –6% for the Netherlands, +13% for Ireland and +27% for Portugal. The international negotiations on climate change have continued since Kyoto. The Kyoto Protocol established a legal framework for delivering emission reductions. But some of the detailed questions, such as how the Kyoto mechanisms will operate and the extent to which sinks can be counted against targets, still need to be resolved. An effective compliance procedure is also required to ensure that countries meet their targets. At the Fourth Conference of the Parties (COP 4) in Buenos Aires, Argentina, in 1998, Ministers agreed to a comprehensive two year work programme (the Buenos Aires Plan of Action) which set COP 6 in November 2000 as the deadline for decisions on most key issues. COP 5 in Bonn in November 1999 reviewed progress and agreed to intensify the negotiating process so that the COP 6 deadline could be met. The Kyoto Protocol will enter into force when 55 countries have ratified it, including developed countries accounting for at least 55% of developed countries' CO_2 emissions in 1990. Most developed countries, however, will want more certainty on some of the key outstanding issues before ratifying. In particular, countries are likely to want to know more about how the Kyoto mechanisms will operate, and what the consequences of non compliance with their Kyoto targets will be.

Recognising the problem of potential global climate change the World Meteorological Organisation (WMO) and the United Nations Environment Programme (UNEP) established the IPCC in 1988. The role of the IPCC is to assess the scientific, technical and socioeconomic information relevant for the understanding of the risk of human induced climate change. It does not, however, carry out new research nor does it monitor climate-related data. It bases its assessment mainly on published and peer reviewed scientific technical literature. The IPCC comprises three working groups and a Task Force:

- Working Group I assesses the scientific aspects of the climate system and climate change;
- Working Group II addresses the vulnerability of socioeconomic and natural systems to climate change, negative and positive consequences of climate change, and options for adapting to it;
- Working Group III assesses options for limiting greenhouse gas emissions and otherwise mitigating climate change.

The Task Force on National Greenhouse Gas Inventories oversees the National Greenhouse Gas Inventories Programme.

The Panel meets in plenary session about once a year. It accepts and/or approves IPCC reports, decides on the mandates and work plans of the working groups, the structure and outline of reports, the IPCC Principles and Procedures and the budget. The IPCC completed its First Assessment Report in 1990. It played an important role in establishing the Intergovernmental Negotiating Committee for a UNFCCC by the United Nations (UN) General Assembly. The UNFCCC was adopted in 1992 and entered into force in 1994. It provides the overall policy framework for addressing the climate change issue.

CHAPTER FOUR

Emission Inventories

4.1 INTRODUCTION

Emission inventories are an important facet of AQM. They identify and quantify emission sources over small or large areas such as an industrial site or for a complete local authority area. Determination of the character of pollutant emission sources (e.g. the nature of the pollutant; emission rate; gas efflux velocity; gas efflux temperature; source morphology, etc.) can aid the quantification of the source's significance and level of potential impact. Inventories can also help to refine air quality monitoring strategies, assess trends in air quality and help identify pollutant sources that may need emission control.

Pollutant emissions are estimated from a knowledge of the process which forms them. For some of the pollutants this is relatively straightforward because emissions are largely dependent on fuel composition (e.g. the sulphur content of the fuel). For other pollutants (e.g. NO_x) emissions depend upon combustion conditions, such as temperature and pressure, and therefore are more difficult to quantify and may therefore be less accurate. Emission estimates from point or area sources generally tend to be better defined than from mobile sources (e.g. motor vehicles, trains, etc.) as more factors affect the type and quantity of the pollutant emitted. For example, motor vehicle emissions depend upon fuel usage; engine design and emission control; maintenance of the vehicle; driver behaviour; traffic conditions; vehicle speeds; vehicle age; the number and the mileage of different types of vehicles; etc.

The compilation of an emission inventory requires data to produce pollution factors and loads. The pollution factor is the amount of a pollutant or a combination of pollutants released by a source (directly or indirectly) per unit of merchandise produced or per unit of raw material consumed, depending upon the type of industry or method of calculation of the pollution factor. Pollution loadings are the total amount of a pollutant or a combination of pollutants released (directly or indirectly) by a source in a given period of time (WHO, 1982a). Essentially therefore an emission estimate is the product of at least two variables:

- An activity statistic and a typical average emission factor for the activity (e.g. annual fuel consumption (in tonnes fuel/year) and an emission factor (e.g. grammes SO_2 emitted/tonne fuel consumed); or
- An emission measurement over a period of time and the number of such periods occurring in the required estimation period (e.g. measured SO_2 emissions (in grammes per hour) and number of operating hours per year).

In practice, however, the calculations tend to be more complicated but the principles remain the same. Emission estimates are collected together into inventories or databases which usually also contain supporting data on, for example, the locations of the sources of emissions; emission measurements where available; emission factors; capacity, production or activity rates in the various source sectors; operating conditions; methods of measurement or estimation; etc. (EMEP/CORINAIR, 1999). Eggleston and McInnes (1987) provide guidance on methods for the compilation of emission inventories. Extensive studies, such as the USEPA AP-42 (USEPA, 1995a) and the EMEP/CORINAIR Atmospheric Emission Inventory Guidebook (EMEP/CORINAIR, 1999) provide a detailed list of source emission factors. Other sources include the DETR (2000d). National emission inventories for many pollutants are currently under development and are updated regularly (e.g. Gillham *et al.*, 1994; UNECE, 1995; London Research Centre, 2000).

4.2 THE PURPOSE OF EMISSION INVENTORIES

The principal purpose of an air emission inventory is often, but not necessarily, regulatory based. Emission regulations or statutes can require air emission inventories to determine the amount of pollutants released to the atmosphere. For example, the US CAA, as amended in 1990, details requirements for specific inventories. International protocols and conventions also necessitate the need for national emission inventories. The reporting of emission inventory data to CLRTAP is required in order to fulfil its obligations regarding strategies and policies in compliance with the implementation of protocols under the Convention. These protocols include for example the Helsinki Sulphur Protocol (1985); the Sofia NO_x Protocol (1988); the Geneva VOC Protocol (1991); the Oslo Sulphur Protocol (1994) and the Aarthus Protocols on Heavy Metals and on Persistent Organic Pollutants (POPs) (Chapter 7, Section 7.9).

Other examples of the purpose of emission inventories include EU Member States who are required to submit annual national emissions for NO_x, SO_2, NMVOC, CH_4, CO, NH_3 and various heavy metals and POPs for 11 main source categories (Level 1 of Selected Nomenclature for Sources of Air Pollution (SNAP)) by 31st December of the following year. Member States are also invited to report emissions of more detailed source sub sectors (SNAP – Level 2). In addition they are required to provide EMEP periodically with emission data within grid elements of 50 km x 50 km. Member States are required to use the draft reporting procedures (EB.AIR/GE.1/1997/5) when compiling their inventories. EC Decision 99/296/EC requires Member States to report to the Commission their anthropogenic CO_2 emissions and removal sinks. They are also required to report their national inventory data on emissions/removal for the six Kyoto greenhouse gases (CO_2, CH_4, N_2O, PFCs, HFCs and SF6) on an annual basis. In accordance with the EC Directive Integrated Pollution Prevention Control (IPPC) (96/61/EC) a Committee has been formed to establish the format and particulars of the 'inventory of principal emissions and sources responsible' (the so-called Polluting Emissions Register (PER)). The inventory will be based on data supplied by Member States to DGXI, who are then required to report this inventory on a 3 yearly basis to Council and Parliament. The first inventory is expected to be reported in 2002.

Under Article 12 of the UNFCCC all parties are required to develop, periodically update, publish and make available to the Conference of Parties national inventories of anthropogenic emissions by sources and removals by sinks of all greenhouse gases not controlled by the Montreal Protocol.

On a smaller scale, inventories have a wide variety of applications. They are used as the basis for drawing up operating permits (i.e. authorisations); determining compliance with existing permit conditions or emission regulations; conducting environmental impact assessments for proposed new sources and for input to human health risk assessment studies. Emission inventories are also being used to predict and delineate local and regional air assessments and management areas. The UK DETR (2000a) have categorised the uses of emission inventories in Local Air Quality Management (LAQM):

- Provision of an overview of the intensity of emissions across an area, and to quantify individual sources such that absolute or relative contributions can be established;
- To guide and refine the design of air quality monitoring networks. An emissions inventory can indicate, for example, where the highest concentrations of pollution are likely to be found, or which areas are the most representative, and can therefore guide the most appropriate siting or monitors; and
- Help in estimating the extent and cost of any necessary emission controls and in identifying the activities that should bear these costs.

Case Study 3 (Chapter 10, Section 10.4) shows the methodology employed to undertake an emission inventory study for a UK Local Authority (Falkirk Council). Case Study 7 (Section 10.8) employs the results from Case Study 3 to predict actual and future air quality levels in the Falkirk locality. Case Study 12 (Section 10.13) shows the methodology used to estimate fugitive VOC emissions from an oil terminal.

4.3 ATMOSPHERIC EMISSION INVENTORY INITIATIVES

There have been several major international initiatives to develop emission inventories, they include (EMEP, 2000):

- Organisation for Economic Co-operation and Development (OECD) Control of Major Air Pollutants (MAP) Project: MAP was designed to assess pollution by large scale photochemical oxidant episodes in Western Europe and evaluate the impact of various emission control strategies for such episodes. The project, which started in 1983 and was reported in 1990, quantified emissions for SO_2, NO_x, and VOCs from point and area source emissions in nine main source sectors from European OECD countries. The source sectors were: mobile; power plant; non-industrial combustion; industry; organic solvent evaporation; waste treatment and disposal; agriculture and food industry; nature and miscellaneous.
- DGXI Inventory: The Council for European Communities (CEC) Environment Directorate (DGXI), in 1985, funded the compilation of an emission inventory for the then EU Member States. The aim of the inventory was to collect data

on emissions from all relevant sources in order to produce a database. The inventory covered four pollutants (i.e. SO_2, NO_x, VOCs and particulates) and recognised 10 main source sectors (i.e. utility power plant; industrial combustion plant; district heating; oil refineries and petrochemical plant; domestic heating; industrial processes; solvent use; transportation; agriculture and nature).

- CORINE Programme and subsequent work by the European Environment Agency Task Force: Council Decision 85/338/EEC established a work programme, in 1986, given the name CORINE (CO-oRdination d'INformation Environnementale), that included a project to gather and organise information on emissions into the air relevant to acid deposition (CORINAIR). The inventory co-ordinated atmospheric emissions from the EU Member States covering three pollutants (SO_2, NO_x, and VOCs) and eight main source sectors (i.e. combustion (including power plant but excluding other industry); oil refineries; industrial combustion; processes; solvent evaporation; road transportation; nature and miscellaneous. The inventory was developed in collaboration with the EU Member States, Eurostat, OECD and UNECE/EMEP and was completed in 1990. In addition the project developed a source sector nomenclature (Nomenclature for Air Pollution Socio-Economic Activity (NAPSEA) and SNAP- for emission source sectors, sub-sectors and activities); a default emission factor handbook and a computer software package for data input and the calculation of sectorial, regional and national emission estimates.

  In 1991 it was agreed to produce an update of CORINAIR 1990. This update was performed in co-operation with EMEP and IPCC-OECD to assist in the preparation of inventories required under the CLRTAP and UNFCCC respectively. The update resulted in a more developed nomenclature (source sector split, i.e. SNAP90) involving over 260 activities grouped into a three level hierarchy of sub-sectors and 11 main sectors (e.g. public power, co-generation and district heating plants; commercial, institutional and residential combustion plants; industrial combustion; production processes; extraction and distribution of fossil fuels; solvent use; road transport; other mobile sources and machinery; waste treatment and disposal; agriculture and nature); a greater list of pollutants (SO_2, NO_x, NMVOC, NH_3, CO, CH_4, N_2O, CO_2); an extended number of sources were considered as point sources; a recognition that the emission inventory needed to be complete, consistent and transparent, and extension of the availability of the CORINAIR system to 30 countries, an increased awareness of CORINAIR and the need to produce an inventory within a reasonable time scale to serve the requirements of the user community (policy makers, researchers, etc). Initial data from CORINAIR90 became available in early 1994 and the project was completed and a series of reports prepared during 1995 and early 1996. The work was finalised and published by the European Environment Agency in 1996 and 1997.
- The Co-operative Programme for Monitoring and Evaluation of the Long-range Transmission of Air Pollutants in Europe (EMEP): EMEP, formed by a Protocol under CLRTAP, has arranged a series of workshops on emission inventory techniques to develop guidelines for the estimation and reporting of emission data (e.g. SO_x, NO_x, NMVOC, CH_4, NH_3 and CO). The 1991

workshop agreed to recommend that a task force on emission inventories be established to review present emission inventories and reporting procedures for the purpose of further improvement and harmonisation, and the EMEP Steering Body should approve the guidelines prepared by the workshop for estimation and reporting for submission to CLRTAP. These guidelines included a recommendation that emission data should be reported as totals and at least for the 11 major sources detailed in the CORINAIR 1990 inventory. The task force had the objective to provide a technical forum to discuss, exchange information and harmonise emission inventories including emission factors, methodologies and guidelines; conduct in-depth evaluation of emission factors and methodologies in current operation and to co-operate with other international organisations working on emission inventories with the aim of harmonising methodologies and avoiding duplication of work. Subsequent meetings of the task force agreed an EMEP/CORINAIR Emission Inventory Guidebook.

- IPCC/OECD Greenhouse Gas Emissions Programme: The OECD, in 1991, held a workshop on greenhouse gas emission inventory methodology to consider the OECD report 'Estimation of Greenhouse Gas Emissions and Sinks' (Background Report). The workshop produced a consensus on a basic methodology document as the most suitable starting point for work on consistent national emission estimates and a proposed plan for a two year programme of work to improve and disseminate the inventory methodology. IPCC subsequently adopted the Work Programme to be carried out by IPCC Working Group 1 (Chapter 3) with support from OECD and the European Environment Agency.

  The Work Programme prepared Draft Guidelines for National Greenhouse Gas Inventories. The Guidelines were revised through a series of expert workshops on agricultural soils, waste, new gases/industrial processes, land use change and fuel combustion followed by a formal review process. This resulted in the 'Revised 1996 IPCC Guidelines for National Greenhouse Gas Inventories'.

  The Guidelines cover the main sources of the three major greenhouse gases (CO_2, CH_4, N_2O) and three additional groups of greenhouse gases (HFCs, PFCs, SF6). They also prompt for emission estimates from three O_3 precursors (NO_x, CO, NMVOC). Furthermore, it is likely that information may be requested on SO_2 and NH_3 (which are important in the formation of aerosols and hence cloud formation which may have a negative effect on global warming) and other greenhouse gases and precursors. The IPCC Guidelines specify six main sectors for reporting emissions and include all energy (combustion and fugitive); industrial processes; solvent and other product use; agriculture; land use change and forestry and waste.

The OECD has developed a guidance document for governments who are considering establishing a national pollutant release and transfer register (PRTR). The Guidance Manual for Governments, published in 1996 (OECD/GD(96)32), was developed through a series of workshops that addressed the key factors countries should consider when developing a PRTR. The OECD and the Environment Agency of Japan hosted an international conference on PRTR (1998) to take stock of the progress and status of PRTRs worldwide and to discuss future directions for its use and design. The conference recommended that OECD countries should continue to set the example in implementing PRTRs and take the lead in sharing their experiences; that OECD should review its PRTR Guidance Manual for Governments and identify areas where supplementary policy and technical guidance might be needed to better share methodologies for estimating pollutant releases, verifying the data, standardising reports and comparing PRTR data across borders and using PRTRs to indicate cleaner technology and technology transfer opportunities; that international organisations should work together to identify how a PRTR could be used to monitor commitments set forth in international environmental agreements; and that all countries without PRTRs should consider the initiation of a national system.

4.4 TYPES OF EMISSION RELEASE AND SOURCES

In any emission inventory there is a necessity to define the type of emission release as well as the type of emission source. Categories of emission release include normal/controlled, abnormal, fugitive or accidental (Environmental Analysis Co-operative, 1999). Normal or controlled releases are emissions that arise from a process running under normal operating conditions. Abnormal releases are other emissions that may occur during operation of a process that is likely to exceed normal emission release rates for a short time (e.g. discharges from safety control devices, such as emergency release valves). The term may also be appropriate for other infrequent releases, which are outside the normal releases, such as process start-up and shutdown and routine maintenance. Fugitive emissions are mainly associated with losses of gases or vapours to air from joints, valves, pipes, etc. in equipment handling volatile substances under pressure. Breathing losses from storage tanks (Chapter 10, Case Study 12) are also classified as fugitive emissions. While each point source may be a tiny release, a complex industrial source may have hundreds of such sources resulting in a significant emission rate when the process is operating normally. Finally, accidental releases are leaks or spills that may escape in uncontrolled ways into the environment. They usually result from equipment failure, operating errors, etc. Table 4.1 shows the various ways that releases to air may occur.

The types of emission sources (i.e. morphology) commonly include line, area and point. Line source emissions are releases arising from transport-related activities taking place along lines of route, such as roads, shipping lanes, aircraft flight paths or railways. Point source emissions are emissions arising from activities at a fixed point. Area source emissions are releases arising from small scale diffuse activities, for which data is usually only available on a small area, as opposed to a site-specific, basis. Area sources include domestic and landfill sources. It is not possible to measure emissions from all of the individual examples of these sources or, in the short-term, from all the different source types. In practice, atmospheric emissions are estimated on the basis of measurements made at selected or representative samples of the (main) sources and source types. Although emission inventories may contain data on the three principal types of sources, in some inventories the data may be transferred on to an area format basis (e.g. regional, country, sub-region, etc) (Case Study 3, Section 10.4).

Table 4.1 Categories of releases to air (Environmental Analysis Co-operative, 1999)

Disposal to another process	*Normal releases*	*Abnormal releases, fugitive emissions, start-up, shut-down, maintenance*	*Accidental releases - minor*	*Accidental releases - major*
Landfill, incinerator or other external treatment	Emissions to air	Flange and gland losses, tank breathing losses, releases and wastes during maintenance	Minor leaks and spills	Loss of containment, fires, etc.
	Planned	Foreseeable	Unforeseeable	Unforeseeable

4.4.1 Industrial emissions

Studies undertaken by the European Environment Agency (1997) show that power generation is a principal source of SO_2, NO_x and CO_2 (Table 4.2). Emissions of SO_2 and NO_x are particularly important at the regional and transboundary level, as they are the precursors to the formation of acid deposition (Chapter 3, Section 3.8.1). Emissions of these pollutants arise from the combustion of fossil fuels in LCP. In Europe there are statutory requirements by Member States to ensure that emissions from these activities make an appropriate contribution to the reductions in emissions of these pollutants required under various international treaties and European legislation (Chapter 7, Section 7.9.4). The most important of these are the UNECE Second Sulphur Protocol and the LCP Directive (88/609/EEC). Stricter operating practices and the use of modern pollution abatement techniques have resulted in a considerable reduction in emissions from power stations. Nevertheless high concentrations still occur in many countries, for example, Eastern European countries, particularly from older power stations and from the use of high sulphur lignite or brown coal.

At a national level industrial emissions can make a significant contribution to total emissions. For example, in Scotland it is estimated that LCP account for around 90% of SO_2 emissions and less than 50% NO_x emissions (SEPA, 2000). The principal source of NO_x emissions are motor vehicles. Overall, SO_2 and NO_x emissions from LCP were considerably less in 1998 than in 1996 with 20% and 10% reductions in SO_2 and NO_x emissions respectively (Table 4.3) (SEPA, 2000). Table 4.4 also shows the principal emissions from coal combustion in the UK (DETR, 2000b). A review of one of Scotland's major power stations (Longannet),

Table 4.2 Percentage (%) contribution of industrial pollution sources for European countries (European Environment Agency, 2000a)

Pollutant	*Power generation*	*Other industry and waste disposal*
SO_2	60.2	24.9
NO_x	19.5	12.8
NMVOC	0.4	31.3
NH_3	0.1	2.4
N_2O	6.1	36.6
CO_2	33.0	24.2
CO	0.8	16.7
CH_4	0.2	52.1

Table 4.3 Emissions of SO_2 and NO_x from large combustion plant in Scotland (kt/a) (SEPA, 2000)

Pollutant	*Source*	*1996*	*1997*	*1998*
SO_2	Electricity supply industry	99.1	90.5	77.0
	Refineries	11.3	14.2	10.0
	Other industries	2.5	3.0	2.2
NO_x	Electricity supply industry	44.7	38.9	38.3
	Refineries	2.9	3.4	3.4
	Other industries	1.5	2.1	1.8

Table 4.4 Contribution (%) to UK emissions from road transport and coal combustion (DETR, 2000b)

Pollutant	*Coal*	*Transport*	*Pollutant*	*Coal*	*Transport*
HCl	99	0	BaP	20	26
1,3-butadiene	0	91	CO_2	23	21
Black smoke	27	55	Cr	42	1
CO	6	73	Se	34	0
SO_2	72	1	PAH	24	8
Benzene	2	65	Zn	7	25
NO_x	18	46	Hg	31	0
Pb	6	57	VOC	1	27
PM_{10}	30	24	Ni	21	1
Cu	46	1	Cd	9	4

demonstrates how SO_2 and NO_x emissions are being reduced significantly through upgrading programmes to meet legislative requirements (SEPA, 2000). Longannet Power Station lies on the northern bank of the Forth estuary close to Kincardine Bridge. The coal-fired power station comprises four 600 MWe units. Total emissions for the plant for 1998/99 were 48.8 and 19.2 kt/a for SO_2 and NO_x respectively and are the most significant power station emissions (Table 4.5) (SEPA, 1999a; 2000). The current sulphur levels in coal, approximately 0.3%-0.6%, presently limit further SO_2 emission reductions. Nevertheless, effective reductions in NO_x and particulate emissions have occurred. The introduction of low NO_x burners has reduced emission concentrations from about 1000-1100 mg/Nm^3 to approximately 500 - 600 mg/Nm^3. Gas re-burn facilities in one unit have further reduced concentrations to 250-300 mg/Nm^3. The most noticeable improvement in emissions, however, has been for particulates. The refurbishment of the electrostatic precipitators (Chapter 8) at the plant in the early 1990s reduced emissions to <50 mg/Nm^3 from more than 800 mg/Nm^3 (SEPA, 1999a; 2000).

Table 4.5 SO_2 and NO_x emissions from principal power stations in Scotland (kt/a) (SEPA, 2000)

Power station	NO_x	SO_2
Longannet	19.2	48.8
Cockenzie	10.2	19.2
Peterhead	8.5	3.1

Table 4.6 Principal industrial emission sources and associated pollutants (Scottish Office, 1998)

Industrial sector	*Sub-section*	*Pollutants emitted*
Fuel production and combustion processes	Gasification, carbonisation and combustion processes	NO_2, SO_2, CO, PM_{10}, VOCs, Pb
	Petroleum processes	PM_{10}, benzene, 1,3-butadiene, VOCs
Metal production and processing	Iron and steel	PM_{10}, Pb
	Non-ferrous	Pb, PM_{10}, SO_2, VOCs
Mineral industries	All sections	NO_2, SO_2, CO, PM_{10}, VOCs, Pb
Chemical industries	All sections	NO_2, SO_2, CO, PM_{10}, VOCs,
Waste disposal and recycling	Incineration and the production of fuel from waste	NO_2, SO_2, CO, PM_{10}, VOCs, Pb
	Recovery	VOCs
Other industries	Pulp and paper	NO_2, SO_2, PM_{10}, VOCs
	Di-isocyanate processes	PM_{10}
	Tar and bitumen	VOCs, PM_{10}
	Coating processes and printing	VOCs, CO, NO_2, PM_{10}
	Manufacture of dyestuffs and coatings	VOCs, CO, NO_2, PM_{10}
	Timber	VOCs, PM_{10}
	Rubber	VOCs, CO, NO_2, PM_{10}
	Animal and vegetable matter	SO_2, NO_2, PM_{10}

Although power stations are a significant source of industrial pollution in many countries, all industry can be a significant local source of a wide range of air pollutants. Table 4.6 shows the sources of principal industrial emissions and their respective pollutants.

4.4.2 Domestic emissions

Table 4.7 shows the percentage contribution of domestic sources to pollution emissions in Europe (European Environment Agency, 2000a). Prior to 1960, the domestic use of coal was the principal source of particles and contributed to elevated concentrations of airborne particles in many European cities, frequently exceeding 1000 $\mu g/m^3$ and annual average concentrations of several hundred $\mu g/m^3$ were commonplace (Chapter 2, Section 3.2.1). Today, annual average concentrations in most western European cities have fallen to less than 30 $\mu g/m^3$. In Eastern Europe much higher concentrations still occur as, to a lesser extent, they do in southern Europe. Lignite is an important source of particles in many parts of Eastern Europe (European Environment Agency, 1997). It is probably the poorest quality in terms of calorific value and contains 67% carbon (compared with the 95% in anthracite) and burns easily, though inefficiently, on an open fire. European countries like the UK switched to smokeless fuels in the 1950-60s. The greater use of renewable energy (wind, solar, tidal wave, etc.) and enhanced energy efficiency measures in homes and offices in the future will also improve air quality.

Table 4.7 Percentage (%) contribution of domestic sources to European air pollution emissions (European Environment Agency, 2000a)

Pollutant	*Percentage contribution*	*Pollutant*	*Percentage contribution*
SO_2	6.5	N_2O	2.4
NO_x	3.4	CO_2	14.9
NMVOC	6.6	CO	9.9
NH_3	0.0	CH_4	1.3

Table 4.8 Percentage (%) contribution of agricultural sources to European air pollution emissions (European Environment Agency, 2000a)

Pollutant	*Percentage contribution*	*Pollutant*	*Percentage contribution*
SO_2	0.0	N_2O	48.4
NO_x	0.3	CO_2	1.0
NMVOC	16.4	CO	0.8
NH_3	96.5	CH_4	45.2

4.4.3 Agricultural emissions

Agricultural practices can be a significant source of nuisance, contributing both to local levels of air pollution and causing odour problems. The main emission sources are the burning of agricultural waste or crops in the field and large intensive livestock units. Sources such as soil type and fertilisation, the nitrogen in the dung and urine of grazing cattle contributes 20-40% of N_2O emissions from agricultural land. Cattle and other ruminants also emit CH_4. Agricultural practices are also a significant emitter of NH_3. N_2O and CH_4 are both greenhouse gases. Table 4.8 shows the percentage contribution of agriculture to pollution emissions for European countries. UK studies (Ministry of Agriculture, Food and Fisheries (MAFF), 1992) have shown that agriculture is an important source of greenhouse gas emissions. Agriculture produces about 1% of UK CO_2 emissions, 30% of CH_4 and 40-50% of N_2O. Findings are comparable to those of the European Environment Agency (2000a).

4.4.4 Motor vehicle emissions

Air quality data and emissions inventories show that major air pollution concerns relate to urban areas. Motor vehicles are, for the most part, the principal source of air pollution although localised industrial processes may also be a concern. The air emitted from motor vehicles can cause a twofold problem - primary and secondary pollution. Petrol and diesel engine motor vehicles emit a wide variety of pollutants, principally CO, NO_x, VOCs and particulates. Whilst improvements in motor exhaust emission controls and fuel technology have resulted in an improvement in air quality (e.g. lead in air (Fig. 8.9) and CO (Fig. 4.1)) concerns still persist about the elevated levels of air pollution encountered in urban areas; including the occurrence of photochemical smogs (or hazes).

The form, type and the total mass of motor vehicle emissions are a function of a series of variables including fuel type, vehicle age, size and type of engine, speed, driving habits, etc. Each variable will have a discernible effect on the resultant pollutant emission. National policies and local actions to ameliorate air quality require accurate projections of emissions in order to evaluate their effectiveness. This requires using best available information on traffic growth, fleet composition changes and the impact of new technologies and legislation on vehicle emissions (Murrells, 2000) (Table 4.9). Table 4.4 shows the 1998 UK emissions from road

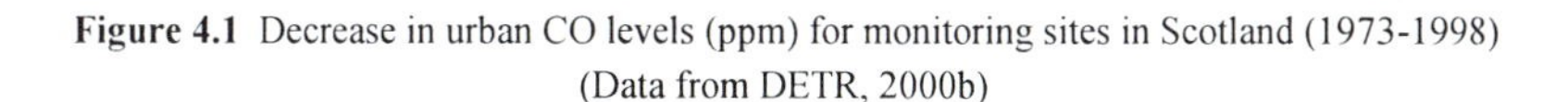
Figure 4.1 Decrease in urban CO levels (ppm) for monitoring sites in Scotland (1973-1998) (Data from DETR, 2000b)

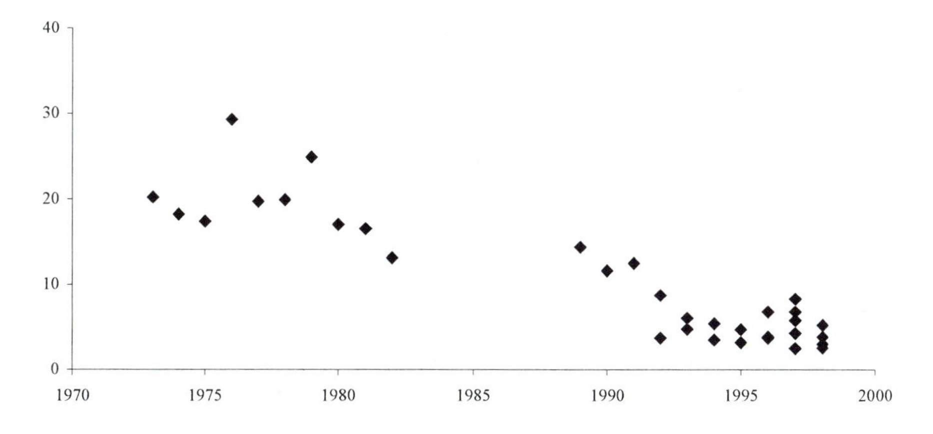

Table 4.9 UK average emission factors (g/km) of different vehicle type and pollutant (1997 base emission factors) (AEA Technology, 2000b)

	NO_x				CO				PM_{10}			
Standard	*Urban*	*Rural single carriageway*	*Rural dual carriageway*	*Motorway*	*Urban*	*Rural single carriageway*	*Rural dual carriageway*	*Motorway*	*Urban*	*Rural single carriageway*	*Rural dual carriageway*	*Motorway*
Petrol cars												
Pre-ECE	1.99	2.30	2.59	2.85	29.4	22.3	17.8	17.2	-	-	-	-
ECE 15.00	1.99	2.30	2.59	2.85	20.5	14.5	15.2	22.9	-	-	-	-
ECE15.01	1.99	2.30	2.59	2.85	20.5	14.5	15.2	22.9	-	-	-	-
ECE 15.02	1.70	1.98	2.52	3.69	17.2	12.1	7.70	10.0	-	-	-	-
ECE 15.03	1.57	1.74	2.07	2.85	18.3	11.7	7.65	9.58	-	-	-	-
ECE 15.04	1.54	1.83	2.33	3.31	11.4	7.81	6.10	8.33	-	-	-	-
Euro 1	0.292	0.279	0.332	0.622	1.68	0.97	1.42	4.60	-	-	-	-
Euro 2	0.129	0.123	0.146	0.274	1.17	0.68	0.99	3.22	-	-	-	-
Diesel cars												
Pre Euro 1	0.653	0.575	0.577	0.757	0.760	0.545	0.437	0.429	0.180	0.151	0.146	0.195
Euro 1	0.451	0.291	0.210	0.310	0.451	0.295	0.214	0.301	0.045	0.026	0.024	0.062
Euro 2	0.325	0.209	0.151	0.223	0.316	0.207	0.150	0.211	0.027	0.016	0.015	0.037
Petrol LGV												
Pre Euro 1	1.76	1.97	2.39	3.32	17.2	9.34	9.64	39.0	-	-	-	-
Euro 1	0.326	0.339	0.414	0.648	3.42	1.71	1.47	4.70	-	-	-	-
Euro 2	0.143	0.149	0.182	0.285	2.40	1.19	1.03	3.29	-	-	-	-

Diesel LGV												
Pre Euro 1	1.26	1.14	1.17	1.57	1.192	0.910	0.845	1.416	0.378	0.310	0.299	0.458
Euro 1	0.618	0.399	0.288	0.423	0.916	0.658	0.560	0.829	0.119	0.084	0.083	0.161
Euro 2	0.469	0.303	0.218	0.321	0.641	0.461	0.392	0.581	0.072	0.050	0.050	0.097
HGV rigid												
Old	11.80	14.40	14.40	14.40	6.00	2.90	2.90	2.90	1.593	0.982	0.899	0.899
Pre Euro 1	9.84	6.10	5.53	5.16	4.32	2.65	2.45	2.04	0.779	0.470	0.427	0.352
Euro 1	6.76	4.41	3.97	3.62	2.42	1.73	1.54	1.05	0.512	0.313	0.279	0.191
Euro 2	4.86	3.41	3.07	2.71	1.95	1.53	1.36	0.99	0.270	0.164	0.147	0.101
HGV artic												
Old	18.20	24.10	24.10	19.80	7.30	3.70	3.70	3.10	1.322	0.792	0.618	0.618
Pre Euro 1	26.19	17.20	15.74	12.20	5.00	3.08	2.83	2.44	1.128	0.694	0.628	0.472
Euro 1	13.73	10.17	9.30	6.33	2.56	1.73	1.54	1.16	0.775	0.479	0.428	0.296
Euro 2	9.99	7.63	6.97	5.18	2.09	1.44	1.29	1.16	0.298	0.184	0.164	0.114
Buses												
Old	16.20	14.80	14.80	13.50	18.80	7.30	7.30	1.76	1.392	1.218	1.218	0.618
Pre Euro 1	12.22	7.89	6.93	5.45	7.15	3.89	3.50	3.50	0.738	0.425	0.378	0.268
Euro 1	13.21	6.71	6.04	4.31	3.17	1.73	1.51	1.05	0.490	0.346	0.307	0.208
Euro 2	9.43	5.03	4.53	3.53	2.54	1.44	1.26	1.05	0.302	0.133	0.118	0.080
Motorcycles												
<50cc	0.03	0.03	0.03	0.03	18.6	18.6	18.6	18.6	0.04	0.04	0.04	0.04
>50cc 2st	0.03	0.03	0.03	0.03	23.1	23.1	23.1	23.1	0.04	0.04	0.04	0.04
>50cc 4st	0.18	0.18	0.18	0.18	18.9	18.9	18.9	18.9	0.12	0.12	0.12	0.12

- not applicable

transport (DETR, 2000b). Principal emissions are shown for 1,3-butadiene (91%), CO (73%), black smoke (55%), benzene (65%), Pb (57%) and NO_x (46%). Case Study 3 (Chapter 10, Section 10.4) details the contribution of motor vehicle emissions to total pollution emissions levels for a Local Authority area (Falkirk).

4.4.5 Aircraft emissions

The main pollutants emitted from aircraft activities are CO, NO_x, hydrocarbons, particulates and odours (Harrop and Daunton, 1987). On a national scale, aircraft emissions may be small compared to other major emission categories. For example, it has been estimated that in the US aircraft engine emissions contribute approximately 1% to national emissions of CO, NO_x and hydrocarbons (Segal and Yamartino, 1981) and up to 3% of total annual emissions on a regional scale (Cirillo, Tschanz and Camaioni, 1975; Jordon, 1977). This percentage may increase where the airport is situated in a rural environment (Yamartino *et al.*, 1980).

4.5 INFORMATION REQUIREMENTS

Emission inventories are essentially an information gathering and data processing exercise. Eggleston (2000) has identified information requirements and principal procedural steps in compiling an emission inventory. These broad steps and data requirements are detailed in Table 4.10. The WHO (1982a) has also previously devised a rapid assessment methodology to compile an emission inventory. The methodology comprises instructions from project initiation to the preparation of the final report. The methodology stresses the need to use information available to crosscheck and improve the accuracy of estimates. The rapid assessment procedure is designed to produce only a preliminary assessment. The fact that every study area is unique in many respects makes advance detailed step-by-step planning difficult. Therefore the study should have the flexibility to collect whatever data and information is felt appropriate and plan visits to data sources and industries, if necessary, as they deem necessary. It is also important that both public health officials and project team members understand the rapid assessment procedure, including its managerial and administrative support requirements. The following sections briefly describe the general points to be considered in making assessment surveys. The procedure consists of the following steps (also see Table 4.10):

- Define area: The definition of the study area depends upon the selection of appropriate boundaries. The boundaries can be physical (such as drainage basins, ridges, or canyons), political (such as city limits or state/provincial limits or even international boundaries), or economic (such as industrial zones or economic development or planning areas).
- Project personnel and support: Rapid assessment studies can be undertaken with as few as one or two properly qualified personnel. If larger groups are to be involved, a project leader should be designated and assigned overall

responsibility. There is a need for close co-operation between the team members.

Table 4.10 Data requirements and procedural steps for compiling an emission inventory

Action	*Description*
Area	Identify extent of area boundaries and sub-divide into smaller units (e.g. 1 km^2) where appropriate.
Study period	Define the period of emission (e.g. 1 year)
Pollutants	Identify pollutants: Generally criteria pollutants (e.g. PM_{10}, SO_2, NO_x, CO, etc.) or pollutants where there is readily available data (Group 1). However, a second group of pollutants should be considered following completion of Group 1 for pollutants where data is not so readily available (e.g. VOC, PAH, benzene, Cd, secondary pollutants (e.g. O_3, PAN, etc.) (Group 2).
Source type	Specific source categories (e.g. industry, domestic and other area sources, traffic)
Data collection	• Fuel consumption (e.g. types of fuel, quantities used by process type (transport, domestic, industry, etc); • Traffic activity (e.g. vehicle class, traffic speed, traffic flows, etc); • Industrial sources (e.g. type, location, production, emissions, emission conditions (stack height, temperature, etc), and so on); • Other sources (e.g. refuse burning; domestic, institutional and commercial combustion, etc); • Other transport (e.g. aircraft, trains, etc); • Population data (e.g. geographical distribution, size, etc); • Emission factors (e.g. amount of pollutant emitted per unit of production, per unit input, per kilometre driven, per unit of fuel burnt, etc); • Consultation.

Construct inventory	Establish a list of sources:
	  • point (stationary) (e.g. waste combustion, electricity combustion, industrial fuel combustion, commercial fuel combustion, domestic fuel combustion, petroleum and solvent sources, other industrial sources, etc)
	  • area (e.g. industrial units, residential areas, etc)
	  • line (e.g. motor vehicles, off-road, aircraft, railways, marine/waterways)
	  • fugitive sources
	• Collect data for each source
	• Compile activity data for area and line sources
	• Review data for suitability for inventory and for developing strategies
	• Process individual source and activity data to provide a spatially disaggregated inventory, etc

Review data sources	Elicit data from: • pollution control agencies • plant operators • regional and local planning authority • census authorities • trade organisations • fuel suppliers • highway authorities • transport operators, etc
Emission factors	Elicit data from measurements or from derived factors (e.g. USEPA-42, etc)
Archive data	Store data on data retrieval system
Audit	Review collected data.

Proper authorisation for access to data and information should be obtained from all government departments and industries from which data are to be collected. A level of co-operation between the project team and those providing the data (government offices, industries, business groups, etc.) is encouraged. Such co-operation will facilitate data collection, improve the completeness and accuracy of the study, and lay the groundwork for future development of pollution control activities. In addition, some administrative support may also be required.

- Data collection should involve:
  - Identification of the type and size of pollution source in study area;
  - Identification of the location of pollution sources in relation to major population centres;
  - Subdivision of the study area and identification of the potential principal sources of pollution;
  - Identification of the data required for establishing pollution factors;
  - Utilisation of the pollution factors to calculate the pollution loads, and indication of the nature of the data required;
  - Determination of those government departments or interested parties that have the data required to complete compilation of the pollution factors and loads;
  - Cross-checking of the data collected from various sources and verification, to the extent possible, of their accuracy. Identification of data of questionable accuracy. If any assumptions are made to complete the data, these should be clearly indicated.
- Reporting: In addition to the findings of the report, the following information may also be included:
  - An interpretation of the environmental impact of estimated emissions in relation to the supplementary geographical, meteorological, environmental quality, and monitoring data collected;
  - A cursory assessment of the impact of major emissions on the population and on valuable natural resources;
  - A summary of possible areas where environmental control measures will be most effective;
  - An assessment of the effectiveness of the existing pollution control programmes and recommendations for improvements as appropriate.

In order to allow emissions from diverse topologies to be aggregated and manipulated, a system of regular grids (usually 1 km^2) is conventionally adopted as the lowest common denominator topology. Using the capabilities of geographical information systems (GIS) software, emissions from line, point and area sources can be allocated to a square in which they fall (Fig. 4.2) (DETR, 2000a).

An example of a screening emission inventory study is that required by the UK Government of individual Local Authorities. The UK Environment Act (1995) requires Local Authorities to assess the air quality in their locality. As part of this review and assessment process it is necessary to estimate the magnitude of polluting emissions to air. Generally in only a small number of cases is it possible to obtain a direct continuous measurement of source emissions. Therefore emission calculations have to rely largely on estimation techniques. As noted (Section 4.1), estimated emission techniques applied by Local Authorities are basically a two-stage process. Emissions are usually proportional to the scale or intensity of an emitting activity (e.g. the activity rate). The first step, therefore, is to quantify the activity rate of the emission source. The data is then related to emission factors, which are empirically derived indices describing the relationship between the type and extent of the activity and the nature and quantity of emissions released (DETR, 2000a). Chapter 10 (Section 10.4) provides a Case Study of a UK Local Authority emission inventory study.

Figure 4.2 A conceptual model of an emissions inventory system (DETR, 2000a) (Crown copyright is reproduced with the permission of the Controller of Her Majesty's Stationery Office)

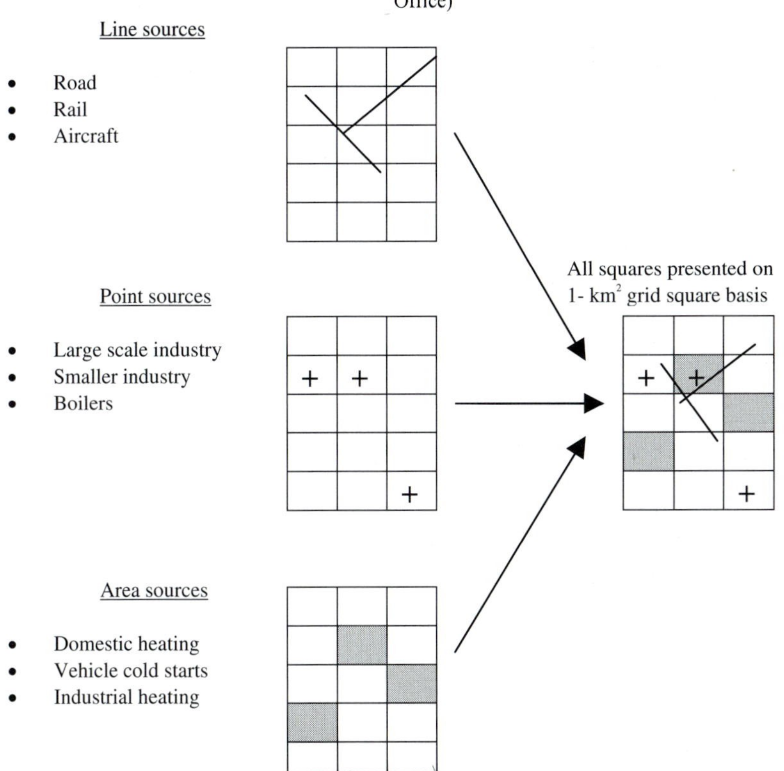

4.6 EXAMPLES OF NATIONAL EMISSION INVENTORIES

Many countries have developed emission inventory databases as part of their own or intergovernmental air quality management programmes. Examples of four national and international emission inventories are given below (UK, European, US and Canada).

4.6.1 UK national atmospheric emissions inventory

The UK National Atmospheric Emissions Inventory (NAEI) has been developed to cover the whole of the UK (Goodwin *et al.*, 1999). The NAEI is a top-down inventory where total emissions for each pollutant are derived from national data and then attributed to 1 km grid square on the basis of population density (major large emission sources i.e. major roads and industrial processes are entered individually). Whilst this type of emission inventory is useful on a small scale basis, it lacks detailed local information (Goodwin *et al.*, 1999). Fig. 4.3 compares the emission levels for CO, NMVOC, NO_x and SO_2 for UK countries. The NAEI provides the UK emission data to various interested parties such as the UNECE for assessment and air dispersion modelling purposes (Section 4.2).

Figure 4.3 Pollutant emissions for UK (Data from NAEI, 1999)

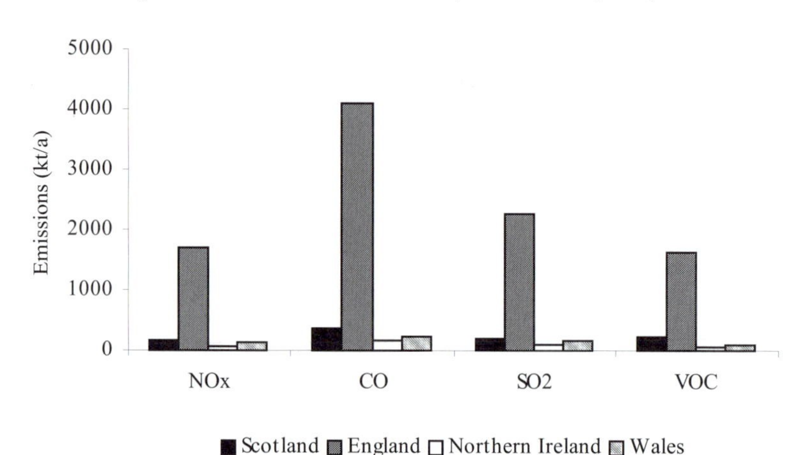

Table 4.11 Contribution of transport emissions (%) to total pollutant emissions for Glasgow and Edinburgh, Scotland (SEPA, 2000; London Research Centre, 1998)

Pollutant	*Edinburgh*	*Glasgow*
NO_x	96.0	76
SO_2	94.0	28
CO	99.8	95
CO_2	-	29
NMVOC	-	56
Benzene	98.0	92
1,3-Butadiene	100.0	98
PM_{10}	62.4	73

- no data

To gain an appreciation of localised emission sources the UK DETR has funded the compilation of more detailed inventories in a number of urban and industrial areas. These inventories are compiled on a bottom-up principle using local detailed inventory data and addresses the limitations of the NAEI. One such inventory was for the City of Glasgow (London Research Centre, 1998). The inventory showed that within the City of Glasgow (owing to the motorways) motor vehicles were the most dominant pollutant source contributing more than 76%, 95%, 92%, 98% and 73% of NO_x, CO, benzene, 1,3-butadiene and PM_{10} emissions respectively (Table 4.11). The contribution of industry to total emission levels is less of a concern in Glasgow than for other UK urban areas. Commercial, residential and industrial combustion sources accounted for 64% and 66% of SO_2 and CO_2 emissions respectively. Industrial processes and solvent use accounted for 36% of NMVOC emissions (SEPA, 2000). Similar comparisons were also found for emission inventory studies undertaken by the City of Edinburgh Council (City of Edinburgh Council, 1999) although a higher proportion of SO_2 emissions were estimated to be from the transport sector. Emissions for Glasgow and Edinburgh accounted for approximately 32%, 1% and 25% of NO_x, SO_2 and CO of total Scottish emissions respectively (Table 4.12).

Table 4.12 Total emissions (t/a) of pollutants for Glasgow and Edinburgh, Scotland (SEPA, 2000)

Pollutant	*Edinburgh*	*Glasgow*
NO_x	10790	11894
SO_2	601	1798
CO	49816	36726
CO_2	-	2721000
NMVOC	-	13514
Benzene	291	308
1,3-Butadiene	88	70
PM_{10}	912.4	1023

- no data

4.6.2 European emission inventories

In 1994, as part of the European Environment Agency's first work plan, European Topic Centres (ETC) were designated to address inland waters, marine and coastal environment, air quality, nature conservation and air emissions respectively for a first 3 yearly work period. This was extended for a second period in 1998. The ETC on Air Emissions (ETC/AE) is led by the German supported Umweltbundesamt and a consortium of partners. The purpose of the ETC/AE is to support the national experts of European Environment Agency countries and ensure the delivery of high quality, reliable, comparable and timely air emissions data in the context of the European Environment Agency's mandate, objectives and work programme. Air emissions data are in particular required by EU legislation and the various international conventions and protocols (Section 4.2). ETC/AE also contributes to the production of the European Environment Agency's state of the environment reports, where air emission estimates are required for climate change, O_3 depletion, acidification, tropospheric O_3, dispersion of hazardous substances and urban air quality.

The main aim of the ETC/AE work programme is to set up an annual European air emission inventory system (CORINAIR), including collecting, managing, maintaining and publishing the information, based on official national inventories. ETC/AE also assists participating countries to report their national emission inventories to the various international obligations mainly by providing

Figure 4.4 Comparison of pollutant emission for selected European countries 1994 (Data from NAEI, 1999; Barret and Berge, 1996a; 1996b)

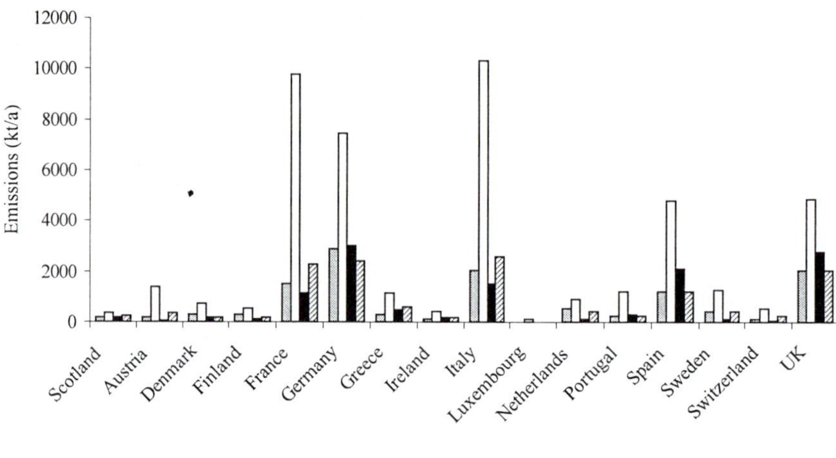

software (CollectER, Collect Emission Register, and ReportER, 1998) and organising regular workshops. The software system makes use of the SNAP97 source nomenclature. In addition, a software package to estimate national emissions from road transport is also available (COPERT2, Computer Programme for estimating Emissions from Road Transport) to participating countries. This was revised in 1999 (COPERT3). COPERT3 includes most results from the DGVII (Directorate for Transport) programmes COST 319, the Estimation of Emissions from Transport, and MEET (Methodologies to Estimate Emissions from Transport). Fig. 4.4 shows the European estimated emission levels of NO_x, CO, SO_2 and VOCs (also see Table 4.17).

4.6.3 Canadian national emission inventory

The 1995 National Emissions Inventory of Criteria Air Contaminants was compiled by Environment Canada and the provincial/ territorial ministries of the environment and energy, through the National Emissions Inventory and Projections Task Group (NEIPTG) of the National Air Issues Co-ordinating Committee (NAICC). The emission inventory contains estimates for more than 60 industrial and non-industrial activities, which were based on activity statistics and plant specific information.

The update to the national emissions inventory contained estimates for PM_{10} and $PM_{2.5}$, which were not previously estimated. The update contained numerous improvements over previous inventories using the latest methodologies and technical information. The emission estimates were improved for numerous sectors such as transportation vehicles (on-road and off-road), the upstream oil and gas industry, mining industry, the residential fuel-wood combustion, road dust, construction activities, forest fires, and the domestic and commercial uses of solvents. Additional improvements were also made in the models used for the spatial disaggregation of the emissions, which is important for the provincial and regional reporting of the emissions, and the various dispersion models used across Canada and in the US. A total of 1100 new industrial sources were added to the emission inventory, bringing the total number of facilities to 4600.

The National Pollutant Release Inventory (NPRI) provides information on 176 listed substances, specifically on their on-site releases to air, water, land and underground injection; off-site transfers in waste; and off-site transfers for recovery, reuse and recycling, and energy recovery. The NPRI is the only legislated, nationwide, publicly accessible inventory of its type. One of the fundamental aspects of the NPRI is to provide Canadians with access to pollutant release information for facilities located in their communities. The NPRI also supports governments and others to identify priorities for action, encourages industry to take voluntary measures to reduce releases, allows tracking of progress in reducing releases, and supports a number of regulatory initiatives across Canada. Table 4.13 details the 1995 emissions for Canada by major source category.

Table 4.13 Emissions (thousand short tons) for Canada (1995) by major source category (Environment Canada, 1999)

Source	*CO*	NO_x	*VOC*	SO_2	*TP*	PM_{10}	$PM_{2.5}$
Industrial	2400	684	1037	2149	685	317	189
Non industrial fuel combustion	1189	367	449	624	248	197	173
Transportation	7394	1422	810	150	108	105	92
Incineration	51	3	7	1	3	2	1
Miscellaneous	16	1	606	0	24	16	10
Open source	7380	239	1033	1	16222	5920	1209

TP – total particulates

4.6.4 US national emission inventory

The USEPA have compiled an extensive inventory of emissions factors to aid the compilation of their emissions inventories. The Compilation of Air Pollutant Emission Factors, AP-42 (USEPA, 1995a), is divided into two volumes. Volume I contains information on over 200 stationary source categories and includes brief descriptions of processes used, potential sources of air emissions from the processes and in many cases common methods used to control these air emissions. Methodologies for estimating the quantity of air pollutant emissions are presented in the form of emission factors. Volume II contains information on emission factors from line sources.

Figure 4.5 Total emission levels (short tons) for the US for 1998 (Data from USEPA, 2000)

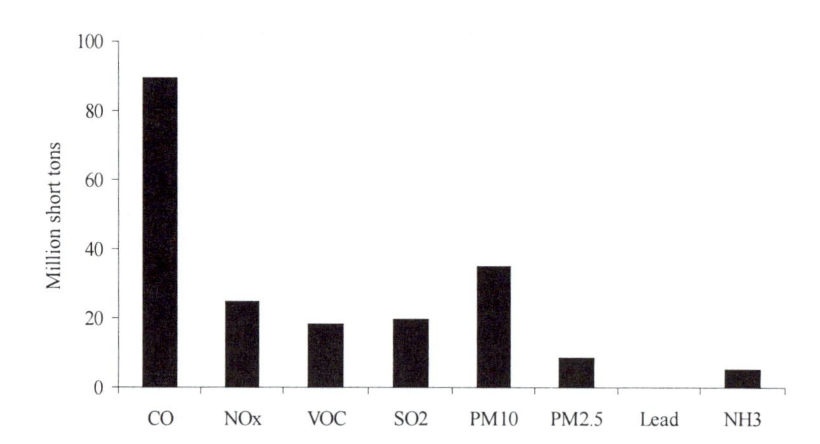

The USEPA provides emission inventory data for a series of sources. Source reports provide information about stationary and line sources and can be generated using data either from the Aerometric Information System (AIRS) (stationary sources only) or the National Emission Trends (NET) Inventory (stationary and line sources). Table 4.14 provides information of USEPA internet sites to retrieve emission inventory source reports. The Emission Factor and Inventory Group of USEPA provides a 'Clearing House for Inventories and Emission Factors (CHIEF)' internet site. The site provides access to tools for estimating air pollutant emissions in various geographic domains (e.g. urban areas, regions, or the entire nation). It serves as the USEPA's central provider for the latest information on air emission inventories and emission factors. Emission estimation databases, newsletters, announcements and guidance on performing inventories are also included in CHIEF. Section 4.8 details the US pollutant emission levels and trends. Fig. 4.5 details total emission levels for the US for 1998.

4.7 TRANSBOUNDARY EMISSIONS

Pollutant emissions are by their nature transboundary. The major air pollutants can lead to a number of environmental problems on global and regional scales (Chapter 3, Section 3.8). To be effective, therefore, AQM studies should be able to evaluate budgetary mass emission levels at a transboundary level for both marine and terrestrial emissions.

4.7.1 Terrestrial

All countries are estimated to export a sizeable fraction of their emissions. For oxidised sulphur over 75% of emissions are exported from most European countries and most countries export more than 80% of their oxidised nitrogen.

Table 4.14 USEPA internet sites to retrieve emission inventory source reports (USEPA, 2000)

	Source report types
AIRS	
Ranking	Lists each source in order of its pollutant emissions, ranking them from largest to smallest (help/hsrcrank.htmhelp/hsrcrank.htm)
Compliance	Indicates whether each source is complying with regulations governing air pollutant emissions (help/hsrccmpl.htm)
Address	The name and address of each source plus additional descriptive information (help/hsrcaddr.htmhelp/hsrcaddr.htm)
Count	The number of sources and total air pollutant emissions for each geographic area (county, state, or USEPA region) (help/hsrccnt.htmhelp/hsrccnt.htm)
SIC	The number of sources and total air pollutant emissions for each SIC(help/hsrcsic.htmhelp/hsrcsic.htm)
Year	The number of sources that submitted emissions estimates for each calendar year (indicates how recent are the data) (help/hsrcyear.htmhelp/hsrcyear.htm)
All Columns	Primarily for exporting data; includes all columns in the Ranking and Address reports (help/hsrccols.htmhelp/hsrccols.htm)
NET	
Ranking	Lists each large point source in order of its pollutant emissions, ranking them from largest to smallest (help/hnetrank.htm help/hnetrank.htm)
Tier	Summarises all NET emissions (point, area, and line) into 14 major categories (Tier 1) or into 75 more detailed categories (Tier 2) (help/hnettier.htmhelp/hnettier.htm)
SIC	The number of sources and total air pollutant emissions for each SIC (help/hnetsic.htmhelp/hnetsic.htm)
Count	The number of sources and total air pollutant emissions for each geographic area (county, state, or USEPA region) (help/hnetcnt.htmhelp/hnetcnt.htm)

SIC = Standard International Classification

Reduced nitrogen is transported less far, however, all countries export more than 40% of their emissions (Barret and Berge, 1996b). Typically across Europe reduced nitrogen shows the most rapid depletion and therefore deposition away from the emitting source/country shows the steepest gradient and restricted field of influence. This pattern is then followed by SO_x with NO_x demonstrating a lesser initial depletion or otherwise they have a lower dispersion gradient and a wider field of influence. These dispersion patterns are reflected in the overall export and import budgets of these pollutants for European countries (SEPA, 2000). Studies have estimated the transboundary budgets (import and export) of emissions for European countries (Table 4.15) (Barret and Berge, 1996b). The UK imports approximately 20% of total SO_x depositions and exports 81% of emissions. With regard to oxidised nitrogen annual import and export levels are approximately 39%

Table 4.15 European transboundary import-export budgets (%) for 1996 (EMEP, 2000)

Country	SO_x				NO_x			
	E	*I*	*To sea*	*In area*	*E*	*I*	*To sea*	*In area*
Albania	88	88	13	54	95	96	11	47
Austria	84	95	10	65	93	93	10	55
Belarus	78	81	7	69	89	89	9	64
Belgium	89	71	30	78	96	85	27	73
Bosnia Herzegovina	85	55	9	54	93	92	10	48
Bulgaria	86	44	13	49	92	79	9	44
Croatia	88	94	20	66	94	96	12	57
Cyprus	91	60	8	39	98	93	11	36
Czech Republic	86	60	17	77	93	84	16	70
Denmark	91	79	44	79	97	87	32	73
Estonia	89	80	15	74	96	96	14	68
Finland	76	88	20	73	87	81	15	66
France	75	67	29	72	82	65	24	67
Germany	73	56	21	77	82	63	22	72
Greece	90	76	24	46	93	77	16	39
Hungary	81	60	10	75	90	86	10	71
Iceland	88	88	43	48	94	96	27	36
Ireland	86	71	57	72	94	89	41	64
Italy	80	64	23	59	86	43	18	51
Latvia	86	94	17	74	95	98	14	67
Lithuania	83	85	15	72	94	95	14	67
Luxembourg	95	94	18	75	98	96	21	71
Macedonia	90	83	5	43	97	96	4	37
Moldova	89	89	8	64	96	96	7	57
Netherlands	88	85	41	80	96	83	31	73
Norway	80	96	35	70	87	88	23	58
Poland	71	46	16	75	83	67	15	69
Portugal	86	50	16	37	92	68	12	34
Romania	74	62	6	60	83	77	6	53
Russian Federation	61	63	7	53	67	67	4	47
Slovakia	85	83	9	73	94	92	10	68
Slovenia	88	77	8	53	96	94	8	45
Spain	81	28	22	47	85	50	16	42
Sweden	79	92	31	74	88	86	23	66
Switzerland	83	93	9	62	93	91	11	51
Turkey	78	77	18	48	84	58	12	39
Ukraine	69	62	9	67	80	81	9	63
Yugoslavia	78	69	7	65	88	94	6	57
Remaining Land Areas	94	72	30	52	91	92	7	34
Baltic Sea	71	96	40	78	88	98	25	71
North Sea	66	89	51	80	85	91	37	71
NE Atlantic Ocean	69	89	33	39	77	88	26	35
Mediterranean Sea	97	100	3	13	95	100	5	15
Black Sea	0	100	0	0	0	100	0	0

E – export
I - import

and 90% respectively. For reduced nitrogen they are 20% and 50% respectively. The percentage of emissions retained in the UK for oxidised sulphur and nitrogen are 75% and 68% respectively. For reduced nitrogen the value is 90% (Barret and Berge, 1996b).

The impact of transboundary emissions is particularly evident for Scotland where studies (Smith *et al.*, 1996) have shown that European sources account for 44% of the total sulphur deposition annually on the Scottish landscape. English and Welsh sources account for 33% of emissions, Northern Ireland 4% and Scottish sources up to 19%. It is possible that some of the sulphur attributed to Scottish low level sources may be from Northern Ireland or the North of England. Results indicate that at least 80% of sulphur deposited in Scotland comes from sources outside the country. Similarly 80% of sulphur emitted in Scotland is exported (Smith *et al.,* 1996). The spatial distribution of sulphur across Scotland is greatest in the central belt area (Fig. 4.6).

4.7.2 Marine

On a global scale oceans are a major source of natural sulphur accounting for ~15% of total emissions (Bates *et al.*, 1992). For the EMEP region (Europe) independent estimates suggest yearly contributions from marine sources of 600-1000 kt/a (Andres, 1993). Accounting for >90% of oceanic reduced sulphur (Cline and Bates, 1983), dimethylsulphide (DMS), produced by phytoplankton, is considered to be the principal biogenic compound contributing to atmospheric marine sulphate (Turner and Liss, 1985). Laboratory studies suggest that 66% of atmospheric DMS is oxidised to SO_2 within hours of release (Yin *et al.*, 1990).

With regard to anthropogenic emission sources studies have previously concentrated on terrestrial emissions as being the predominant emission source, but, it is now becoming increasingly recognised that they have not adequately addressed the contribution from marine sources (ship emissions) (Corbett *et al.*, 1999). Emissions of SO_2 and NO_x from ships may now represent a growing and significant proportion of the remaining emissions, other than terrestrial, contributing to acidification concerns in Europe. This concern is highlighted by the higher estimates of SO_x and NO_x that ship emissions assumed when the Second UNECE Sulphur Protocol was negotiated. Shipping is now becoming increasingly recognised as a significant source of SO_x and NO_x (Lowles and ApSimon, 1996).

Table 4.16 Total NO_x and SO_x emissions (kt/a) from international shipping in European waters

Sea area	SO_x	NO_x
NE Atlantic	641	911
North Sea	439	639
Total	1080	1550

Recent studies have found total NO_x and SO_x emissions from ships to be comparable in magnitude with the largest energy-consuming nations (Corbett *et al.*, 1999). The total emissions of SO_2 and NO_x for the North Sea and N E Atlantic Ocean are 1080 and 1550 kt/a (Tsyro and Berge, 1997) (Table 4.16). North Sea international shipping contributes 5-10% to the total NO_x deposition in the UK (Tsyro and Berge, 1997). The relative contribution for oxidised sulphur from international shipping is 2.5%. Fig. 4.7 shows a comparison of emissions from shipping on a global and regional perspective and Fig. 4.8 shows the distribution of shipping emissions around European waters.

Figure 4.6 Distribution of sulphur deposition in Scotland (Smith *et al.*, 1996) (Permission from The Centre for Ecology and Hydrology, Edinburgh)

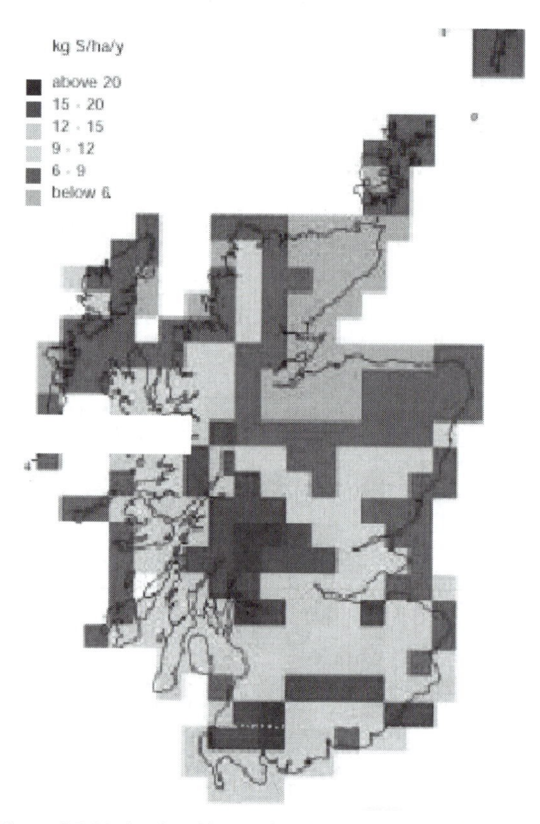

Figure 4.7 National and international emissions from shipping (Data from BMT Murray, Fenton, Edon, Liddiard, Vince, 2000)

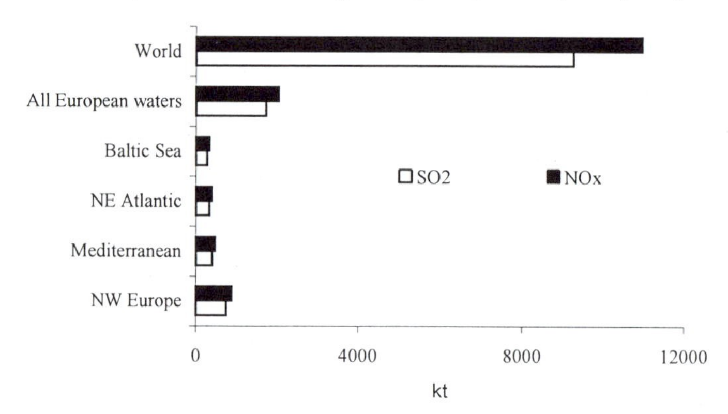

4.8 POLLUTION EMISSION TRENDS

Extensive emission inventories have lead to a greater appreciation of pollution emission trends. For example, Fig. 4.9 shows for the US that CO, VOC, SO_2 peaked in or around 1970 with a general downward trend during 1970 to 1998, whilst total NO_x emissions has risen slightly in recent years. Table 4.17 shows the temporal variation of European and Canadian emissions from 1980 to 1997.

Figure 4.8 Distribution of NOx shipping emissions (t) in the European waters (EMEP, 2000) (courtesy of EMEP/MSC-W based on calculations made at Lloyd's Register (Ref. EMEP/MSC-W Note 5/00, 'Effects of International Shipping on European Pollution Levels', by Jonson *et al.*)

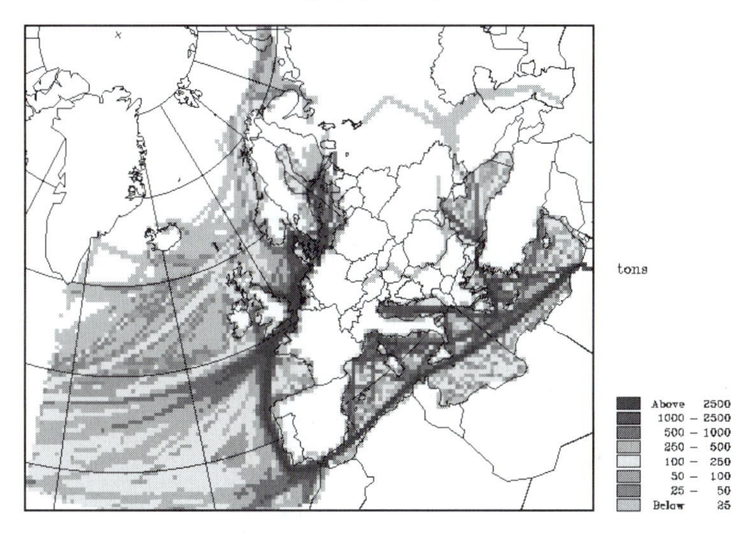

Figure 4.9 Total US national emissions (million short tons) 1940 – 1998 (USEPA, 2000) (Data from the USEPA, 2000)

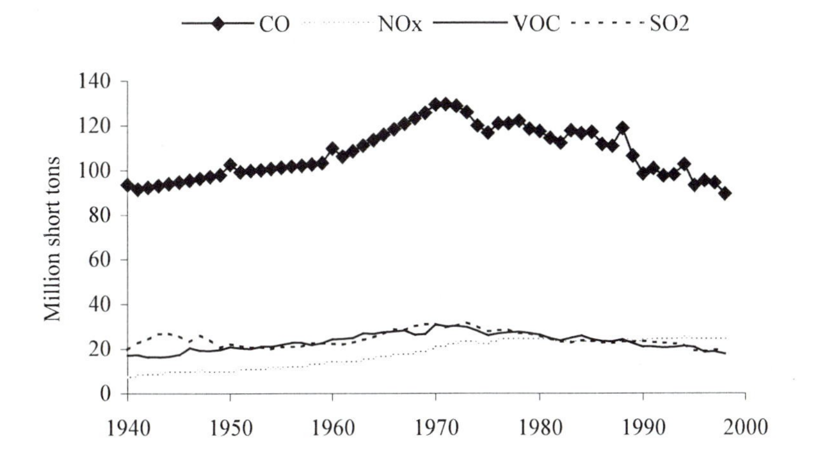

Table 4.17 SO_2 and NO_x emissions (thousands of tonnes) (1980-97) in the ECE region (EMEP, 2000)

SO_2	*1980*	*1982*	*1984*	*1986*	*1988*	*1990*	*1991*	*1992*	*1993*	*1994*	*1995*	*1996*	*1997*
Austria	400	329	218	176	115	91	83	63	60	56	52	52	57
Belarus	740	710	690	690	720	637	652	458	382	324	275	246	208
Belgium	828	694	500	377	354	372	334	318	297	253	246	241	216
Canada	4643	3612	3955	3627	3838	3236	3245	3117	3008	2651	2681	2722	-
Czech Rep.	2257	2387	2305	2177	2066	1876	1776	1538	1419	1270	1091	946	701
Denmark	450	369	295	282	242	182	243	190	156	155	150	186	109
Finland	584	484	368	331	302	260	194	141	124	112	96	105	100
France	3338	2490	1866	1342	1226	1250	1369	1201	1061	1009	958	947	-
Germany	3164	2841	2576	2228	1215	5313	3996	3299	2938	2466	2102	1543	1468
Hungary	1633	1545	1440	1362	1218	1010	913	827	762	741	705	673	657
Iceland	18	18	19	18	18	24	23	24	25	24	24	24	25
Ireland	222	158	142	162	152	178	179	161	157	177	161	147	165
Italy	3757	2850	2114	1929	1963	1651	1539	1394	1333	1271	1322	-	-
Lithuania	311	304	303	316	300	222	234	139	125	117	94	93	77
Moldova	308	287	270	297	273	265	184	126	36	23	27	19	17
Netherlands	490	404	299	264	250	202	173	172	164	146	147	135	124
Norway	137	111	96	91	68	53	44	36	35	34	34	33	30
Poland	4100	-	-	4200	4180	3210	2995	2820	2725	2605	2376	2368	2181
Russian Fed.	7161	7090	6503	5707	5145	4460	4392	3839	3456	2983	2838	2685	2449
Slovenia	234	256	250	247	210	194	181	190	183	177	119	110	120
Spain	3072	2938	2714	2456	2016	2266	2223	2195	2060	2065	1927	-	-
Sweden	491	371	296	272	224	119	96	88	87	82	79	83	69
Switzerland	116	-	84	68	56	43	41	38	34	31	34	30	26
Ukraine	3849	3427	3470	3393	3211	2782	2538	2376	2194	1715	1639	1293	1132
Yugoslavia	406	409	456	470	502	508	446	396	401	424	462	434	522
NOx													
Austria	231	221	218	217	203	194	198	188	176	184	170	163	172
Belarus	234	235	240	258	262	285	281	224	207	203	195	173	189
Belgium	442	-	-	317	345	339	335	343	341	342	336	325	310
Canada	1959	1897	1871	2043	2204	2104	2003	1997	2006	2026	2032	2011	-
Czech Rep.	937	818	844	826	858	742	725	698	574	435	412	432	423
Denmark	282	267	273	320	302	282	321	276	274	272	250	288	248
Finland	295	271	257	277	293	300	290	283	282	282	259	267	260
France	1823	1688	1632	1618	1615	1879	1941	1876	1773	1745	1726	1695	-
Germany	2617	2525	2598	2546	2306	2693	2521	2308	2151	2020	1946	1887	1803
Hungary	273	268	264	264	258	238	203	183	184	188	190	196	198
Iceland	21	21	22	22	25	26	27	28	29	29	28	30	29
Ireland	73	86	84	100	122	115	119	125	122	117	115	121	124
Italy	1638	1605	1596	1690	1854	1938	1984	2010	1990	1789	1768	-	-
Lithuania	152	156	162	169	172	158	166	98	78	77	65	65	57
Moldova	58	50	44	72	74	100	67	39	15	18	24	11	30
Netherlands	583	562	573	587	602	580	568	556	535	510	498	501	470
Norway	188	181	200	226	221	218	208	207	215	212	212	220	222
Poland	1229	-	-	1510	1550	1280	1205	1130	1120	1105	1120	1154	1114
Russian Fed.	1734	2002	187	1871	2358	3600	3325	3093	3054	2685	2570	2467	2379
Slovenia	51	52	52	58	59	62	54	55	61	66	67	70	71
Spain	1056	980	985	978	1052	1177	1227	1251	1223	1241	1243	-	-
Sweden	404	412	411	432	432	338	339	329	324	331	301	302	280
Switzerland	170	-	177	177	172	166	160	153	145	139	136	130	125
Ukraine	1145	1153	1102	1112	1090	1097	989	830	700	568	531	467	455
Yugoslavia	47	50	58	58	63	66	57	49	54	52	59	57	66

European emissions for SO_2 and NO_x show a similar trend to that of the US with an overall general decline in emissions since the 1980s.

4.9 EXERCISE

Calculate the annual mass emissions from a point source that emits 10 mg SO_2/Nm^3. The water (H_2O) and oxygen ($O_{2(d)}$) content of the flue gas are 6.4% and 11% respectively. The flue gas temperature is 270°C ($T_s = 270 + 273 = 543K$) and the standard $O_{2(s)}$ content for the process is 10%. The actual gas flow volume of the process is 14.8 m^3/sec (v_v).

To convert the normalised emission concentration (C_s) to the actual discharge concentration (C_d) from the process the following equation is applied:

$$C_d = C_s (273/T_s) ((100\text{-}H_2O)/100) ((20.9 - O_{2(d)})/(20.9\text{-}O_{2(s)})) \tag{4.1}$$

$$C_d = (10)(0.503)(0.936)(0.908) = 4.28 \text{ mg/m}^3$$

$$\text{Discharge rate } (Q_a) \text{ (g/sec)} = ((v_v\, C_d)/1000 \tag{4.2}$$

$Q_a = (14.8 \times 4.28)/1000 = 0.063$ g/sec. Therefore the annual emission rate of SO_2 equates to 1.99 t SO_2/a.

CHAPTER FIVE

Air Pollution Monitoring

5.1 INTRODUCTION

Air quality monitoring may be undertaken for a number of reasons, but essentially it is to determine the occurrence, distribution and concentration of pollutants. Monitoring can also assist in demonstrating whether or not there is a significant risk of an AQM strategy objective being exceeded in a relevant location, and to enable potential air quality management areas (AQMA) to be identified. Whatever the purpose of the exercise the over-riding consideration is to ensure that the samples and measurements obtained provide adequate and accurate data for the intended purpose.

5.2 FORMS OF MONITORING

Monitoring strategies can be divided into two forms. Those that measure targets, which may show changes in distribution and performance (i.e. target monitoring), and those that measure factors, which may cause changes in the environment (i.e. factor monitoring) (Holgate, 1979). For many baseline air quality studies it is environmental monitoring that is frequently required, which is a form of factor monitoring. Factor monitoring is largely concerned with measuring pollution levels, and may be carried out either:

- At source (emission monitoring);
- In the environment (environmental monitoring);
- At the point of exposure (exposure monitoring); or
- Within the target/receptor (internal monitoring).

Baseline monitoring, a form of environmental monitoring, is the repeated measurement of a substance considered to be important to monitoring objectives. Baseline studies can be used for impact prediction and assessment, and modified as necessary in order to focus on key impacts and issues as the process proceeds through its construction, commissioning, operation and eventual decommission. Post-operational monitoring provides invaluable information on the verification of impact predictions (Chapter 6).

5.3 SITE SELECTION

An important consideration in any monitoring study is site selection. To be purposeful the sample should be representative of conditions prevailing in the environment at the time and place of collection. Thus not only must the sampling location(s) be carefully chosen but also the sampling position at the chosen location. Site selection requires the need to:

- Identify the purpose to be served by monitoring;
- Identify the monitoring site type(s) that will best serve the purpose;
- Identify the general location where the site is placed;
- Identify specific monitoring sites.

A number of issues need to be taken into account in the site selection process. The rationale used to identify appropriate monitoring sites should be fully documented. Generally sites are located where there is a likelihood of human exposure although this can differ depending on the study objectives. The selection process must take into account the spatial distribution and variability of the monitored pollutant. For example, concentrations of traffic pollutants such as CO are often highest at roadside locations, whereas SO_2 concentrations may be highest at urban background or rural locations as a result of domestic or point source emissions. For this reason, it is usually not possible to optimise measurements for all pollutants at any one location (DETR, 2000c). Therefore in such circumstances, some degree of compromise may be required or, several sites monitoring individual pollutants may be the chosen option. Air dispersion modelling studies (Chapter 6) can aid the practitioner in the monitoring site location process. They can help to determine likely 'hot spots' locations where air quality levels may not meet the required AQS. Similarly, modelling studies can help refine the selection of monitoring locations.

Monitoring sites may be classified according to the type of environment in which they are located, in order to permit a more meaningful evaluation of collected data (e.g. comparison of data for kerbside sites, background sites, etc). Generally the site description reflects the influence of either the particular pollutant source or the overall land use. Typical monitoring location types include (DETR, 2000c):

- City/urban centre - an urban location representative of typical population exposure in towns or city centres (e.g. pedestrian precincts and shopping areas);
- Urban background - an urban location distanced from sources and therefore broadly representative of city-wide background conditions (e.g. urban residential areas);
- Suburban - a location type situated in a residential area on the outskirts of a town or city;
- Roadside - a site sampling within 1–5 m of a busy road;
- Kerbside - a site sampling within 1 m of a busy road;
- Industrial - an area where industrial sources make an important contribution to the total pollution burden;
- Rural - an open countryside location, in an area of low population density distanced as far as possible from roads, populated and industrial areas;
- Other - any special source-orientated or location category covering monitoring undertaken in relation to specific emission sources such as power stations, car-parks, airports, tunnels, etc.

Case Studies 1, 4, 7 and 14 provide examples of air quality monitoring studies. Case Study 1 (Section 10.2) details a screening monitoring study using passive diffusion tubes. Case Study 4 (Section 10.5) outlines a baseline air quality data review for Belfast City Harbour and a subsequent air quality monitoring study to supplement existing data. Case Study 7 (Section 10.8) details the automatic monitoring (static and mobile monitoring) undertaken by a Local Authority (Falkirk Council) as part of their LAQM studies. Finally Case Study 14 (Section 10.15) details a baseline monitoring study undertaken to evaluate the accuracy of a predictive air dispersion modelling exercise.

5.4 MONITORING STRATEGIES

Before embarking on a monitoring programme, it is important to have a clear understanding of what the proposed monitoring will achieve and how it will aid the AQM process. Given below are some of the basic points and questions that the practitioner should consider (DETR, 2000c):

- What pollutants need to be monitored?
- What monitoring methods are appropriate?
- What monitoring equipment is required and how much will it cost to purchase and to operate?
- How long to monitor for?
- Where to monitor and how many monitoring sites are required?
- What data quality is required and how to process and evaluate the data?
- How does the monitoring data relate to the requirements of the strategy?

It is important that the financial implications of embarking on a monitoring programme are fully understood before any action is taken (DETR, 2000c). An obvious concern in the design of an air quality monitoring strategy is that each requires different answers. Typical queries, as seen above, include the number and location of sampling sites, the duration of the survey and the time resolution of sampling. All unfortunately will vary according to the use to which the collected data are to be put (Harrop, 1999). Decisions on what to monitor, and how to monitor often become easier once the purpose of monitoring is clearly defined. Therefore the crucial first step in the design of a monitoring programme is to set out the study objectives. Having done this the programme may be designed by consideration of a number of steps in a systematic way such that the generated data is suitable for its intended purpose. A systematic approach could be that given by Hewitt and Harrison (1986) (Table 5.1). Not all surveys will require all of these tasks, nevertheless all surveys should have a logical sequence of auditable events.

Monitoring strategies are inevitably limited in their duration, frequently owing to cost, and therefore caution should always be given to the interpretation of the results as air pollution can vary both spatially and temporally. A survey undertaken for strategic purposes may be restrictive in its contribution to understanding the complexities of air quality at a localised level. The intensity and level of air quality monitoring is variable and dependent upon local needs, available resources and existing baseline data records.

Table 5.1 Monitoring strategy (Hewitt and Harrison, 1986)

Step	*Strategy*
1.	Set out the study's objectives
2.	Review the number of monitoring sites, parameters and duration of the survey
3.	Appropriate sampling methods
4.	Equipment selection
5.	Appropriate analytical techniques
6.	Calibration methods
7.	Data recording and analysis facilities
8.	Data presentation

Most forms of monitoring are generally useful, but their level of usage will be dependent upon the harmony of existing and proposed monitoring networks, methods and techniques in the study area. Should baseline data already exist for the locality it is important that monitoring methods and techniques be compatible, where possible, to ensure a level of consistency in the data-gathering process. Similarly additional monitoring should compliment existing data-gathering practices and protocols where possible. Whilst it may be useful from a prestige view point to have an automated monitoring system, it may not necessarily be essential or economic. Existing data may be appropriate for the purposes of the study. It is therefore important to undertake a thorough baseline air quality review study to see if additional monitoring is actually warranted (Case Study 4, Section 10.5). The choice of sampling methods and techniques should also be considered against the AQS guideline or objective requirements (Chapter 7). It should be remembered that site-specific ambient air quality data is always preferable in any AQA study, although it is frequently the most difficult to obtain. The cost limitations frequently imposed upon projects will often determine the extent of baseline monitoring.

Other than the initial purchase of equipment there are other costs associated with a monitoring programme. These include (DETR, 2000c):

- Staff and site operation costs - frequent documented site visits are an essential component and cost of air monitoring. The frequency of visits required will depend on the type of monitor being used. For instance, the daily volumetric apparatus (smoke) which uses a gravimetric sampler may require daily site visits. Passive samplers may need to be replaced at 1, 2 or 4 weekly intervals, etc. Telemetry systems can provide an efficient and cost-effective method for data acquisition from automatic sites, but their adoption does not obviate the need for regular visits by operators for equipment maintenance and calibration purposes.
- Staffing and data processing costs - automatic analysers may produce large quantities of data, requiring collation and archiving for subsequent analysis. Although passive and active sampler methods produce much less data, extensive surveys can soon accumulate large data sets which may become difficult to manage if not efficiently processed and archived.

- Equipment maintenance costs - to ensure best performance all equipment should be regularly maintained. In addition to routine servicing, provision should be made for emergency breakdown and repair visits, in order to minimise equipment downtime. Equipment spare parts should be kept for these eventualities.
- QA/QC - a documented quality assurance and control programme should be developed in order to ensure reliable and credible measurements (Section 5.5). Typical QA/QC programmes include an established schedule of regular site calibrations, validation of data, monitoring methods and techniques, documentation of all procedures, etc.

Monitoring results should be continuously appraised against the study objectives. Limitations in the design, organisation, or execution of the study should be identified at an early stage and where possible remedied (Bisset, 1991).

5.5 MONITORING STANDARDS AND ACCREDITATION

Standards for sampling, analysis and on-line measurement of air quality have an important role to play in ensuring the credibility of measurements. However, the mere existence of a suitable standard will not ensure that monitoring results are valid. There will be a need to build an infrastructure in which measurements are made to ensure that practitioners use properly validated methods and techniques; adopt internal analytical quality assurance procedures; participate where possible in proficiency schemes and utilise third party audit and assessment as a mechanism for ensuring recognition of a laboratory's quality assurance protocols (Her Majesty's Inspectorate of Pollution (HMIP), 1995b). To this end, organisations such as the National Measurement Accreditation Service (NAMAS), the Community Bureau of Reference (BCR), the US National Institute of Standards and Technology (NIST) and the UK Laboratory of the Government Chemist (LGC), International Standards Organisation (ISO), Comite Europeen de Normalisation (CEN), etc are involved in the creating and extending of QA/QC quality infrastructures.

5.6 MONITORING METHODS AND TECHNIQUES

Careful attention should always be given to the methods and techniques available for monitoring and the resources required. Particular concerns are the general methods for recording very low concentrations; compatibility of measurements and reference methods; and reliability and calibration of instrumentation. Simple methods that meet monitoring objectives should always be considered. However, only proven and generally accepted measurement methods and techniques should be adopted. The UK DETR (2000c) have identified some points to be considered when choosing monitoring equipment. These include issues relating to:

- The data's quality objectives for the monitor;
- The required time resolution of measurement;

- The need for continuous monitoring;
- The availability of financial resources. It is important to establish the full cost of any monitoring programme including operation and support as well as equipment purchase;
- Equipment reliability. Practitioners should ask other users, and in addition equipment suppliers for details;
- The availability of independent type approval/designation of equipment.

Since monitoring instrumentation covers a wide range in capital and running costs, it is therefore often advisable to choose a simple method to meet the monitoring strategy's specification. Many baseline monitoring, spatial screening and indicative surveys can be perfectly well served by inexpensive active or passive sampling methods. Simple monitoring equipment (such as indicator tubes, diffusion tubes, passive samples (e.g. moss bags)) will provide an indication of pollution levels for a moment in time. Real-time monitoring is of more value in identifying the principal air quality trends and short- and long-term pollution episodes, however, its use will precipitate greater costs.

5.6.1 Methods of measurement

Air monitoring methodologies can be divided into four main types, covering a wide range of costs and performance levels. The methods and their relative merits and disadvantages are shown in Table 5.2 (DETR, 2000c).

5.6.1.1 Passive sampling methods

Diffusion tubes (or badges) provide a simple and inexpensive technique for screening air quality to provide a general indication of average pollution concentrations over a period of several weeks or months. Diffusion tubes are simple plastic tubes or discs, open at one end to the atmosphere and with a chemical absorbent at the other end (Fig. 5.1). Exposed tubes are then analysed in the laboratory. The method is, therefore, particularly useful for assessment against annual (long-term) mean objectives. The low cost per tube permits sampling at a number of points of interest; this is useful in highlighting 'hotspots' of high pollutant concentrations such as alongside major roads. Diffusion tubes surveys are simple to undertake and minimal operator training is required. The method is frequently used in the UK to measure NO_2 levels (Fig. 5.1), although the method is also used to monitor other pollutants (e.g. SO_2, NH_3, 1,3-butadiene, BTX (e.g. benzene, toluene and xylene)) (DETR, 2000c). Fig. 5.2 shows the variation of annual NO_2 levels from 1993-1999 in Glasgow, Scotland using diffusion tubes for different monitoring sites.

Case Study 1 (Section 10.2) provides an example of diffusion tube air quality monitoring for NO_2 and SO_2. Diffusion tubes were deployed over a large area for a year to gain an appreciation of likely air quality levels. Tubes were placed in both urban and rural locations.

Figure 5.1 NO_2 and benzene diffusion tubes at a streetside location (courtesy of (a) Scottish Environment Protection Agency and (b) Environmental Health Department, Falkirk Council)

Table 5.2 Methods and their relative merits for measuring air quality (DETR, 2000c)

Method	*Advantages*	*Disadvantages*
Passive sampling	Low cost and simple Useful for screening and pollutants base-line studies	Unproven for some pollutants. In general, only provides weekly or longer averages
Sensor-based systems	Portable	Low sensitivity
Active (semi-automatic) sampling	Low cost, easy to operate and reliable Historical data sets available from UK networks	Provide daily averages Labour intensive Laboratory analysis required
Automatic point monitoring	Provides short-term data On-line data collection possible	Relatively expensive Trained operator required Regular service and maintenance costs
Remote optical/long-path	Provide path or range-resolved monitoring data Useful near sources Multi component measurements with point possible measurements	Relatively expensive Trained operator required Data not readily comparable

Figure 5.2 NO_2 levels (ppb) for the City of Glasgow measured using diffusion tubes for 1993 to 1999 (Data from DETR, 2000b)

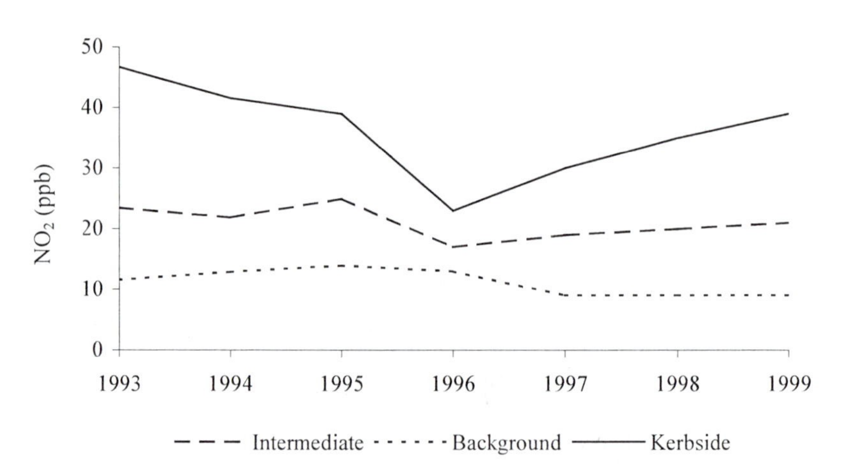

5.6.1.2 Active sampling methods

These methods collect pollutant samples either by physical or chemical means for subsequent analysis in a laboratory. A known volume of air is pumped through a collector such as a filter or chemical solution for a set period of time, which is then removed for analysis. Samples can be taken daily (24 hour), thereby providing detailed information about pollutant levels. There is a long history of using active sampling methods in Europe, providing valuable baseline data for trend analyses

Figure 5.3 SO_2 and smoke active monitoring apparatus (daily volumetric apparatus) (courtesy of the Environmental Health Department, Falkirk Council)

and comparison (e.g. SO_2 (Fig 1.2) and smoke). The type of equipment most widely used is the daily volumetric apparatus (Fig. 5.3) (DETR, 2000c).

5.6.1.3 Automatic methods

These methods produce high resolution measurements (typically hourly averages or better) at a single point for pollutants such as O_3, NO_x, SO_2, CO and PM_{10} (or $PM_{2.5}$) (DETR, 2000c). Methods used include a variety of sophisticated techniques such as infrared or UV absorption, UV fluorescence, chemiluminescence, or for particulates, a variety of sophisticated filtration techniques (Case Study 7 (Chapter 10, Section 10.8) details results from automatic mobile and mobile monitoring undertaken in a Local Authority area). Gas chromatography (GC) analysers also provide high resolution data on benzene, 1,3-butadiene and other speciated hydrocarbon concentrations (DETR, 2000c). To ensure that the recorded data is accurate and reliable, a high standard of maintenance, calibration, operational and QA/QC procedures is required as well as sophisticated data recording devices to record, process, analyse and present the information. Fig. 5.4 shows an automatic monitoring station at a busy roadside site monitoring NO_2, CO and PM_{10} (also see Section 10.8).

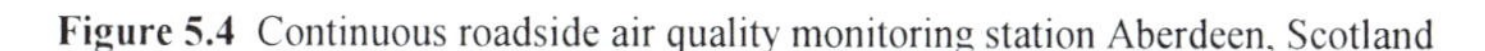

Figure 5.4 Continuous roadside air quality monitoring station Aberdeen, Scotland

5.6.1.4 Optical/long path methods

These are instruments that use a long path spectroscopic technique to make real time measurements of the concentration of a range of pollutants integrated along a path between a light source and a detector. Instruments using the Differential Optical Absorption Spectroscopy (DOAS) system can be used to monitor NO_2 and SO_2 (DETR, 2000c). In order to ensure that the data produced are accurate and reliable, again a high standard of maintenance, calibration, operational, QA/QC and data recording procedures are required (DETR, 2000c).

A range of relatively low cost automatic analysers have been developed specifically for portable and personal exposure-monitoring applications. These are either battery or mains operated electrochemical or solid state sensor-based systems, which can continuously monitor a range of pollutants (e.g. CO, NO_2 and SO_2) at variable time averages. These sensors are of relative low sensitivity and are mostly suitable for identifying areas of elevated pollution at roadsides and near point sources locations (DETR, 2000c). Whilst several different analysers may be suitable for the intended purpose, equipment should where possible have some independent verification of performance such as the USEPA or German TUV designation.

5.6.2 Measurement techniques

Many measurement techniques have been used in air monitoring. In common with most analytical techniques they consist of the measurement of some physical property of the substance which is being monitored. The sample may require some prior treatment to ensure that it is in the correct chemical or physical state for the measurement to be made (Thain, 1980). Thain (1980) identifies the principal monitoring techniques in greater detail.

5.6.2.1 Ionisation techniques

When an organic compound is burnt in a hydrogen flame, ions are produced and collected by an electrode. The electric current which results is proportional to the ion concentration. Ionisation techniques include photo-ionisation, which is caused by irradiation with UV light and thermionic detection where ionisation is induced thermally, either by flame or by an electrically heated filament.

In mass spectrometry (MS) a sample is bombarded by a beam of electrons. Under these circumstances substances break up in a reproducible manner and give rise to a unique pattern of fragments of different masses, which are related to the molecular structure of each substance. The observed pattern and the abundance of the fragments of each mass can be measured by separating the fragments according to mass. This is achieved by accelerating the fragments electrically and separating them in a magnetic field or in a mass filter, after which the abundance of fragments of each mass is measured in turn by a photo-multiplier.

Electron capture is a technique which is allied to ionisation methods. The device consists of a cell in which are located two electrodes, one of which contains a radioactive source. The radiation ionises the air within the cell and maintaining a potential difference between the electrodes can produce an ionisation current. When an electromagnetic species is introduced into the detector it captures electrons and reduces the standing current. The extent of this decrease is dependent upon the number of electron-capturing species present and their electro-negativity. The detector measures the quantity present.

5.6.2.2 Thermal techniques

Combustion heat is a technique that only measures combustible gases. A sample is passed over a heated catalytic sensor to above the ignition temperature of the measured gas. A filament within the sensor is connected in an electrical bridge circuit. The heat of combustion changes the resistance of the filament and unbalances the bridge by an amount proportional to the concentration of the analyte in the sample.

In thermal conductivity a sample is passed through a cell which contains a heated wire filament. The loss of heat from the filament, which depends on the thermal conductivity of the sample components, is measured electrically. The technique is non-selective and is affected by normal atmospheric components such as water and CO_2.

5.6.2.3 Chemical and electrochemical reaction techniques

Such techniques include chemiluminescence, electrolytic conductivity and coulometry. Light is emitted when organic substances undergo certain chemical reactions with an intensity which cannot be attributed to thermal radiation. This is called chemiluminescence. The most common reaction which gives rise to this effect is the reaction of O_3 with unsaturated compounds. The wavelength of light which is emitted is dependent on the nature of the substances involved, and the intensity of the light is proportional to their concentrations. The technique is commonly used for monitoring O_3 and NO_x.

Electrolytic conductivity operates by treating the sample in such a manner that the pollutant is converted to an ionic species, which is dissolved in de-ionised water. The change in electrical conductivity of the water is a measure of the concentration of the ionic species, from which the concentration of the original substance can be calculated.

Coulometry occurs when a measured substance is allowed to react stoichiometrically with a reagent that is generated electrolytically. The current passed to the coulometric generator to maintain equilibrium is a measure of the substance undergoing reaction. Compounds containing sulphur, nitrogen or chlorine lend themselves to this technique.

5.6.2.4 Optical measurement techniques

Many substances can be detected and measured at low concentrations by their reaction with reagents that produce coloured products. This is called colorimetry. Interferometry occurs when the refractive index of a mixture of gases depends on the refractive indices of its various components. A contaminant, which has a refractive index different from that of air, can be detected and measured in air by measuring the difference between the refractive index of the sample and that of uncontaminated air.

Spectral absorption occurs when radiation is passed through a sample that contains organic gases or vapours, absorption of radiation will occur at certain wavelengths. The sampled substance can be determined by the wavelengths at which absorption takes place, and the quantity of the substance is proportional to the level of absorbed radiation.

Certain substances are readily detected by the light that they emit when their compounds are heated or undergo combustion. The light emitted can be measured by a photomultiplier tube and an optical interference filter. This technique is flame photometry.

5.6.2.5 Techniques based on the effects on semi-conductors

These devices are based on non-stoichiometric metallic oxides whose electrical properties depend on the atmosphere that surrounds them. In simple forms they are sensitive but unspecific, and respond to almost all organic substances with which they come in contact.

5.6.2.6 Techniques based on separation

Many of the techniques of measurement described above are non-specific and so can be used only if a correction can be made for the effects of other interfering substances. If the substance, however, can be separated from another substance which is present, the measurement can be made by a non-specific technique. The most common technique used for separation is GC. Where GC cannot make separation it may be necessary to use other techniques. The most common of these is liquid chromatography.

Table 5.3 shows some of the different techniques commonly available to measure selected air pollutants (Barrowcliffe, 1992; DoE, 1993). Leithe (1971), Thain (1980) and Ruch (1970) provide a comprehensive listing of analytical methods and the techniques for determining and measuring air pollutants.

5.6.3 Particulate measurement

Particulates can be measured as either a concentration, deposition or as a soiling/presence factor. The former is frequently related to assessing health effects. Deposition and soiling measurements tend to provide a quantification of a level of nuisance.

5.6.3.1 Particulate deposition

The deposition of particulate matter is generally expressed as a mass per unit area per unit of time (e.g. $mg/m^2/day$). In selecting the actual site for the deposit gauge, care should be taken so that the position chosen is such that the dustfall from the atmosphere will be typical of the surrounding neighbourhood. To prevent any shielding effect, the site should be such that the distance of the gauge from any object in the vicinity is not less than twice the height of the object above the top of the funnel (Palmer, 1974).

Table 5.3 Air quality monitoring techniques (Barrowcliffe, 1992; DoE, 1993)

Pollutant	*Monitoring technique*
SO_2	Hydrogen peroxide acidemetric and colorimetric
	Colorimetric para-rosaniline
	Gas phase fluorescence
	DOAS
	Flame photometry
	Diffusion tube
NO_2	Chemiluminescence
	Christie arsenite
	Diffusion tube
	DOAS
O_3	UV photometry
	Chemiluminescence
	DOAS
CO	Infra-red absorption
	Electrochemical cell
Particles	Hi-volume sampler
	Smoke shade reflectance
	Beta gauge
	Tapered elemental oscillating microbalance (TEOM)
PAN	GC/electron capture
	GC/chemiluminescence
Benzene	GC
	DOAS
	Diffusion tube
VOCs	Total hydrocarbon/non-methane hydrocarbon analyser
	GC
PAH	High performance liquid chromatography (HPLC)
TOMPS	Polyurethane filter
Metals in air	Atomic absorption spectrometry
	X-ray fluorescence
	Filter

The measurement of the rate of atmospheric particulate deposition is not a precise determination due to the vacuities and accuracy of the equipment (Parker, 1978). The types of deposit gauge in use differ a great deal in design and orifice height above the ground owing to the compromises necessary to meet different and sometimes conflicting requirements. As a result, the measurements made with different types of gauge cannot be expected to give results in agreement with one another. Even gauges of the same design may give widely varying results when placed in adjacent positions. Two types of deposit gauge are the American and BS Deposit Gauges. The gauges are essentially similar. The British deposit gauge (Fig. 5.5) has a collection bowl with vertical sides fixed in a stand so that the orifice of the funnel is 1.2 m above ground level. The aerodynamics of the equipment are affected to some extent by the open cylinder. A wire mesh bird guard, which projects about 100 mm above the bowl, prevents birds from perching on the edge of the funnel. The funnel shaped bowl drains into a collecting vessel. After 30 days (a calendar month) the bowl is washed down with distilled water and the contents are analysed for soluble matter, insoluble matter, and for any other constituents required. Other techniques utilised include the use of inverted 'Frisbees' as a collecting vessels. A variation of the British technique is the directional deposit gauge where the nuisance from dust may be more closely related to the amount carried horizontally, or at a low angle to the horizontal, by the wind than the amount falling vertically, or near vertically. The gauge comprises four cylinders, each with a vertical slit in them, mounted on a stake. The slits face the four points of the compass.

Figure 5.5 BS deposit gauge (British Standards Institution, 1969a) (From Palmer, 1974).

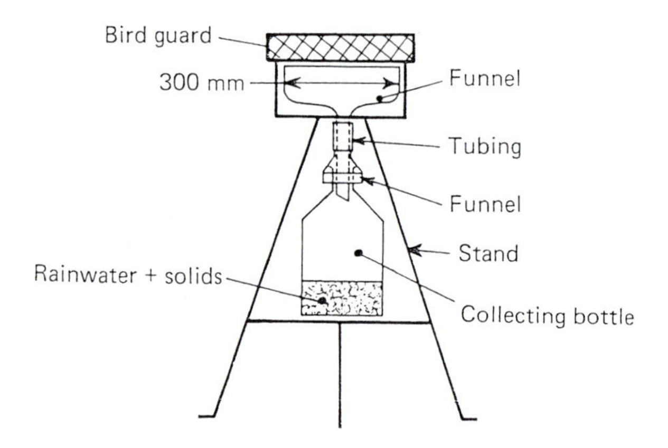

Another type of deposit gauge is that used in the People's Republic of China. These deposit gauges consist of ceramic pots about 10 cm in diameter and 30-40 cm high, which are deployed 3 to 15m above ground level. In cities, like Shenyang, one gauge and sulphation plate is located approximately in each 1km^2 of the central part of the City and the deposition is expressed as ton/km^2/month (AEA Technology, 2000a). A simpler and cost effective method for the measurement of particulate deposition is the use of petri dishes. The method is useful for periods of 24 to 48 hours in fairly dry and calm conditions. Should wind speeds exceed 25 km/hr the survey should be abandoned (Palmer, 1974).

Today the use of the deposit gauge is becoming obsolete in developed countries. Nevertheless, owing to its simplicity the technique has historically gained wide use. The method provides an indication of air quality trends of particulate deposition if applied over an extended period. For example, Fig. 5.6 shows the decrease in annual deposition levels for Glasgow, Scotland from 1918 to 1990. Deposition levels have decreased from more than 400 to 150 mg/m^2/day over a 70 year period (City of Glasgow Council, 1999; SEPA, 2000).

Figure 5.6 Annual average total deposition levels of particulates in Glasgow, Scotland 1918 – 1990 (Hampton *et al.*, 1993; SEPA, 2000) (courtesy of the National Society for Clean Air and Environmental Protection)

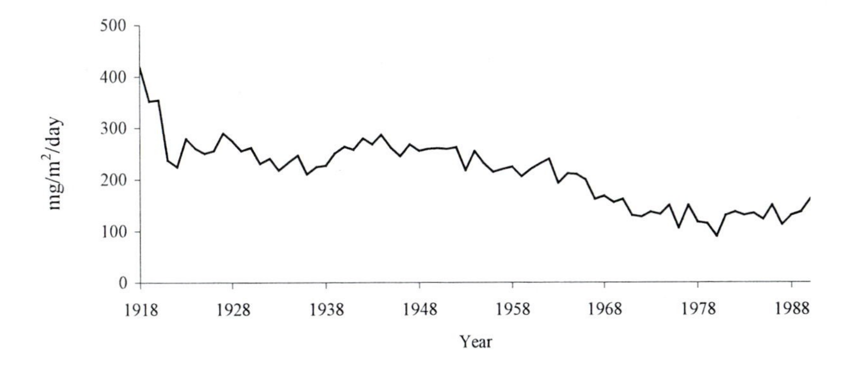

5.6.3.2 Suspended particulates

Fine particulates in the atmosphere are generally measured as a concentration (i.e. mass per unit volume of air (e.g. $\mu g/m^3$)). More suitable methods for measurement are therefore required, as deposition gauges cannot measure the SPM. The only method for measuring the concentration of SPM in air in absolute terms is to filter a sample of air and determine the weight of material collected on the filter and the total amount of particles in the ambient air given as total suspended particulate matter (TSPM). A rough conversion factor to determine particles smaller than 10 μm in diameter (i.e. PM_{10}) assumes that 80% of the TSPM is PM_{10} (PM_{10} = 0.8*TSPM).

The American high volume sampler (e.g. Grazeby-Anderson) is the simplest of the direct weighing methods and has been adopted by the USEPA (Fig. 5.7). A glass fibre filter paper is held horizontally in a holder. Air is drawn through the filter paper at a determined flow rate (i.e. 1.13-1.70 m^3/min). A corresponding BS method uses a sampling rate approximately 20 times less than the American high volume sampler (British Standards Institution, 1969a; 1969b). Another simple gravitational method is the Warren Spring Laboratory (WSL) 'M' Type sampler (Ralph *et al.*, 1982). The sampler comprises a sampling head that conforms to the specifications given in the Annex to the EC Directive (82/884) on Pb concentrations in ambient air (Council for European Communities, 1982). The sampler has been subsequently revised for directional sampling (Barnett *et al.*, 1987).

Historically the most widespread form of particulate measurement has been smoke (Fig. 5.3). The principal method used for estimating smoke (black smoke) concentrations in air is the BS method (British Standards Institution 1969b; WSL, 1966) and is in widespread use in the UK National Survey of Air Pollution. It is also used in other countries and is, in effect, the method adopted by OECD for such surveys (OECD, 1964). The sampled air is drawn through the apparatus at a rate of about 2 m^3/day by a low volume suction pump and enters through an inverted

Figure 5.7 American high volume sampler (From Palmer, 1974)

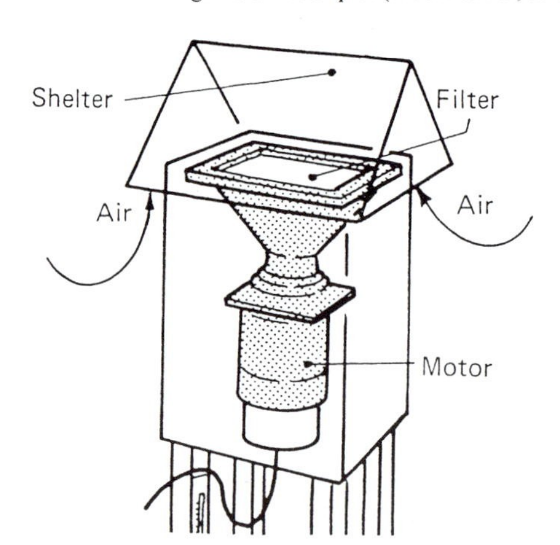

funnel. The air passes through a sheet of Whatman No. 1 filter paper, held in a filter clamp, followed by a suction pump and gas meter. After 24 hours the filter paper is removed and the darkness of the smoke stain on it is determined by a photoelectric reflectometer. The concentration of smoke is then read from a calibration curve relating the darkness of the stain to the concentration of equivalent standard smoke in microgrammes per cubic metre. The following samplers are available for PM_{10} monitoring:

- Gravimetric;
- TEOM;
- Beta-gauge absorption;
- Light scattering.

5.6.3.3 Particulate soiling

A number of factors influence the human perception of particulate soiling (e.g. dust nuisance). The contrast in colour between the deposited dust and the receiving surface is of principal importance. It has been demonstrated that the perception of soiling is markedly influenced by the comparison with a clean surface (Chapter 3, Section 3.6). A clean surface will act as a reference point to distinguish between the clean and soiled areas. The nuisance value (Chapter 7, Section 7.5.1) is associated with the rate of soiling, i.e. the length of time that it takes an unacceptable level of dust to accumulate. A given level of soiling which takes only one day to occur will probably be considered less acceptable than the same level of soiling which occurs over a period of one week (Moorcroft and Eyre, 1989).

Methods used to measure soiling include the glass slide deposit gauge technique (Brooks and Schwar, 1987). A glass microscope slide is used as the dust collector surface, and measurements of the soiling level determined by the loss in surface reflectance using a standard 45° incident/45° reflected light glass meter. A clean slide will give a 'gloss' reading of 100%, or a soiling level of zero. A dirty slide will give a 'gloss' reading of <100% and a soiling level of greater than zero.

5.6.3.4 Short-term surveys

Short-term particulate surveys are often more effective than monthly collections for the purpose of identifying a nuisance source, because they can be confined to periods when the wind is in a given direction. For these surveys, several kinds of collector can be used and include sticky plates, petri dishes, glass jars with or without a funnel, plastic bowls, glass lens slides, tiles, etc. The basic procedure for investigating a complaint for grit or dust nuisance is given by Sanderson (1963):

Figure 5.8 Illustration of a hypothetical particulate survey (after Sanderson, 1963)

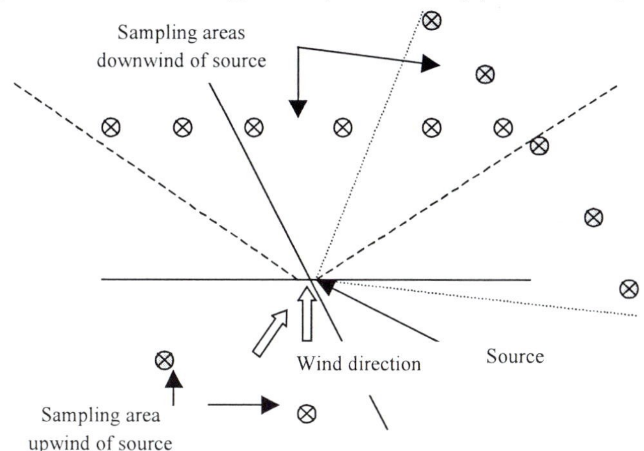

- Determine the wind direction under which the nuisance occurs;
- Samples of deposited matter should be collected in the affected area when the wind is from this direction;
- A second or third sample may have to be taken downwind of the suspected source when the wind directions are altogether different from those in the first test. If the particles found in these further tests are similar in composition to those in the first sample, the area of origin can be more closely located;
- An extra collector or two upwind side of the suspected source at the time of the downwind exposure should be undertaken;
- It would be helpful to collect a sample of material from the source suspected of causing a nuisance for comparison with collected samples;
- Careful handling of the samples is most important. Individual particles are easily crushed beyond recognition;
- To collect samples 3 or 5 collectors should be placed about 100 m apart, one where the nuisance has occurred and one or two on either side at right angles to the wind direction from the suspected source (Fig. 5.8);
- The collecting vessels should be washed thoroughly before each use and covered until placed in position;

- The collecting vessels should be level and sited all at the same height above ground (e.g. >1.25 m and preferably 3 m);
- For good exposure the collecting vessel should be at least twice the height of that object or building above the vessel;
- The vessel should be correctly labelled. A note of the direction of the wind should be made.

Best exposure conditions are for 6 to 8 hours in wind speeds of 5 and 15 mph (8 and 24 kph). If rain occurs the survey should be curtailed.

5.6.3.5 Composition of particulates

Microscopic analysis provides enlarged images of objects such as small particles and permits detailed analysis of smoke, dusts, fogs and other aerosols. Microscopic analysis not only reveals the shape and size of small objects but also in some cases the proportions of various constituents of a mixture. Determination of the composition of the particles will help to ascertain the source. Hamilton and Jarvis (1963) and Palmer (1974) provide detailed information for identifying the type and the character of deposited material. Optical microscopy is also used in the determination of the presence and type of fibrous material present in the atmosphere (e.g. asbestos) (Purdy and Williams, 1977; Moorcroft, 1985). If the dust source is not distinguishable by simple microscopic investigation then a more detailed crystalline composition analysis may be required (e.g. scanning electron microscope).

5.6.4 Carbon monoxide

There are no passive sampling devices available to monitor CO. There are, however, a number of portable sensor-based systems. These have low sensitivity, typically 0.5 ppm resolution but would be suitable to identify pollution hotspots at roadsides, etc. Automatic non-dispersive infra-red analysers are also available. A common analytical methodology for the determination of CO includes the use of iodine pentoxide (Grant *et al.*, 1951).

5.6.5 Sulphur dioxide

One of the most common methods of SO_2 determination is the daily volumetric apparatus (Fig. 5.3). Determination of SO_2 and smoke is carried out on a daily basis (i.e. 24 hours) and involves passing air through a filter paper to remove any particulates (Section 5.6.3.2). The air sample on passing through a filter paper is bubbled through a 0.05% solution of hydrogen peroxide (H_2O_2) in distilled water. The reaction that occurs in the bubbler is $H_2O_2 + SO_2 \rightarrow H_2SO_4$. In order to prevent CO_2 in air from dissolving in the H_2O_2, a known amount of H_2SO_4 is also added to the absorbing solution. The air leaving the bubbler passes through a gas meter, to record the actual volume of air sampled and then to atmosphere through a pump. The H_2SO_4 formed in the bubbler is then titrated with a standard alkali (e.g.

sodium tetra borate) using a BDH 4.5 indicator. The titration figure obtained is equivalent to the SO_2 present in the atmosphere. The methodology has been adopted as a BS method (WSL, 1966) and by the OECD (OECD, 1964).

The USEPA have previously adopted the pararosaniline method of determination (USEPA, 1971) where SO_2 is absorbed from sampled air by a solution of potassium tetrachloromercurate and a dichlorosulphitomercurate complex is formed. The complex is treated with pararosaniline and formaldehyde, when the intensely coloured pararosaniline methyl sulphonic acid is formed. The absorbance of the coloured product is measured in a spectrometer and the concentration of SO_2 is related to absorbance.

The wet chemical methods for the measurement for SO_2 are of limited use for continuous monitoring of peak short-term concentrations. For screening monitoring, active samplers and portable samplers can be used. For more advanced monitoring automatic analysers are required. Both UV fluorescent analysers and DOAS systems are available.

5.6.6 Oxides of nitrogen

Various methods are available for the detection and determination of NO_2. Analytical methods include using the reagents N-(1-naphthyl)ethylenediamine dihydrochloride and acetic acid (Saltzman, 1954). For screening monitoring, diffusion tubes, active samplers and portable samplers can be used. For more advanced monitoring automatic systems are required. Both chemiluminescent analysers and DOAS systems are available.

Whilst diffusion tubes have many advantages, a principal shortcoming is that they are not convenient for routine short-term measurements, which may be required for the protection of human health (Association of London Chief Environmental Health Officers and Greater London Council Scientific Services Branch, 1985/86). Nonetheless, it has been shown that there is a reasonably consistent relationship between long-term averages of NO_2, and short-term concentrations (Association of London Chief Environmental Health Officers and Greater London Council Scientific Services Branch, 1985/86). For example, the annual mean has been found to be a good predictor of the 98th percentile of hourly values (Chapter 7, Section 7.2.1), the relationship being:

$$\text{98th percentile} = 2.4 \times \text{annual mean} \qquad (5.1)$$

The error in the multiplier (2.4) is $\pm$ 10%. Similarly, the 50th percentile of hourly values, also a required parameter of the EC NO_2 Directive, is given by:

$$\text{50th percentile} = 0.93 \times \text{annual mean} \qquad (5.2)$$

5.6.7 Ozone

Various methods are available for the detection and determination of O_3. Analytical methods include the use of a potassium iodide solution (Smith and Diamond, 1952), sodium diphenylaminesulfonate (Bovee and Robinson, 1961), etc. However, several continuous monitoring techniques are available. The principal techniques include UV photometry, chemiluminescence and DOAS.

5.6.8 Volatile organic compounds

Techniques available to measure VOCs include:

- Indicator tubes (e.g. Drager tubes) are a widely used method used for the measurement of organic vapours. An analytical tube is inserted into a hand held pump, and a measured volume of air is drawn through the tube. A chemical reaction between the pollutant and the tube contents causes a colour change in the crystals and the concentration can be briefly read from the graduated markings along the side of the tube. Indicator tubes can be used to measure the concentration of a wide range of pollutants.
- Thermal desorption offers a considerable number of advantages to the analysis of VOCs. Tubes packed with a suitable adsorbent, such as Tenax TA are used to sample air by drawing a measured volume of air through the tube, thus concentrating organic materials in the tube to the level at which they can be detected. Samples are thermally desorbed in the laboratory and the vapours are injected into a GC.
- Infra-red spectrophotometry is based upon the absorption of infra-red light, which occurs at a specific wavelength for individual compounds. By the selection of different wavelengths, it is possible to monitor for a wide range of VOCs.
- GC is based on the separation of individual compounds, which are then measured using a variety of detection devices. If the GC is also interfaced with a MS, then it is possible to obtain unequivocal identification for an extremely wide range of organic compounds. GC/MS can be operated *in situ* or a grab sample is collected and analysed in the laboratory.

Diffusion tubes are available for BTX (Fig. 5.1). The only other suitable methods for benzene are based on GC analysis, which generally requires expensive and complex instrumentation. A reference method is currently being developed by CEN for an EC Directive. This requires sampling of benzene using a pumped sampling method on a sorbent cartridge followed by GC determination. Diffusion tubes have also been developed for monitoring 1,3-butadiene, which are similar to those used for benzene, and are suitable for screening studies. 1,3-butadiene also can be monitored using a GC.

USEPA methods T014 and T02 detail methods and techniques for the measurement of ambient VOCs levels (USEPA, 1984; 1988).

5.6.9 Lead

Lead is monitored using samplers that capture fine ambient particulate matter on a filter for subsequent analysis. In Europe, the 'M Type' sampler has been widely employed. A gravimetric sampler with PM_{10} sampling inlet is likely to be the method recommended for the new EC Directive on lead, but not until 2005. For screening surveys, other gravimetric type samplers can be used.

5.7 ODOUR MEASUREMENT

The strength of an odour is expressed in terms of its odour threshold (OT) concentration (Chapter 7, Section 7.5.2). This is the lowest concentration at which the odour is detectable by 50% of a group of people (i.e. an odour panel). An airborne release is likely to be regarded as a nuisance when its concentration is greater than 5 times the OT. This concentration is referred to as the odour nuisance threshold (ONT) (Environmental Analysis Co-operative, 1999). Gibson (1994) identifies the following sequence of events for undertaking an olfactometry test:

- Sampling strategy:
  - Identify the relevant odour-producing process or processes;
  - Assess the toxicity and potential risk to the panel members of any emissions;
  - Identify location(s) of odour emission points;
  - Determine likely fluctuations in odour emission with time;
  - Locate and install odour sampling points.
- If the odour emission is considered to be constant then point samples will be adequate. If fluctuations occur in the emissions longer sampler times should be considered to cater for the odour's variations.
- Sufficient samples should be taken to ensure that the odour emission is properly quantified. Replicate sampling may be required.
- The most commonly used method of sampling is to use a sample bag placed in a rigid container. The air is removed from the container using a vacuum pump and the reduced pressure causes the bag to fill. The container should be free from odour, smooth surfaced, have minimal physical or chemical interaction with the sample pollutant and be leak proof. Common container materials include glass, stainless steel, teflon, PVDF, PET, tedlar, PVF.

Following collection the sample can be used for olfactometry purposes. The olfactometer is a dilution apparatus that mixes the odorant with dilution air, which is odour free, in a certain ratio. One or more sets of sniffing ports are attached to the olfactometer, through which either reference air (odour free) or the diluted odorant can pass through. The composition of the odour panel should be a minimum of 6 and they should be at least 16 years old. In addition it is recommended that the panel should not (Gibson, 1994):

- Smoke for half an hour before the start of the measurement;
- Eat or drink (except water) immediately before measurement;

- Use cosmetics, perfumes etc.;
- Participate whilst infected with a cold;
- Communicate, either verbally or by expression, with other panel members about the study findings during the measurement.

The odour mass flow rate (E; ou/sec) is the product of the odour concentration (DTT; ou/m^3) and the volume discharge rate at the exit point (v_v; m^3/sec):

$$E = v_v \,.\, DTT \tag{5.3}$$

5.8 OTHER EQUIPMENT

In addition to analysers, other site infrastructure may be required to set up an air quality monitoring station. Additional equipment may include:

- Equipment housing
- Air conditioning/heating
- Gas cylinder storage facilities
- Air sample inlet system
- Auto-calibration facilities
- Electrical systems
- Telephone lines/modem
- Data logging and acquisition facilities
- Security systems
- Safety systems

Active samplers and diffusion tubes require a much lower level of site infrastructure than for automatic sampling equipment. However, access to appropriate analytical facilities at a central laboratory may be required. A frequent problem encountered by practitioners is an appreciation of the costs for running, maintaining and calibrating equipment.

5.9 MONITORING NETWORKS

The transboundary nature of air pollution has necessitated the need for international air quality monitoring networks. These have included the WHO GEMS (Global Environmental Monitoring System) and EMEP networks amongst others. A 1984 protocol ratified by the EU in 1986 provides funding for EMEP and comprises some 100 monitoring stations (Table 5.4). National air quality monitoring programmes are now relatively common. Table 5.5 shows the number of air quality monitoring stations in selected countries for criteria pollutants (UNECE, 1995).

5.9.1 GEMS

GEMS/AIR was a global programme for urban air quality management operated jointly by WHO and the United Nations Environment Programme (UNEP) from 1975 to 1996. GEMS/AIR was a component of the UN GEMS, which is a component of the UN Earthwatch System. Technical collaboration with developing countries was a prime focus of the programme with the provision to those countries with monitoring devices. The programme closed down in 1996.

Table 5.4 EMEP monitoring sites in Europe (Active sites (non-active sites)) (EMEP, 2000)

Sites	*Sites*	*Sites*
Austria 3 (1)	Greece 1	Romania (6)
Belarus (1)	Hungary 1	Russia 3 (2)
Belgium (5)	Iceland 1(1)	Slovakia 4
Bosnia Hercegovina (1)	Ireland 5	Slovenia 4 (1)
Croatia 2	Italy 1(3)	Spain 6(1)
Czech Republic 2	Latvia 2	Sweden 8(5)
Denmark 5(2)	Lithuania 1(1)	Switzerland 5(1)
Estonia 2(1)	Moldova (1)	Turkey 1
Finland 5(1)	Netherlands 2 (5)	UK 20(4)
FYROM (1)	Norway 13(5)	Ukraine (3)
France 8(3)	Poland 4(1)	Yugoslavia 2
Germany 11(12)	Portugal 3(2)	

FYROM = Former Yugoslavian Republic of Macedonia

The Air Management Information System (AMIS) is a programme developed by the WHO under the umbrella of the Healthy Cities Programme and has followed on from GEMS. The objective of AMIS is to transfer information between countries and cities. In this context AMIS acts as a global air quality information exchange system. The AMIS programme includes:

- Co-ordinating databases with information on air quality issues in cities;
- Acting as an information broker and provider between countries;
- Providing and distributing technical documents on air quality management;
- Publishing and distributing air quality data reviews;
- Provision of air quality monitoring and management training courses;
- Facilitating research studies of non-commercial institutions;
- Running Regional Collaborative Centres to support data transfer activities.

AMIS's core database contains summary statistics of air pollution data and the number of days on which WHO guidelines are exceeded (Chapter 7). Any compound for which WHO air quality guidelines exist can be entered into the open-ended database. In the existing version data (mostly from 1986 to 1996) from about 100 cities in 40 countries are represented. To continue updating and expanding the system, in particular to cover more participating sites, WHO has sought the assistance of Alam Sekitar Malaysia Sdn Bhd (ASMA). ASMA has been awarded a 20 year concession by the Government of Malaysia to build, own and operate among others a network of 50 continuous air quality monitoring stations in Malaysia. ASMA also manages an environmental database for Malaysia. Table 5.6 summarises historically the number of WHO air quality monitoring stations. Table 5.7 lists the monitoring sites that currently form the network of ASMA sites on behalf of the WHO.

Table 5.5 Number of air quality monitoring sites in selected countries (UNECE, 1995)

Country	SO_2	NO_x	O_3	*CO*	*Particulates*	
Austria	204	182	135	77	144	
Belgium	95	22	10	1	19	(80)
Bulgaria	103	101	7	10	95	
Canada	78	88	126	58	94	(39)
Croatia	68	22			4	(54)
Cyprus	6	6	6	6	6	
Czech Republic	93	93	25	30	93	
Denmark	16	11	9	6	16	(5)
EC	2208	979	459	524	1000	(1050)
Finland	71	29	10	6	63	
France	420	135	55	45	120	(80)
Germany	31	31	31	18	31	
Greece	22	22	21	20		(21)
Hungary	47	47	26	40	37	
Italy	512	435	202	340	452	
Netherlands	39	45	38	21	18	(14)
Norway	25	23	14	0	0	(8)
Portugal	50	27	15	14	40	
Romania	143	143	3	3	35	
Slovakia	33	33	17	10	28	
Slovenia	20	8	7	2	2	
Spain	571	78	25	38	109	(455)
Sweden	45	45	20	10	5	(45)
Switzerland	90	99	116	40	56	(40)
Yugoslavia	29	3				(29)

Note: Number of black smoke monitors is given in brackets
EC – European Community

Table 5.6 WHO-ASMA monitoring sites 1986-1999 (WHO - ASMA, 2000)

	1986	*1987*	*1988*	*1989*	*1990*	*1991*	*1992*	*1993*	*1994*	*1995*	*1996*	*1997*	*1998*	*1999*
SO_2	172	178	158	167	205	206	221	224	232	211	180	89	94	34
NO_2	132	141	117	127	166	172	181	181	196	201	192	105	81	29
CO	71	74	79	85	95	101	111	112	121	119	96	65	71	24
O_3	38	46	48	56	62	61	73	74	80	102	105	75	64	24
Pb	12	12	15	15	23	24	32	29	35	34	37	31	27	9
SPM	74	73	51	61	85	86	89	92	101	75	68	32	28	7
PM_{10}	9	8	11	13	13	35	36	38	40	48	61	46	49	22
BS	15	15	17	13	16	14	18	15	16	17	17	11	6	4

Table 5.7 Monitoring sites comprising the WHO-ASMA air quality monitoring network (WHO - ASMA, 2000)

Country	*City*	*Years*	*Country*	*City*	*Years*
Argentina	Buenos Aires	1986-96[1]		Duisburg	1986-98[16]
	Cordoba City	1987-91[2]		Frankfurt	1986-98[6]
	Mendoza	1986-97[3]		Hamburg	1986-98[34]
	Santa Fe	1986-97[4]		Munich	1986-98[37]
Australia	Melbourne	1986-95[6]		Stuttgart	1986-98[34]
	Perth	1986-99[6]	Ghana	Accra	1986-90[14]
	Sydney	1986-95[7]	Greece	Athens	1986-97[36]
Austria	Vienna	1988-96[8]	Guatemala	Guatemala City	1994-97[38]
Belarus	Minsk	1986-95[9]	Honduras	Tegucigalpa	1995-97[38]
Belgium	Brussels	1986-99[6]	Hungary	Budapest	1991-99[37]
	Liege	1996-99[10]		Eger	1995-99[29]
Brazil	Sao Paulo	1986-95[11]		Gyor	1995-99[29]
	Osasco	1986-95[12]		Miskolc	1995-99[29]
	Porto Alegre	1986-96[13]		Pecs	1995-99[29]
	Rio De Janeiro	1986-96[14]		Szeged	1997-99[8]
	Victoria	1992-96[13]	India	Ahmedabad	1986-96[28]
Bulgaria	Sofia	1986-98[15]		Calcutta	1986-96[28]
	Plovdiv	1986-98[15]		Chennai	1986-96[28]
Canada	Hamilton	1987-97[16]		Hyderabad	1986-96[28]
	Montreal	1986-97[17]		Jaipur	1986-96[28]
	Toronto	1994-97[17]		Kanpur	1986-96[28]
	Vancouver	1986-97[16]		Kochi	1986-96[28]
Chile	Santiago	1988-95[18]		Mumbai	1986-96[28]
China	Beijing	1981-94[13]		Nagpur	1986-96[28]
	Chongqing	1983-94[19]		New Delhi	1986-96[28]
	Guangzhou	1981-94[13]	Japan	Osaka	1986-94[31]
	Shanghai	1981-94[13]		Tokyo	1986-95[31]
	Shengyang	1981-94[13]	Jordan	Amman	1994-95[39]
	Xi'an	1981-94[13]	Kuwait	Kuwait City	1986-96[35]
	Hong Kong	1986-98[6]	Latvia	Daugavpils	1986-96[17]
Costa Rica	Heredia	1996[6]		Liepaja	1986-96[17]
	San Jose	1993-97[6]	Lithuania	Kaunas	1986-96[19]
Croatia	Zagreb	1986-99[20]		Klaipeda	1986-96[19]
Cuba	Havana	1990-95[21]		Siauliai	1986-96[19]
Denmark	Copenhagen	1986-94[22]	Mexico	Mexico City	1986-96[6]
Ecuador	Ambato	1986-95[4]	New Zealand	Auckland	1986-99[42]
	Cuenca	1986-95[4]		Christchurch	1989-99[41]
	Esmeraldas	1986-88[4]	Nicaragua	Managua	1996[43]
	Guayaquil	1986-95[23]	Panama	Panama City	1996-97[44]
	Quito	1986-95[24]	Philippines	Batangas	1987-95[13]
El Salvador	San Salvador	1996[25]		Manila	1987-95[13]
Estonia	Tallinn	1996-98[19]		Quezon City	1987-95[13]
Finland	Helsinki	1986-95[6]	Portugal	Lisbon	1986-95[42]
France	Aix en Provence	1992-96[26]	Romania	Bucharest	1986-96[15]
	Gardanne	1992-99[26]	South Africa	Capetown	1986-94[45]
	La Penne sur Huveaune	1999[26]		Durban	1986-95[46]
	Marseille	1992-99[5]		Johannesburg	1986-95[40]
	Paris	1994-98[35]	Spain	Madrid	1986-95[5]
	Strasbourg	1990-96[31]	Switzerland	Basel	1986-99[32]
Germany	Berlin	1986-98[33]		Geneva	1986-96[16]
	Cologne	1986-98[6]		Lugano	1986-99[32]
	Dortmund	1986-98[16]		Zurich	1986-99[32]

Thailand	Bangkok	1989-95[47]		Liverpool	1986-99[30]
Turkey	Ankara	1986-99[4]		London	1986-99[31]
	Bursa	1986-99[27]		Manchester	1986-99[40]
	Erzurum	1986-99[27]		Middlesbrough	1996-99[8]
	Istanbul	1986-99[27]		Sheffield	1986-99[48]
	Izmir	1986-99[27]	US	Denver	1986-96[31]
	Kutahya	1986-99[27]		Los Angeles	1986-95[31]
UK	Belfast	1996-99[30]		Lynwood	1998-99[31]
	Birmingham	1986-99[30]		Rubidoux	1998-99[31]
	Edinburgh	1986-99[30]	Venezuela	Caracas	1986-95[15]
	Glasgow	1986-99[40]			

1. NO_2, *BS, CO, SPM, Pb*
2. NO_2, *SPM*
3. SO_2, NO_2, *BS, Pb*, PM_{10}
4. SO_2, *BS*
5. SO_2, NO_2, *BS*
6. SO_2, NO_2, O_3, *CO, SPM*, PM_{10}, *Pb*
7. NO_2, O_3, *CO, SPM*, PM_{10}, *Pb*
8. SO_2, NO_2, O_3, *CO*, PM_{10}
9. NO_2, *CO, SPM, Pb*
10. SO_2, NO_2, O_3, *BS*, PM_{10}, *Pb*
11. SO_2, NO_2, O_3, *CO, SPM*, PM_{10}
12. SO_2, *SPM*, PM_{10}
13. SO_2, *SPM*
14. *SPM*
15. SO_2, NO_2, *SPM, Lead*
16. SO_2, NO_2, O_3, *CO, SPM, Pb*
17. SO_2, NO_2, *CO, SPM, Pb*
18. SO_2, NO_2, O_3, *CO*, PM_{10}, $PM_{2.5}$
19. SO_2, NO_2, *CO, SPM*
20. SO_2, NO_2, *SPM, BS, Pb, Cd*
21. SO_2, NO_2, NH_3
22. SO_2, NO_2, O_3, *CO, SPM, BS, Pb*
23. SO_2, *SPM, BS*
24. SO_2, *SPM*, PM_{10}, *BS, Pb*
25. NO_2, O_3, PM_{10}, *Pb*
26. SO_2, NO_2, O_3
27. SO_2, PM_{10}
28. SO_2, NO_2, *SPM*, PM_{10}, *Pb*
29. SO_2, *CO*, O_3
30. SO_2, NO_2, O_3, *CO*, PM_{10}, *BS*
31. SO_2, NO_2, O_3, *CO*, PM_{10}, *Pb*
32. SO_2, NO_2, O_3, *CO, SPM, Pb, Cd*
33. SO_2, NO_2, *CO*, O_3, *SPM, NO*, NO_x
34. SO_2, NO_2, *CO*, O_3, *SPM, NO*
35. SO_2, NO_2, O_3, *CO*
36. SO_2, NO_2, O_3, *CO, BS*
37. SO_2, NO_2, O_3, *CO, SPM*
38. NO_2, O_3, *SPM*, PM_{10}, *Pb*
39. *SPM, Pb*
40. SO_2, NO_2, O_3, *CO, BS, Pb*
41. SO_2, NO_2, *CO, SPM*, PM_{10}, *Pb*
42. SO_2, NO_2, *SPM, BS, Pb*
43. NO_2, O_3, PM_{10}
44. NO_2, PM_{10}, *Pb*
45. SO_2, NO_2
46. SO_2, *Pb*
47. SO_2, NO_2, *SPM*
48. SO_2, NO_2, *CO, BS*

BS – black smoke

5.9.2 National and municipal air quality monitoring networks

Individual countries, territories and cities/municipalities have developed their own air pollution monitoring networks. Selected examples of these networks are detailed below.

5.9.2.1 UK air quality monitoring network

The UK DETR funds a network of air quality monitoring sites, collectively known as the UK Air Monitoring Network (UKAMN). The network is organised into three automatic and six sampler-based programmes, and is operated by the National Environmental Technology Centre (NETCEN). In addition to the DETR/NETCEN monitoring sites, individual Local Authorities (i.e. municipalities) also carry out their own air quality monitoring programmes as part of their local air quality management responsibilities under the UK Environment Act 1995. The extent of Local Authority monitoring is variable, but is generally operated and co-ordinated by the Environmental Health Departments.

In addition to the UKAMN and Local Authority monitoring networks, air quality monitoring may also be carried out independently by various other interested bodies. These include industrial operators as part of their authorisations or permitting requirements; universities and educational institutions for research purposes; research organisations and institutes; and local action/environment groups may commission their own air quality monitoring if sources of emissions give rise to local concerns. Table 5.8 details the UK monitoring network. Fig. 5.9

Figure 5.9 Number of automatic monitoring stations in the UK since 1986 (Data from DETR, 2000b)

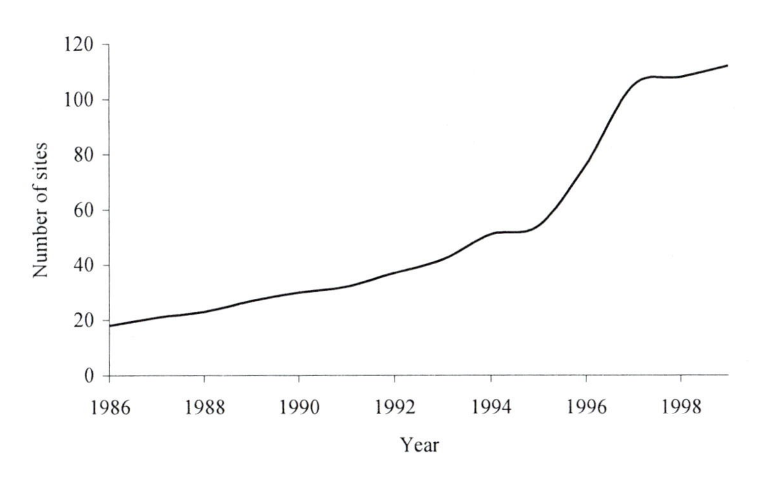

shows the growth in the number of continuous monitoring stations in the UK since 1986. The current level of automatic monitoring undertaken in the UK comprises more than 112 sites. The number of sites for specific pollutants includes (DETR, 2000b):

- O_3 is monitored at 52 urban sites and 19 rural sites;
- NO_x is monitored at 80 urban sites and 7 rural sites;
- CO is monitored at 65 urban sites only;
- SO_2 is monitored at 60 urban sites and 7 rural sites;
- PM_{10} is monitored at 52 urban sites and 3 rural sites;
- Hydrocarbons are monitored at 12 urban sites and 1 rural site.

5.9.2.2 US ambient air monitoring network

The USEPA's ambient air quality monitoring programme (Fig. 5.10) is carried out by State and local agencies and consists of three major categories of monitoring stations, State and Local Air Monitoring Stations (SLAMS), National Air Monitoring Stations (NAMS) and Special Purpose Monitoring Stations (SPMS) that measure the criteria pollutants (e.g. particulate matter, CO, SO_2, NO_2, O_3 and Pb). Additionally, a fourth category of monitoring station, the Photochemical Assessment Monitoring Stations (PAMS), which measures O_3 precursors (approximately 60 volatile hydrocarbons and carbonyl), has been required by the 1990 amendments to the CAA.

SLAMS consist of a network of approximately 4000 monitoring stations whose size and distribution is largely determined by the needs of State and local air pollution control agencies to meet their State Implementation Plans (SIP) requirements. NAMS (1080 stations) are a subset of the SLAMS network with emphasis being given to urban and multi-source areas. In effect, they are key sites under SLAMS, with emphasis on areas of maximum concentrations and high population density. SPMS provide for special studies needed by the State and local agencies to support SIP and other air program activities. The SPMS are not permanently established and can be adjusted easily to accommodate changing needs and priorities. SPMS are used to supplement the fixed monitoring network as circumstances require and resources permit. If the data from SPMS are used for SIP purposes, they must meet all QA and methodology requirements for SLAMS monitoring. A PAMS network is required in each O_3 non-attainment area that is designated serious, severe, or extreme. The required networks have from two to five sites, depending on the population of the area. There has been a phase-in period starting in 1994 and has exceeded 90 sites by the end of 5 years.

5.9.2.3 The Danish air quality monitoring network

The third Danish air quality monitoring programme (LMPIII) was started in 1992. The programme was revised considerably during 1992 compared to the previous phase (LMPII) (Palmgren, Kemp and Manscher, 1992). The programme comprises an urban monitoring network with stations in three Danish cities. The programme is carried out in a co-operation between the National Agency of Environmental Protection, the National Environmental Research Institute (NERI), the Greater Copenhagen Air Monitoring Unit and the municipal authorities in the cities of Odense and Aalborg.

NERI is responsible for the practical programme together with the Agency of Environmental Protection, Copenhagen, the Environmental and Food Control Agency, Funen and the Department for the Environment and Urban Affairs, Aalborg. The measuring programmes at the stations in operation during the major part of 1997 are shown in Table 5.9. Sites and measuring methods are described in Kemp (1993).

Figure 5.10 USEPA currently active air quality monitoring sites for CO, NO_2, O_3, Pb, PM_{10}, particulates and SO_2 (USEPA, 2000) (courtesy of the USEPA internet site (www.epa.gov))

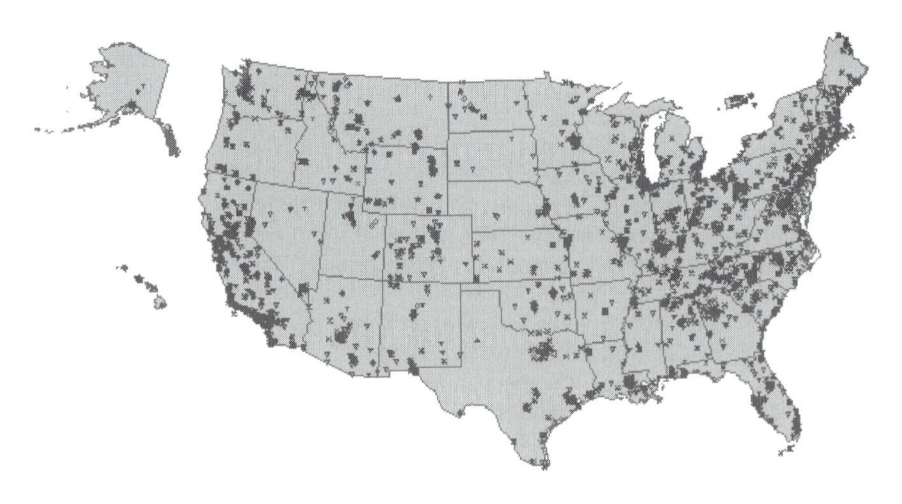

Table 5.8 UK Air quality monitoring network (UNECE 1995, DoE, 1997c)

UK Network	*Pollutants*	*Sites*
Urban	O_3, NO_x, SO_2, CO, PM_{10}	46
Hydrocarbons	25 species	12
Rural	O_3, NO_x, SO_2	16
Diffusion tubes	NO_2	1190
Smoke and SO_2	Smoke, SO_2 (rural SO_2)	222 (29)
Metals in air	Pb, metals	28
Acid deposition	Anions and cations, NO_2	32
TOMP	PAH, PCB, dioxins	4

Table 5.9 Air quality monitoring stations forming part of the third Danish programme

Site name/number	*½ hour average*	*24 hour average*
Copenhagen/1257	NO, NO_2, SO_2,, CO, O_3	SO_2, TSP, Elements
Copenhagen/1259	O_3, NO, NO_2, CO, meteorology	-
Odense/9155	NO, NO_2, SO_2, CO	SO_2, TSP, Elements
Odense/9159	O_3, NO, NO_2, meteorology	-
Odense/9154	-	SO_2, TSP, Elements
Aalborg/8151	NO, NO_2, SO_2, CO	SO_2, TSP, Elements
Aalborg/8159	O_3, NO, NO_2, meteorology	-
Lille Valby/2090	NO, NO_2, SO_2, O_3	SO_2, TSP, Elements
Keldsnor/9055	NO, NO_2, O_3	-

Figure 5.11 Singapore air monitoring network (Singapore Government, 2000) (courtesy of the Public Affairs Department Singaporean Government)

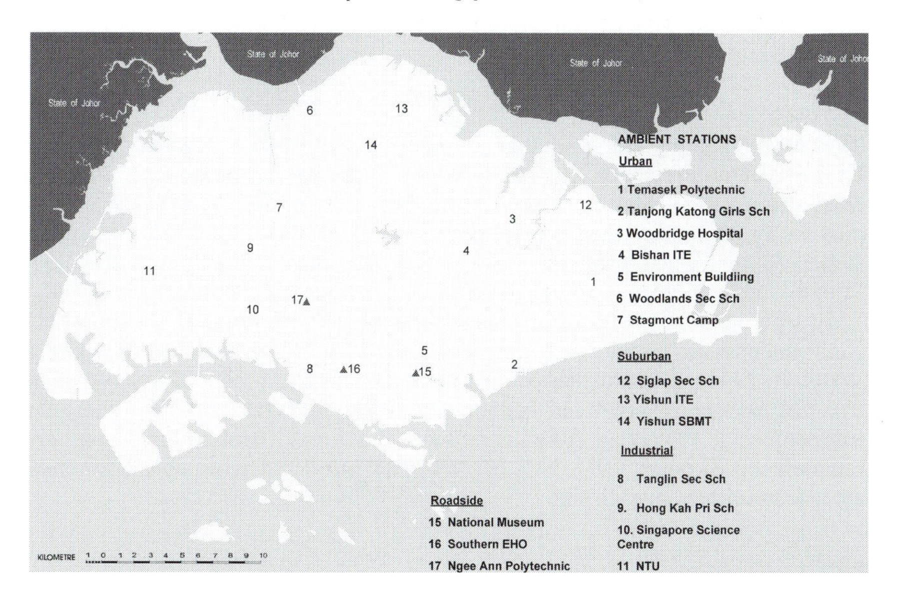

5.9.2.4 India

The nationwide programme was initiated in 1984. In 1995 the network comprised 290 stations covering over 90 towns and cities distributed over 24 States and four Union Territories. The National Ambient Air Quality Monitoring Network is operated through various agencies that include the respective States Pollution Control Boards and the National Environmental Engineering Research Institute, Nagpur. The pollutants monitored are SO_2, NO_2 and SPM. In addition to the three conventional parameters, National Environmental Engineering Research Institute monitors special parameters, including NH_3, H_2S, RSP and PAH.

5.9.2.5 Singapore

The Strategic Planning and Research Department through the Telemetric Air Quality Monitoring and Management System routinely monitor the ambient air quality in Singapore. The system comprises 15 remote air monitoring stations (Fig. 5.11) linked to a central control system via dial-up telephone lines. Of the 15 monitoring sites, 12 monitor general ambient air quality and three monitor roadside air quality. Automatic analysers measure SO_2, NO_2, CO, Pb, O_3, hydrocarbons, PM_{10} and $PM_{2.5}$. Owing to rapid development programmes in the north eastern and north western areas of Singapore four new stations were added to the network in 1999.

5.9.2.6 Tehran, Iran

The Air Quality Control Company, a subsidiary of Municipality of Tehran, is in charge of environmental management plans within the Greater Tehran Area. The Air Quality Control Company also acts as technical and supervisory board to the Municipality of Tehran on all environmental pollution affairs which includes air quality management. The Air Quality Control Company has four fully automated ambient air monitoring stations continuously measuring air quality.

5.9.2.7 Shenyang, People's Republic of China

The Municipality's air pollution monitoring network is operated by State Environmental Protection Bureau (SEPB). Shenyang Environmental Protection Monitoring Centre Station, which is a subdivision of SEPB, is responsible for monitoring air quality (and other environmental media). Air quality monitoring started in the Municipality in the 1970s using manual monitoring techniques. Initially measurements were undertaken at 15 sites with a monitoring frequency of 5 days per season. The number of sites was then rationalised in an optimisation study undertaken by the State Environmental Protection Administration and the WHO. In 1994 the Centre started automatic sampling at a reduced number of sites (5). The pollutants monitored include SO_2, NO_x, CO and TSP. The latest State Environmental Protection Administration regulations extend ambient monitoring requirements to cover PM_{10}, NO_2, O_3, Pb, BaP and F. Meteorological parameters are also monitored at each site.

Table 5.10 Air quality monitoring undertaken in Shenyang (AEA Technology, 2000a)

Type of site	*Sites*
Automatic sites operating >144 days/yr for SO_2, CO, TSP and NO_x	5
Manual sites operating 48 days/yr for SO_2, NO_x, TSP, CO	1
Manual sites for dust-fall rate	81
Manual sites for sulphating rate	81

In 1998 the State Environmental Protection Administration required the key monitoring stations of the national air monitoring network to implement weekly air quality reporting, which necessitated daily reporting and forecasting. Shenyang was one of the cities required to comply with the request. Existing manually operated sites could not meet these reporting requirements; therefore SEPB embarked in 1998 upon improving their air quality monitoring system (AEA Technology, 2000a). In 1999 the sites were converted to continuous automatic sites operating all year. A rural manual operated site is located outside the city at Huishan. Monitoring sites are located in differing land use zones (e.g. industrial, domestic, etc.). Table 5.10 summarises the air quality monitoring undertaken in Shenyang (Chapter 10, Case Study 13, Section 10.14 details an assessment of data for the Shenyang air quality monitoring network).

In addition the monitoring network also comprises 81 passive monitoring sites measuring dust-fall, sulphation rates and NO_x. In 1999 there were 57 precipitation samples collected and analysed for pH and conductivity. In addition, about 20 samples of rainfall and snowfall were analysed for selected anions and cations.

5.9.2.8 Helsinki, Finland

Helsinki Metropolitan Area Council is the joint municipal organisation of the capital area of Finland. It consists of four cities, i.e. Helsinki, Espoo, Vantaa and Kauniainen. The Helsinki Metropolitan Area Council is responsible for matters of their common interest, especially regional public transport, waste management, air quality and development planning. Air quality monitoring is a responsibility of the Helsinki Metropolitan Area Council Environmental Office. Air quality is measured at five stations. The two urban monitoring stations are located in Helsinki at Töölö and Vallila, while the two suburban stations are located in Espoo at Leppävaara and in Vantaa at Tikkurila. Urban background levels are monitored at the Luukki station in Espoo, approximately 20km north-west of the centre of Helsinki. Furthermore, there are two additional stations in Helsinki: one at Kaisaniemi monitoring only SPM and a meteorological station at Kallio giving data on wind direction and speed, temperature, precipitation, and relative humidity. The Helsinki Metropolitan Area Council Environmental Office also has three mobile monitoring stations, which are generally used for air quality monitoring in environments with busy traffic, the typical measuring periods being from six months to one year. The pollutants monitored at these stations depend on the local interests and requirements. The concentrations measured include those of SO_2, NO, NO_2, CO, O_3, non methane hydrocarbons, TSP and PM_{10}. Air quality data obtained from the monitoring stations is collected in computers located in the Helsinki Metropolitan Area Council Environmental Office before being distributed to terminals at the Environmental Offices of the cities of Helsinki, Espoo, Kauniainen, and Vantaa. Air quality data and trends are analysed in monthly and annual reports as well as in a summary report at five-year intervals.

Table 5.11 provides a summary of selected national air quality monitoring programmes (Murley, 1995).

5.10 SOURCE MONITORING

Source monitoring may be undertaken for both mobile and stationary sources. The latter are generally easier to measure, although all sources exhibit their own intrinsic measurement problems. Source monitoring may be undertaken for a number of reasons. These include (Hewitt and Harrison, 1986; SEPA, 2000):

- Determination of the mass emission rates of pollutants from a particular source, and assessment of how these are affected by process variations;
- Evaluation of the effectiveness of abatement techniques for pollution control;
- Concern over the results from operator monitoring;
- Concern over the sample or analytical methods used by the operator;
- Lack of confidence in operator compliance monitoring;
- A release to air is significant;
- Concerns about the impact or potential of a release into the air;
- Process is under complaint;
- Provide independent data on a discharge;
- Evaluation of compliance with emission consent levels.

To be representative and purposeful the sample should accurately reflect the true magnitude of the pollutant emission at a specific location in the process at a given time. This requirement is met by adequate sampling and instrument design (Hewitt and Harrison, 1986). In addition enough measurements should be taken over a period of time and space so that their combined result will accurately and adequately represent the source's emission characteristics. This will require consideration of the emission both in time and space.

5.10.1 Point source monitoring

Where possible all samples should be taken iso-kinetically. The word iso-kinetic means equal speed or velocity. Therefore the velocity of the sample gas entering the sampling nozzle should be the same as that of the gas in the emission exhaust stream at the sampling point. For small emission sources *in situ* design constraints may make it difficult or impracticable to achieve this requirement. In order to take a representative sample from a stack certain other requirements must be satisfied, such as the selection of a suitable sampling position. Therefore it is important that consideration is given to the following criteria:

Table 5.11 National air quality monitoring programmes (Murley, 1995)

Country	*Pollutant and number of sites monitored*
Austria	From April 1994 SO_2 (202), PM (145), NO_x (184), CO (80), O_3 (137), HC (12), H_2S (11), NH_4 (2) and wet deposition (13)
Belgium	SO_2 (60), automatic sites (e.g. SO_2 and occasional PM, NO_x, NMHC), O_3 (72), heavy metals (60), deposition network for heavy metals (60)
Croatia	Continental and coastal areas urban (8), industrial (3) and tourist sites (6) (e.g. SO_2, dust deposition, smoke at all sites and NOx, SPM at some sites). In industrial areas specific pollutants from local industries measured. The State Hydrometeorological Institute runs the background monitoring network, which consists of SO_2 and smoke (7), NO_2 (10), NH_3 (1) and rainfall (13)
Czech Republic	Modern ambient monitoring system in Prague and northern Bohemia, >90 multi-component measuring stations, a number of additional monitoring stations are operated by municipalities and by industry. The Czech Hydrometeorological Institute monitors the levels of main pollutants (SO_2, NO_x, acidity, O_3, HC, and heavy metals)
Finland	At both a municipal and regional national level at background stations. Meteorological Institute responsible for measurements at background sites. Deposition quality monitored by Environment Agency at 40 background sites.
France	In 1992 SO_2 and acidity (446), NO_x (120), particulates and black smoke (227), CO (34), lead (26). VOCs and micro-pollutants now increasingly monitored.
Israel	In 1993 there were 63 monitoring stations.
Netherlands	26 suburban stations of which 15 are macro stations where a large variety of pollutants can be monitored. 7 city roadside stations (<2750 vehicles/day) and 13 sites >10000 vehicles/hour
South Korea	In 1994 there were 74 sites (SPM, SO_2, NO_2, O_3, HC). Scheduled to increase to >100. In 1996 67 acid rain sites

Number of monitoring sites in brackets

- Safety of location for operators;
- Sample plane selection;
- Sample points;
- Sampling platforms;
- Service requirements;
- Access to sampling platforms.

It is preferable to select a sampling position in a straight length of flue and as far as possible from any obstruction or construction which may cause disturbance. The distance of the sampling point should be at least one flue diameter upstream from the obstruction. The further the sampling point from a source of disturbance, the more representative will be the sample. It is necessary to sample at a number of suitably distributed points within each sampling position in order to obtain a good estimate of the effluent material. Fig. 5.12 shows a sampling platform on a stack and Fig. 5.13 shows continuous online stack sampling apparatus.

Figure 5.12 Sampling platform (courtesy of SGL Technic Ltd)

Guidance on the requirements for sampling of releases to atmosphere from a point source is given by the UK's former Her Majesty's Inspectorate of Pollution (HMIP) (1993a). Immediately before each sample measurement at a sampling point, a standard pitot tube (Fig. 5.14) is inserted into the flue to measure the gas velocity. From a calibration chart the pressure drop required across the sampling probe to ensure sampling at the same velocity as in the flue is obtained. The pitot tube is then removed and replaced by a sampling probe, to which suction is applied from a pump. A valve to give the correct pressure drop regulates the suction. Temperature of the flue gas is measured using a thermometer or thermocouple.

When sampling within a stack, it is necessary to sample at a number of suitably distributed points within each sampling position in order to obtain an adequate estimate of the concentration of effluent passing the sampling position. These points should lie along two sampling lines and the greater the number of sampling points the more reliable will be the result. Variations observed in the pitot static pressure differences determine the minimum number of sampling points at which samples are to be taken. An iso-kinetic sampling procedure would include the following steps (Palmer, 1974):

- Pitot static pressure difference readings should be made at the selected sampling points;
- Attach a nozzle of suitable size to the sampling probe to allow the sample to be taken iso-kinetically;

Figure 5.13 Continuous sampling apparatus (courtesy of SGL Technic Ltd)

- Insert the probe through one of the access holes with the nozzle facing downstream at the first sampling point and allow the probe assembly to attain the temperature of the flue gases;
- Turn the nozzle to face upstream and commence sampling at a rate corresponding to iso-kinetic sampling for the prescribed time;
- Progressively move the sampling probe from point to point, sampling at each for the same length of time and adjusting the sampling rate as necessary;
- Transfer to the next sampling line and repeat the procedure until the sampling has been completed at all points;
- If sampling for particulates, detach the bag or thimble containing the collected solids and weigh.

It is good practice when stack monitoring to carry out duplicate tests. To minimise the possibility of errors occurring in the sampling collection process the following should be observed:

- The test apparatus show be allowed to reach the temperature of the flue gas to avoid condensation and caking of particulates occurring in or on the sampling probe;
- The sample nozzle should not touch the surfaces (e.g. wall of stack) whilst sampling (i.e. to avoid dislodging particulates, etc);
- The nozzle must face squarely upstream at all times during the sampling period;

Figure 5.14 Pitot-static tube (after Palmer, 1974)

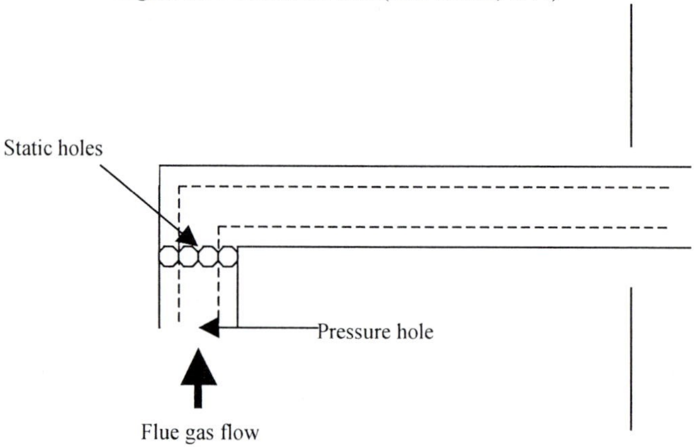

Static holes

Pressure hole

Flue gas flow

- The flow rate through the sampling probe should not be impeded.

5.10.2 Mobile sources

Motor vehicle emissions (Chapter 4) are dependent upon a series of factors that effect the mass and type of pollutant emissions. These factors include:

- Type of engine (e.g. diesel, petrol, LPG, etc.)
- Speed of vehicle;
- Fuel impurities (e.g. sulphur content, lead content);
- Operator behaviour.

Figure 5.15 Motor vehicle drive cycle for emission sampling purposes (after Hewitt and Harrison, 1986) (Reproduced by permission of The Royal Society of Chemistry)

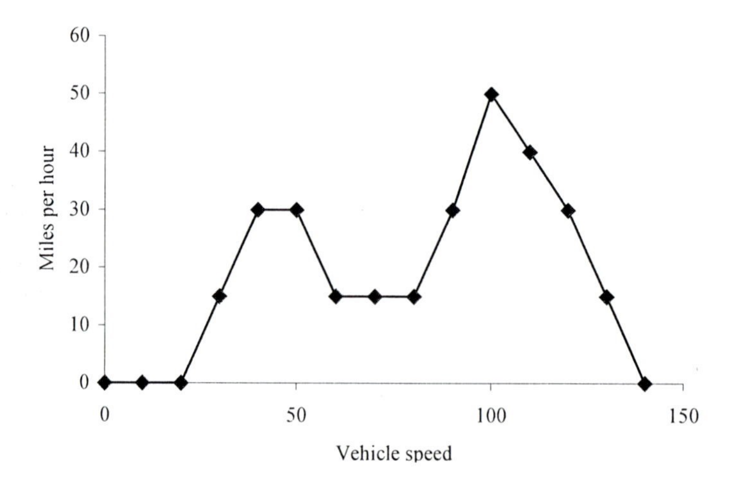

Consequently emission levels are variable and are considered specific to the type of operating cycle. Emission tests therefore reflect these cycles (Fig. 5.15) where sampling is a function of time and vehicle speed.

CHAPTER SIX

Impact Prediction

6.1 INTRODUCTION

Many pollutants show extremely complex dispersion patterns, especially in environments such as towns and cities where there are a large number of emission sources and variations in environmental conditions. This complexity means that it is often very difficult to model or measure pollutant patterns and trends, and thus to predict levels of human exposure. Nevertheless, modelling can compliment monitoring programmes and be an effective AQM tool. The types of model available to predict the impact of source emissions on ambient air quality or air sensitive receptors (ASR) depends on the user's needs and the morphology of the emitting source. Models can be simple or complex, the latter generally requiring onerous data sets. Models may also be point, area, volume or linear in nature as well as being either statistical, mathematical or physical. Mathematical modelling represents the real world in mathematical terms. Statistical models utilise relationships between two or more variables. Physical models are scaled down representations of the actual modelled feature. Modelling seeks a more precise understanding of the significant behaviour of air pollution measurements. It gives the means to check for consistency in the emissions inventories, observed meteorology and measured concentrations of the air pollutant (Middleton, 1996).

6.2 MECHANISMS FOR DISPERSION

Air pollution concentrations are influenced by prevailing weather conditions and vary considerably with time. These variations are largely determined by meteorological factors. The extent to which pollutants are dispersed and diluted is dependent on wind speed, turbulence, mixing depth, topography, etc. Air pollution can be dispersed, diffused and diluted by a number of factors. It is these factors that must be considered in any air dispersion modelling study. The principal factors of concern include:

- Meteorological conditions;
    a) wind speed and direction,
    b) atmospheric stability,
    c) temperature.
- Topography;
- Structures and buildings.

Separation or accumulation of pollutants occurs on the basis of their physical characteristics. Chemical reactions occur, breaking down the original pollutant or converting it into new compounds. Some pollutants can also be removed from the transporting medium through deposition, for example, by settling out under the

Figure 6.1 Meteorological monitoring station (photograph courtesy of Cordah Limited, 2000)

effects of gravity, by washout or by interception (scavenging) by plants and other obstructions.

Whether the modelling method used is short-term or long-term determines the meteorological data requirements for the study. Site-specific data are often difficult to obtain and therefore on-site monitoring is often the most suitable remedy to the problem (Fig. 6.1). This is a costly exercise and therefore may restrict the parameters monitored. Consequently, studies often make use of either assumed meteorological conditions to simulate worst case conditions or meteorological data from the nearest available monitoring site, which inevitably has been recorded by a governmental department (i.e. the UK Meteorological Office) or other interested bodies (i.e. research institute).

6.2.1 Wind speed and direction

Wind speed is determined by atmospheric pressure gradients, which are shown on weather charts by isobar lines (i.e. lines of equal pressure). Isobars close together indicate the wind speed will be high. Under such conditions pollutant concentrations are likely to be relatively low (i.e. dispersed and diluted). Low wind speeds tend to result in the accumulation of pollutants. Pollution concentrations are inevitably inversely proportional to the wind speed (Fig. 6.2a). Table 6.1 lists typical terms given to describe wind speed.

Wind speed also varies with height over the lowest few hundred meters of the earth's atmosphere due to the frictional effect of the earth's surface (Fig. 6.2b). The variation is greatest over rough surfaces (e.g. cities) when the effect could be a

Table 6.1 Beaufort scale of wind velocities (from Parker, 1978)

Force	*Beaufort scale description*	*Velocity(mph)*
0	Calm	0
1	Light	1-3
2	Light	4-7
3	Light	8-12
4	Moderate	13-18
5	Fresh	19-24
6	Strong	25-31
7	Strong	32-38
8	Fresh gale	39-46
9	Strong gale	47-54
10	Whole gale	55-63
11	Storm gale	64-72
12	Hurricane	73-82

Figure 6.2 The effect of wind speed on air pollution dispersion and dilution

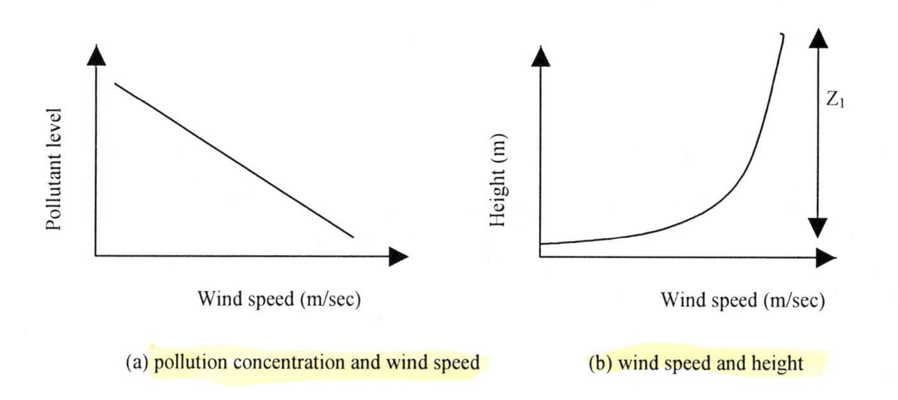

(a) pollution concentration and wind speed (b) wind speed and height

reduction in wind speed of 40%. Over smooth surfaces (e.g. sea) the effect is less and the reduction may only be 20% (Parker, 1978). The frictional drag retards motion close to the ground and therefore gives rise to a sharp decrease of mean horizontal wind speed (u) as the surface is approached (Oke, 1978). The depth of the boundary layer (Z) above which u is approximately constant with height is a function of the roughness of the surface. The force exerted on the surface by the air being dragged over it is called the surface shearing stress (τ) and is expressed as a pressure (Pa, force per unit surface area) (Oke, 1978). A simple power law (equation 6.1) can represent the vertical wind speed profile within the turbulent boundary layer (Turner, 1994).

$$u_z = u_a(Z_1/Z_a)^{\alpha} \tag{6.1}$$

The ASRs most affected by air pollution are those situated downwind of the emission source. Knowledge of the prevailing wind direction is therefore important in predicting the likely impact of the source. A more detailed presentation of the information can be made using a wind rose (i.e. circular histogram) in which the length of the line is proportional to the frequency of occurrence of the wind in each sector (e.g. 45°, 30°, etc) (Fig. 6.3a). The actual fluctuation of wind direction during any amount of time will dilute the concentrated pollutant owing to the greater volume of air into which it is mixed (Fig. 6.3b). Over a few minutes the dilution will be minor and therefore peak pollutant concentrations will remain. Over several hours the dilution effect will be greater and therefore it will reduce pollutant concentrations.

6.2.2 Atmospheric turbulence

Turbulence in the atmosphere is created by two principal actors, the minor of which is surface roughness. This occurs when wind blowing over a surface creates mechanical mixing in the atmosphere. Greater surface roughness, due to buildings and topography, generally acts to increase turbulence, and therefore encourages dispersion. The exception to this may occur during periods of very light winds, when air may stagnate close to the ground, especially in areas like valleys and street canyons. For elevated emission sources, such as stacks, the influence of turbulence can have the opposite effect, as it may bring the plume to ground more rapidly, giving rise to greater concentrations than would be found around a similar source in a less turbulent zone (Section 6.2.4).

The major cause of atmospheric turbulence, however, is incoming solar radiation. Fig. 6.4 shows that if a parcel of air moves upwards in the atmosphere it expands and cools owing to a reduction in air pressure. Assuming that there is no exchange of heat with the surrounding air, the parcel cools at a rate of about 9.8°C/km (dried adiabatic lapse rate (DALR)). However, such a lapse rate of temperature rarely occurs in the atmosphere, since although mechanical turbulence tends to maintain a DALR, atmospheric daytime convection and outgoing radiation at night breaks up the neutrality of the atmosphere (Oke, 1978).

The environmental lapse rate (ELR), which is the actual temperature profile of the atmosphere, varies with different air masses and at different times of the day. If the ELR is equivalent to the DALR the atmosphere is said to be neutral. The stability of the atmosphere can be defined with reference to this neutrally stable state. For example, on days with strong solar heating of the ground the decrease in air temperature with height often exceeds that of the DALR. A parcel of air which rises will always find itself in surroundings cooler than itself and therefore continues to rise, favouring convective instability. Correspondingly, an atmosphere in which the temperature decreases more gradually than the DALR or even increases with height is stable. In cases of a temperature inversion all vertical motion and turbulence are suppressed and any air pollution trapped in an inversion cannot disperse. On the other hand, emissions released above the inversion layer do not ordinarily reach ground level (Oke, 1978).

Figure 6.3 The effect of wind direction on pollution dispersion and dilution

(a) A wind rose, Kinloss, Scotland 1987 – 1996 (Data from Meteorological Office, 2000)

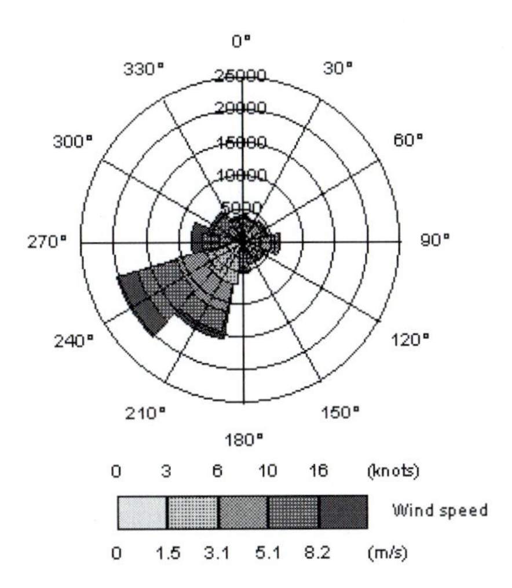

(b) The dilution of pollution concentration with a variation in wind direction (after Oke, 1978) (courtesy of ITPS)

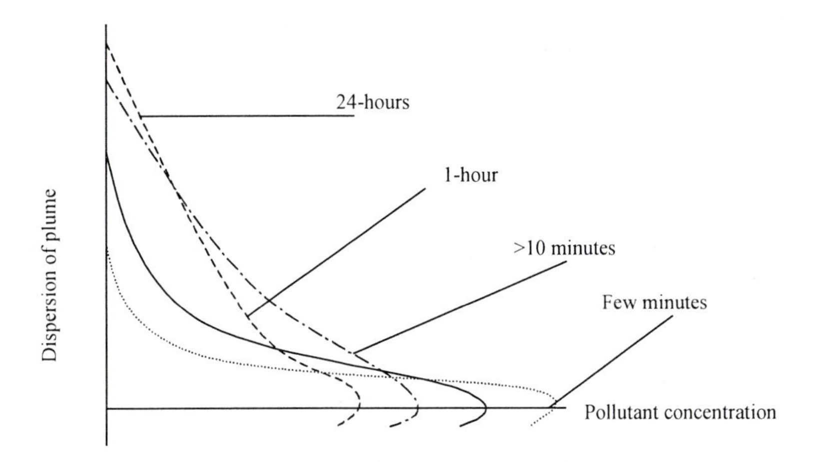

Atmospheric stability is often divided into 7 categories from A (very unstable), through D (neutral) to G (very stable). In addition, hybrid categories A/B, B/C and C/D which are intermediate between A and B, B and C, and C and D, are also defined. The atmospheric stability classes are generally referred to as the Pasquill-Gifford categories. Category A is often associated with a more rapid dispersion of pollution and category G with temperature inversions and the possible accumulation of pollution. The stability categories are estimated from the total cloud amount, wind speed and time of year. During the day an estimate is made of the incident solar radiation and this is combined with wind speed to estimate the stability category. At night, stability is a simple function of cloud cover and wind speed (Table 6.2).

Table 6.2 Pasquill stability categories (Pasquill and Smith, 1982)

Wind speed (m/sec)	*Daytime incoming solar radiation*				*Within 1 hr of sunrise or sunset*	*Night time cloud amount (oktas)*		
	Strong >59	*Moderate 30-59*	*Slight <29*	*Over-cast*		*0-3*	*4-7*	*8*
<2	A	A-B	B	C	D	F	F	D
2-3	A-B	B	C	C	D	F	E	D
3-5	B	B-C	D	D	D	D	D	D
5-6	C	C-D	D	D	D	D	D	D
>6	C	D	D	D	D	D	D	D

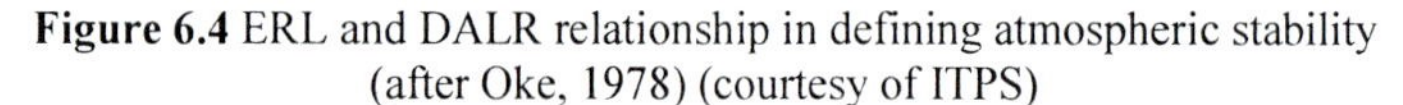

Figure 6.4 ERL and DALR relationship in defining atmospheric stability (after Oke, 1978) (courtesy of ITPS)

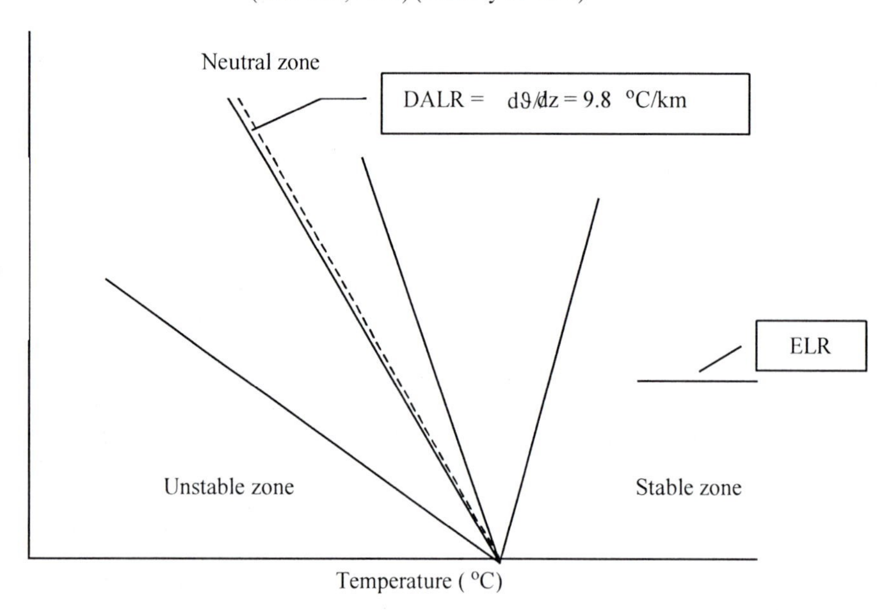

Figure 6.5 Formation of advective temperature inversions (from Oke, 1978) (courtesy of ITPS)

Frontal inversions
a) cold air wedging under warmer air (cold front)
b) warm air overriding colder air (warm front)

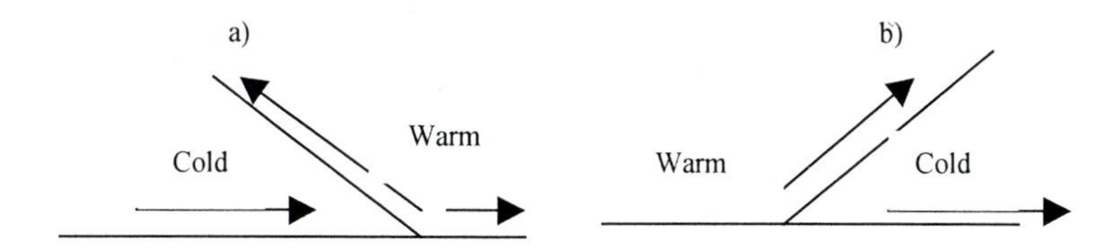

6.2.3 Temperature inversions

By definition an inversion exists when warm air overlies cooler air (i.e. air temperature increases with height). The formation of an inversion can be due to a number of reasons. Typically a surface temperature inversion is normally formed at night by radiative cooling of the ground and subsequently of the air near to the ground. Another mechanism for the formation of a temperature inversion is when air sinks it encounters greater pressure, it is compressed and warms (usually adiabatically). This results in a characteristic elevated bulge in the air temperature profile so that the lowest portion of the warm layer exhibits a subsidence inversion which forms a lid to the underlying mixed layer over an extensive area. Such a meteorological condition is common in large anticyclonic conditions. Inversion conditions may also occur by the advection of warmer or cooler air (Fig. 6.5). Weather fronts are the boundary zones between contrasting air masses. The full frontal surface extends from the surface up into the atmosphere as a sloping plane with the warmer less dense air invariably overlying the colder air. Fronts are therefore always characterised by an inversion (Oke, 1978). Rosenberg (1974) and Oke (1978) provide further examples of the formation of temperature inversions.

6.2.4 Topography

Topographic features can modify the general pattern of wind speed and direction and they can also cause microclimates to develop. Topographically induced radiation variations may cause energy balance differences across the landscape. For example, a south-facing slope will receive greater net radiation than that of a north-facing slope. The south-facing slope is therefore likely to produce greater levels of sensible heat into its surrounding lower atmosphere. Such strong differential heating will produce local slope winds and will produce a wide spectrum of microclimates. The nature of these wind systems will depend on the valley's orientation and geometry. The most pronounced systems are found in a deep and straight valley with a north-south axis. In valleys with other orientations or possessing more complex geometries the airflow system may be either asymmetric or incomplete. Terrain with a mean slope greater than 10% will affect the modelled dispersion of emissions (Environmental Resources Management Ltd., 2000b). Fig. 6.6 shows how the process may occur. During the day air above the valley will be heated by the underlying surface to above that over the centre of the valley. As a result anabatic flow arises and to maintain continuity a closed circulation develops across the valley involving air sinking in the valley centre. At night the valley surfaces cool by the emission of long wave radiation. The lower air layers cool and slide down the slope under the influence of gravity (katabatic winds). The cold and dense air settles to the bottom of the valley and a temperature inversion is created (Oke, 1978).

Figure 6.6 The effect of topography on wind movement (after Oke, 1978) (courtesy of ITPS)

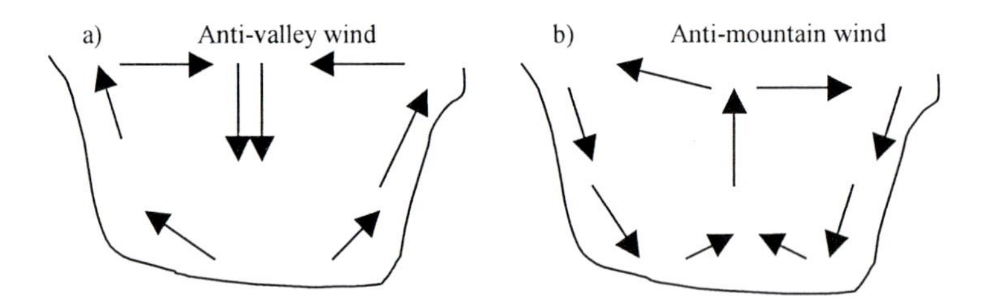

a) During the day winds are anabatic, and the valley wind fills the valley and moves upstream, with the anti-valley wind coming downstream
b) During the night the slope winds are katabatic and reinforce the valley wind which flows downstream, with the anti-mountain wind flowing in the opposite direction

Airflow over non-uniform terrain is not easy to generalise. Nevertheless, an obstacle will create an alteration in the flow pattern, so that the detailed wind regime of every landscape is unique. Oke (1978) identifies some general observations about flow patterns around selected features:

- For flow over a break of slope (i.e. across a valley) an air stream will enter a zone of low pressure. At this point the flow is more turbulent and ground level winds are typically opposite to the direction prior to entering the valley.
- For flow through a topographic constriction wind speeds increase. Upon leaving the constriction the flow diverges and decelerates.
- Foı flow around an isolated object (e.g. building), if it is laminar it is accelerated over the upstream portion of the obstacle and decelerated there after. Beyond the obstacle the flow is retarded by friction. Flow is brought to a standstill where flow separation takes place. Behind the obstacle within the separated flow the motion is considerably more turbulent. This zone is called the wake and includes a lee eddy or vortex immediately behind the object where some of the air becomes trapped in a zone of relatively low pressure. At regular intervals a vortex detaches itself and moves downwind. The vortices are shed alternately with clockwise and counter clockwise rotation (Fig. 6.7).

There are no rules of any consequence relating to the effects of irregular terrain on the dispersion of air pollution. Maps provide most topographical data for study purposes. However, concerns for topographical influences on air dispersion and meteorology arise when an air dispersion model used may not necessarily be

Figure 6.7 The effects of building/structures on wind movement (Cambridge Environmental Research Consultants Ltd, 2000) (courtesy of Cambridge Environmental Research Consultants Ltd)

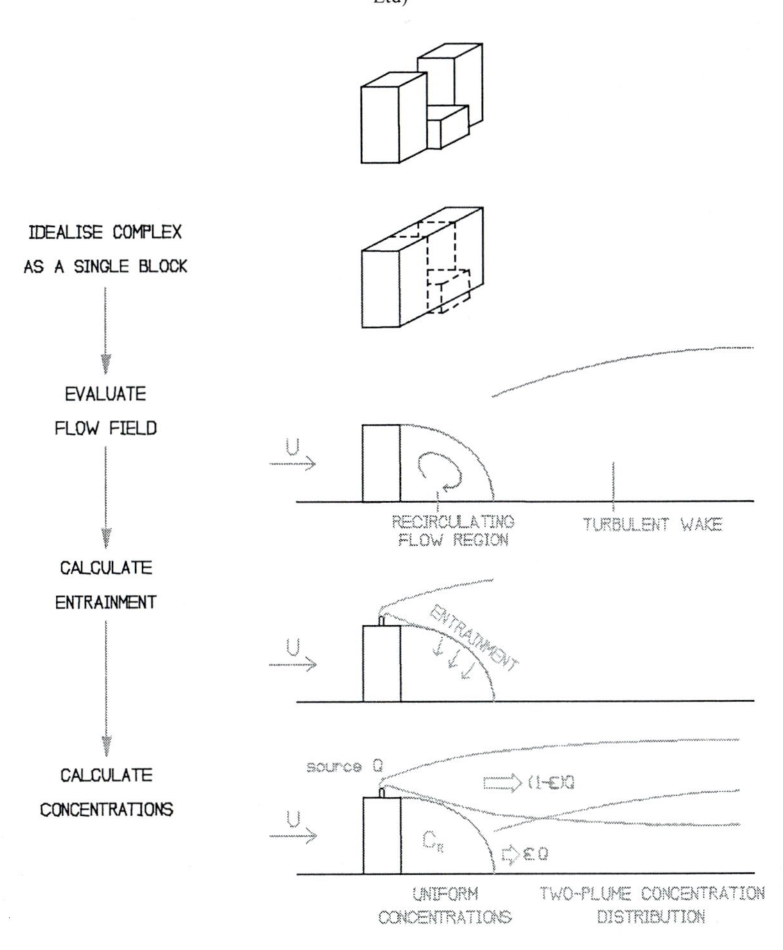

totally applicable to a scenario of diverse terrain. Micro- and macro-meteorological data (i.e. temperature inversions) are seldom available for study areas. The idiosyncrasies of topography and obstacles affecting local meteorology and climate are often beyond the resources of most studies to identify and therefore are frequently neglected or simplified.

6.2.5 Plume formation

After the initial rise of the plume from an emitting source its formation downwind will be largely governed by prevailing meteorological conditions. Using atmospheric stability as a surrogate for turbulence it is possible to classify the most characteristic plume patterns into five basic types (Fig. 6.8) (Oke, 1978). A looping plume is typical of summer daytime conditions with strong atmospheric instability (stability class A). The atmosphere is characterised by large eddy currents associated with free convection that transport the plume in an up and down motion. The eddies can bring the undiluted plume into contact with the ground close to the emission source causing elevated pollution concentrations. A coning plume can occur during the day or night and is characteristic of windy and/or cloudy conditions with atmospheric stability conditions close to neutral (stability class D). Due to the absence of thermal buoyancy in the boundary layer the turbulence comprises smaller frictionally-generated eddies of forced convection and therefore the vertical and lateral spreading of the plume is about equal so that it forms a cone shape symmetric about the plume centre line, because the plume diameter only grows by diffusion, and there is little vertical transport. A fanning plume is characteristic of a strong atmospherically stable situation (stability class F). Stable air suppresses any buoyant movement and with very light winds forced convection is unlikely to be significant at the height of an elevated plume. The lack of vertical diffusion therefore keeps the plume thin in that direction, but the erratic behaviour of wind direction in stable conditions may allow a fan shape to develop when viewed in a plan perspective. At other times the plume may appear as a straight or meandering line. A lofting plume is found in the early evening during the period when nocturnal radiation inversion is building up from the earth's surface. The atmospheric stable layer beneath the plume hinders its transport downward towards the earth's surface but the unstable layer aloft allows the plume to disperse upwards. Fumigation occurs when an inversion lid exists above the plume that obstructs its upward dispersion, but there is a lapse temperature profile beneath so that there is ample buoyant mixing capable of bringing the plume contents to the surface of the earth.

6.3 AIR DISPERSION MODELLING

Many air pollutant problems are often best evaluated by monitoring. This, however, may be expensive, both in terms of staff time, equipment and laboratory costs. Monitoring is inhibitive in that it is not predictive in nature (Chapter 5). One relatively inexpensive and increasingly used alternative is to employ computer-based models, which can simulate the dispersion of air pollution into the atmosphere. However, it should be remembered that monitoring and modelling when applied together compliment each other and have a greater influence on gauging the level of the impact than if the techniques are used separately.

Models have been developed for a variety of pollutants, time scales and operational scenarios. Short-term models are used to calculate concentrations of pollutants over a few minutes, hours or days and can be employed to predict worst case conditions (i.e. high pollution episodes). Long-term models are designed to predict seasonal or annual average concentrations, which may prove more useful in studying health effects and impacts on vegetation, materials and structures.

The purpose of modelling is to provide a more precise understanding of the behaviour of air pollution dilution and dispersion patterns. Models, whether physical, statistical or mathematical, provide a scientific means of relating

Figure 6.8 Plume formation under varying atmospheric stability conditions (after Oke, 1978) (courtesy of ITPS)

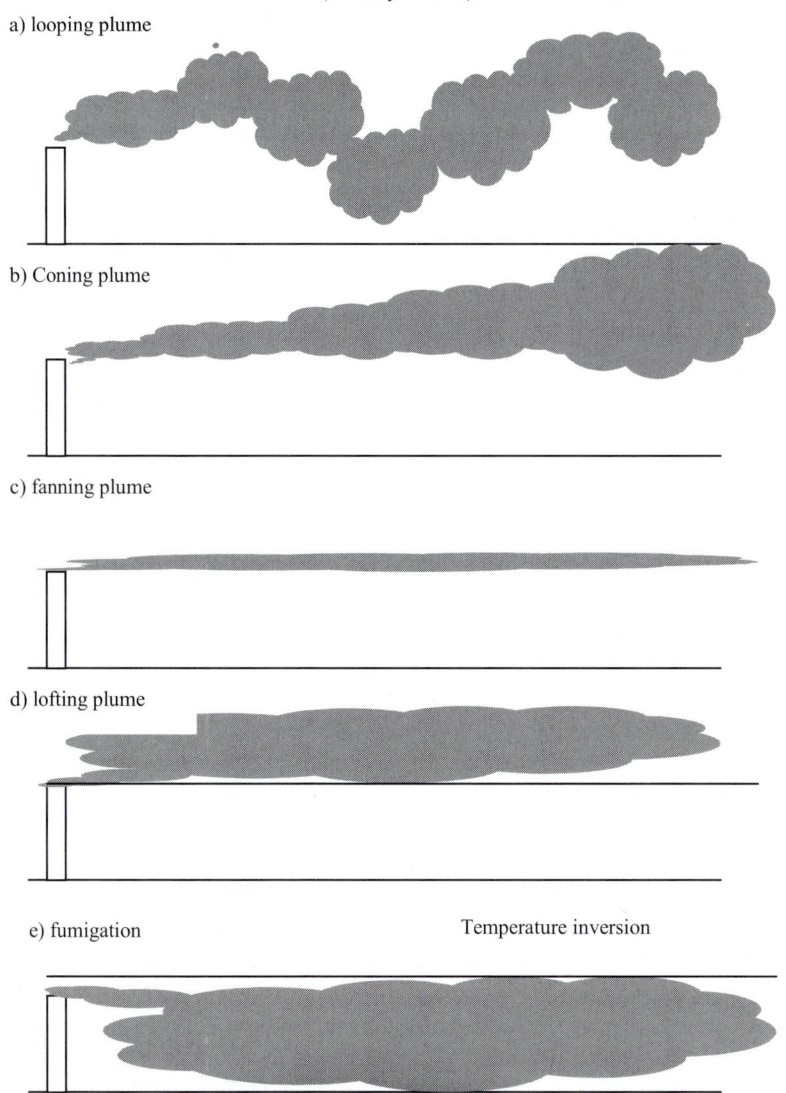

emissions and atmospheric processes to provide estimates of ambient air quality values (Szepesi, 1989). Computer-based models consist of mathematical equations relating the release of emissions into the atmosphere to expected concentrations in ambient air. Air quality models, therefore, can be used to help to identify and to evaluate the level of controls required to remedy air pollution problems. Source emissions, meteorological observations and air quality models are three of the basic components of assessing air quality impacts. The applications of models are numerous and have previously been categorised as follows (Harrop, 1986):

- Historical air quality trend analysis;
- Characterisation of existing air quality in multi-source areas:
  a) temporal and spatial air quality patterns;
  b) identification of source contributions;
  c) likelihood of high pollution episodes;
- Chimney height optimisation;
- Support for impact and economic studies;
- Environmental impact statements;
- Attainment of air quality standards and guidelines:
  a) formulation of control policies;
  b) regulation of polluters;
  c) urban/city alert system;
  d) accidental releases of toxic substances.

Case Studies 5, 7, 10, 11 and 14 (Chapter 10) provide examples of the application of air dispersion modelling techniques for line, point and area sources.

6.3.1 What is a model?

Modelling in the present context is a process to calculate concentrations based on physics where concentration (mass/volume) is the rate of emission (mass/time) divided by the rate of mixing (volume/time) (Middleton, 1996). A model or paradigm is a representation of an existing object or system. It represents the real world in mathematical terms based on physical principles (Middleton, 1996) and is as old as applied mathematics (Scorer, 1998). Turner (1994) has stated that solving a single equation for plume rise to estimate the effective height of the plume from a single source and using a dispersion equation to estimate the concentration from the source on a single receptor would not generally be considered a model. It is usually when a computer is used for repetitious solutions of these equations that we refer to the calculations as being a dispersion model.

Szepesi (1989) has categorised dispersion models into eight classes (Table 6.3). Most current models are deterministic, some are stochastic and a few are adaptive. In fact all air quality models representing specific real situations have features that are either explicitly or implicitly adaptive. The reasons for adaptive systems are the uncertainty in the meteorological data and knowledge of atmospheric turbulence; chemical reactions and reaction rates in the atmosphere; knowledge of the chemical species from, and the emission characteristics of, natural sources; data and knowledge of gas to particle and particle/particle interactions. Szepesi (1989) has identified the procedure for model and database selection. The suitability of an air dispersion model to the evaluation of source impact and control strategies depends upon several factors, these include:

- Detail and accuracy of the database (i.e. emission inventory, meteorological data, air quality data, etc);
- Meteorological and topographical complexities of the area;
- Technical competence of those undertaking the simulation modelling;
- Resource availability.

Table 6.3 Dispersion model categories (Szepesi, 1989)

Type of model	*Description of model*
Simple deterministic	Based on empirical data and are formulated in terms of algebraic relationships (e.g. air pollution indices, simple area source models, etc)
Statistical	Given a set of variables a statistical relationship among them is possible by using regression and spectral analysis, or some other statistical technique. Types of model include averaging time models, time series analysis, etc
Local plume/puff	Simple models for uniform wind and thermal conditions possess the following properties: • concentrations from a continuous steady source vary in proportion with the source strength or rate of emission; • concentrations vary inversely with the mean wind speed at the source; • distance from the source to the receptor and intensity of atmospheric turbulence determine the height and width of a plume at the receptor; • concentrations may decrease because of chemical reactions, radioactive decay and removal. Maximum practical time scales are 1 to 2 hours with maximum space scales of 10 to 30 km; • extra provisions are made to account for calm winds, complex terrain and low inversion levels
Integral, box, column, moving cell and multi-box	Based on the integral form of the diffusion equation over a volume (e.g. region over an urban area, volume in a valley or basin, etc).
Finite difference and grid	Based on solving the transport and diffusion equations by finite difference approximations
Particle	Mixed Lagrangian/Eulerian models, which follow a pollutant passing through an Eulerian grid. In this method spatial distribution of the pollutant is represented by a large number of Lagrangian particles of constant mass that are advected in a fictitious velocity field consisting of the true velocity field plus a turbulent flux velocity field
Physical	Wind tunnel, liquid flume and towing tank are physical modelling techniques. The greatest asset is the ability to investigate dispersion of pollutants for configurations too complicated to be economically simulated by mathematical modelling
Regional	Models usually relate to horizontal scales of several hundred kilometres. (e.g. grid, particle and trajectory models)

6.4 MODEL CHARACTERISTICS

The air pathway process that controls the fate of pollutants from source to receptor is transport, diffusion, transformation and removal. Owing to the complexity of these processes, there exists a large and diverse suite of air models, each having their own dispersion characteristics, some of the most important are detailed below (Szepesi, 1989).

6.4.1 Time and space scales

Air pollution decisions can be characterised in terms of geographical scales. Szepesi and Fekete (1987) have applied a scheme of scales from global to the local level. Table 6.4 shows the relationship between time and space scales where L_{min} and L_{max} represent the minimum and maximum length scales of the phenomena in question.

Models are either steady state or time dependent. If the system of equations governing the phenomena being studied in the model depends on time, the model is time varying. If the system represents the average state of phenomena over a period of time, the model is in steady state. Generally steady state models are applicable when the time and spaces are sufficiently small or when the desired output is sufficiently coarse that variability in the effects of pollutants, emissions and meteorology can be ignored or averaged out. For example the steady state Guassian plume can be used over site-specific scales if the winds and atmospheric thermal structure are nearly uniform over the period of time of interest (e.g. 1 hour).

Table 6.4 Scheme of air pollution scales (Szepesi, 1989)

Scale	*Description of area*	*L_{min}/L_{max} (km)*	*Time*
Global	Continents on the hemisphere or globe	800/40000	3 month
Continental	Regions on a continent	200/10000	3 d
Regional	Cities in a region	40/2000	3 hr
Local	Urban area	2/100	1 hr
Local	Site-specific		

6.4.2 Frame of reference

Air dispersion models, except for some empirical models, are related to a co-ordinate system or reference frame. Reference frames may be fixed at the earth's surface (e.g. ordnance grid reference point) or on a puff of pollutant as it moves downwind from the source. Reference frames fixed at the earth's surface or on the source are called Eulerian while frames fixed to a puff of pollutant are called Lagrangian.

6.4.3 Pollutants and reaction mechanisms

Air quality models describe the fate of airborne gases and particles. As these pollutants travel over their pathways, physical and chemical reactions may occur. The general modelling categories for these reaction mechanisms include:

- Non-reactive such as CO emissions from fossil fuel combustion;
- Reactive such as the formation and deposition of NO_2 from NO_x in the presence of O_3;
- Gas to particle conversions such as the production of particles directly from gases via gaseous reactions or via condensation;
- Gas/particle processes such as particle growth by condensation or by absorption of gases;
- Particle/particle processes (aerosol models).

6.4.4 Treatment of turbulence

Atmospheric turbulence (Section 6.2.2) is the mechanism that dilutes and mixes both gaseous and particulate emissions as they are transported in the atmosphere. Turbulence is produced when certain atmospheric gradients in the wind, temperature and humidity fields occur. For these reasons the formation of turbulence in air dispersion models can range from the simple (a well-mixed situation) to a complex one.

6.4.5 Topography

Surface conditions and topographic features of the earth may generate a field of turbulence, modify vertical and horizontal winds and change the temperature and humidity distributions in the boundary layer (Section 6.2.4), which may modify the transport and diffusion of pollutants. An important characteristic of models is the way in which structures (e.g. buildings, etc.) and topography are dealt with. Szepesi (1989) has characterised topography into four groups for modelling purposes:

- Homogeneous flat terrain such as plains;
- Non-homogeneous flat terrain such as land/water or land use interfaces;
- Simple terrain such as simple hills and valleys;
- Complex such as mountain ranges.

6.4.6 Plume additivity (treatment of multiple sources)

Plume additivity assumes that modelled predictions for non-reactive pollutant plumes from multiple sources can be combined for common spatial and time positions. The validity of this assumption depends on physical and chemical non-interaction between the pollutants in the interacting plumes.

6.4.7 Model accuracy and limitations

The accuracy or uncertainty of modelling is a fundamental consideration. No air dispersion model is extremely precise, owing to the complexity of atmospheric processes and an inability to measure them completely. Any observation of an emission plume shows that the resulting concentration will vary greatly along its length. This only compounds the problems of accurately simulating such a complex process. Most models are subjected to validation exercise where the details of the meteorological situation have been known so that they can be fitted roughly to the pollution actually measured (Scorer, 1998). A factor of two over or under forecast limits on accuracy are quite common (Middleton, 1996). Uncertainties in air dispersion modelling predictions can be caused by (Turner, 1994):

- Errors in emission estimates due to incorrect source emission and location data, unaccounted for time variability in emission rates, uncertainties in source parameters and incorrect plume rise calculations.
- Errors in meteorology due to incorrect or inappropriate meteorological data sets, poorly specified dispersion parameters and incorrect determination of the atmospheric thermal structure.
- Inappropriate dispersion model structure due to unrepresentative choice of model for the problem in question owing to incomplete knowledge of chemical and physical interactions of emitted gases and particles, incorrect formulations of removal processes and poorly specified boundary conditions.

The NSCA (2000b) have identified two further types of error found in resultant predictions from modelling practices:

- Systematic errors occurring when the model shows the same error trend at all times resulting in an over or under prediction;
- Random error occurring when a model shows values sometimes higher and sometimes lower than measured values, even after any systematic error might be allowed for.

It should always be remembered that an over-riding consideration when using models is that their level of accuracy will inevitably be undermined by the level of accuracy of the data used, as seen in the exercise and the expertise of the practitioner. Indeed, Scorer (1998) has wisely observed that in attempting to predict the dispersion of air pollution there is a need for a forecast of the wind structure and several other parameters of the weather and then to compute pollution concentrations according to several further dubious assumptions about the mechanism of dispersion. This makes predicting the concentration of the pollution orders of magnitude more difficult and unreliable than the prediction of the air motion. The knowledge of the vertical velocity of the air carrying the pollution is neither easy to predict nor measure, so that much of the parameterising of the nature of the details of air motion are only roughly and somewhat arbitrarily guessed. To make a model several parameters have to be invented, however, they will restrict the range of possible outcomes.

Table 6.5 Assessing uncertainty (NSCA, 2000b)

Step	*Action*
1	Assure that data of suitable quality are available from at least four air quality monitoring stations. To tabulate the measured and predicted concentrations
2	Plot the data points on a scatter graph with observed data (monitored data) on the horizontal axis and modelled data on the vertical axis (Fig. 6.9)
3	Plot the line of best fit (by regression of y (modelled) on x (observed) through the scattered points (Fig. 6.9). The origin may, or may not, intersect zero, depending on whether an offset for background concentrations, etc. is included. The formulae for the line is in the form: $y = mx+c$ (6.2)
4	Calculate the horizontal difference of the points from the line, i.e., the deviation of the modelled data. This can be done using the formula from the line of best fit: Modelled deviation = ((M.observed) + C) – modelled (6.3) For example for a data point of 38 the modelling deviation = ((0.6148 x 38) + 12.336) – 40 = -4.3016. Repeat this calculation for each data point and tabulate.
5	Use the deviation values to calculate the standard deviation.
6	Calculate the U value from the data using the following expression: U = SD/mean of observed data
7	Calculate the standard deviation for the model (SDM) using the following formula: SDM = Ux Co, where Co is the concentration of the air quality objective under consideration (6.4)
8	Plot contours at locations where the model predicts Co+2.SDM, Co+SDM, Co, Co-SDM, Co-2.SDM

Methods available for assessing the level of the accuracy of predictions include comparison of predicted values with observed measurements at the same location and time period. A regression analysis can then be used to obtain a value of uncertainty (Fig. 6.9). Table 6.5 details the systematic steps to assess the level of uncertainty of model predictions (NSCA, 2000b). The output from the information shown in Table 6.5 is a contour plot of the modelled area with effectively a series of regions, with the line of exceedence roughly in the middle (i.e. at Co (Table 6.5, point 7)). At the centre is a region where the air quality is 'almost certain' (+2.SDM) to exceed the air quality objective and around the outside, the region where air quality is 'almost certain' (-2.SDM) to comply with the objective.

Case Study 7 (Chapter 10, Section 10.8) provides an assessment of the accuracy of modelling predictions in relation to a LAQM study using the UK Air Dispersion Modelling System (ADMS) Urban model. Case Study 14 (Section 10.15) also provides an assessment of the accuracy of predicted levels from a line source model (the UK Design Manual for Roads and Bridges (DMRB)) against baseline air quality data.

Figure 6.9 Comparison of predicted and observed air pollution concentrations (NSCA, 2000b) (courtesy of the National Society for Clean Air and Environmental Protection)

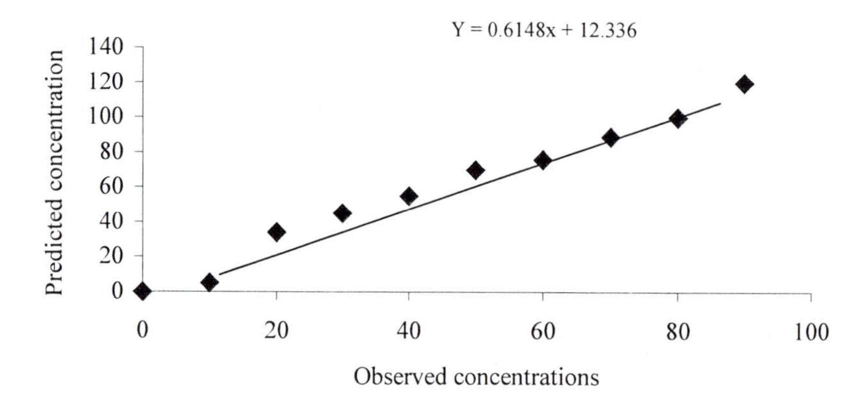

6.5 DATA REQUIREMENTS

It is essential that the appropriate source, transmission (meteorological) and air quality data be used in any air quality simulation model (Szepesi, 1989). Failings in the integrity of the model input data may compromise the study findings (Section 6.4). The information requirements to run a model can be quite onerous. Fig. 6.10 details the principal informational requirements/input to a point source model. The information requirements will begin with a need to know the existing situation. This information will include the ambient air pollution concentrations; pollutant sources and their specific location; meteorology; local topography; physical conditions affecting pollutant dispersion and sensitive receptors and their specific location. The aim is to know what air pollutants are present in the area under consideration and in what quantities, where the pollutants came from, how they will be dispersed and where they are destined to impact upon a ASR. The second step is to determine the specific characteristics of the emissions source (e.g. the nature of the pollutant; emission rate; efflux velocity; efflux temperature; source morphology; etc) (Chapter 4, Section 4.4).

Figure 6.10 Model information requirements for a point source (e.g. stack)

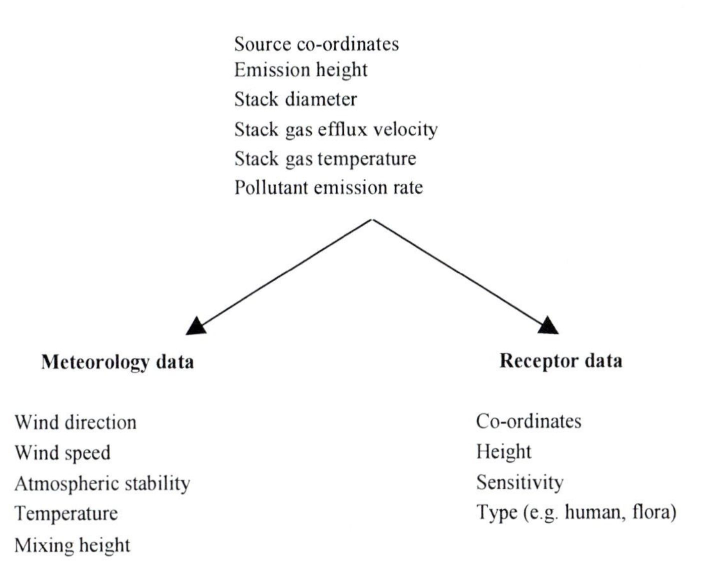

6.6 AIR DISPERSION MODELLING PROCEDURES

To promote best practice in the use of mathematical models for atmospheric dispersion, guidelines have been produced (Royal Meteorological Society, 1995) which emphasize the principle of fitness for purpose in the selection of modelling procedures and effective communication in the documentation of reported results. Table 6.6 identifies 10 criteria that should be considered when assessing the environmental impact of discharges to atmosphere using models.

Air dispersion modelling can be inherently difficult. It should not be the purpose of the AQA to confuse, the objective should be to aid in the decision-making process. It is therefore necessary to make the assessment process as transparent as possible. The use of appendices in an AQA report for supporting technical information may help in the communication process. The presentation of the report is extremely important in the conveyance of the message. It is all too easy to lose the reader in a plethora of technical terms; this should be avoided. The temptation to lend bias to the interpretation of results should always be avoided.

6.7 STACK HEIGHT DETERMINATION

Following the pollution abatement process (Chapter 8) it is important that the final height of the emission source is sufficient to ensure that emissions will not be detrimental to the environment. Stack height determination may be undertaken by air dispersion modelling techniques or independently by established proforma.

Table 6.6 Guidelines for the justification of choice and use of models, and the communication and reporting of results (Royal Meteorological Society, 1995)

Step	*Guidance*
1	Statement of context and objectives: To explain the situation being modelled and the purpose of the dispersion calculations, giving clear account of the relationship between the objectives and the modelling procedures adopted to achieve them
2	Justification of choice of modelling procedure: to demonstrate the fitness for the modelling procedure
3	Use of software implementations of modelling procedures: to provide a fully documented account of the details of the model and its conversion into valid software
4	Input data: to show how the data requirements of the model have been met, and to explore the implications on the assessment in cases where there are deficiencies in the available data
5	Presentation of results and conclusions: to ensure that the findings of the exercise are successfully communicated
6	Explicit quantification: to ensure that best use is made of the opportunity to express results in quantitative terms
7	Sensitivity analysis: to expose how the results depend on choices and assumptions made in respect of variables whose values may be debatable
8	Uncertainty and variability: to ensure that these issues are addressed in respect of uncertainties in model parameters, the inherent variability of dispersion behaviour, and variations that are likely to be displayed between the results of one model and another
9	QA of models: to demonstrate that the model used has been subjected to an evaluation procedure establishing its suitability for a specified range of tasks
10	Auditability: to ensure that there is a clear and transparent account of the exercise for inspection by interested parties

Frequently an integral part of a modelling study is an optimisation study to help determine an appropriate stack height of a process. The stack can be modelled for an array of stack heights as part of the decision-making process (Case Study 5 (Chapter 10, Section 10.6) provides an example of such a study).

Another method is to use relatively simple non-specific techniques of approximating the height of a discharge stack (HMIP, 1993b; Environmental Analysis Co-operative, 1999). The calculation assumes that the discharge stack height is governed by the need to limit local ground level pollutant concentrations below a maximum level that might occur for short periods. The methodology for determining a stack height is given below (Environmental Analysis Co-operative, 1999; HMIP, 1993b):

- Stage 1- calculating the pollution index (P_i)
    - Identify released substances and release rates;

Figure 6.11 Estimation of uncorrected discharge height from P_i and heat release (Q_h) at stack exit (HMIP, 1993b; Environmental Analysis Co-operative, 1999) (Crown copyright is reproduced with the permission of the Controller of Her Majesty's Stationery Office and Institute of Chemical Engineers)

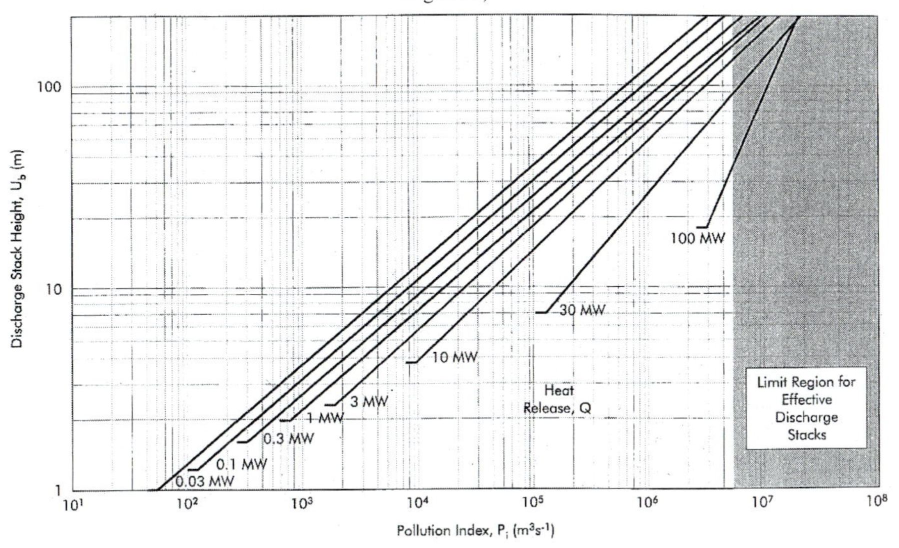

The nonogram, taken from Fig. 2 of the Technical Guidance Note D1, shows the uncorrected discharge stack height plotted as a function of the pollution index and the heat release. It gives a minimum stack height assuming no complex effects from buildings or topography. The shaded area shows the values of the pollution index of around 10^7 and beyond

- Determine the appropriate environmental quality criterion;
- Correct for any ambient concentrations of the same substances;
- Calculate the Pi for each released substance (equation 6.5).

$$P_i = (D/(G_d - B_c)) \times 1000 \qquad (6.5)$$

- Identify the released substance with the highest value of Pi.

- Stage 2 – calculating the discharge stack height.
  - Calculate the plume rise due to heat release (Q_h) of the emission (equation 6.6)

$$Q_h = (v_v(1 - (283/T_s))/(2.9) \qquad (6.6)$$

  - Calculate plume rise due to emission exit velocity.

$$M = (283/T_s).v_v.v \qquad (6.7)$$

$$U_m = 0.82M^{0.32} \qquad (6.8)$$

- Select the higher of the two plume rises as the determining factor.
- Calculate required stack height, based on the substance with the highest P_i value, correcting for the presence of nearby tall structures if necessary. The nomogram in Fig. 6.11 can be used to estimate the required stack height for the calculated P_i and Q_h values. Equation 6.9 can also be employed.

$$C_h=H_b+0.6\{u_c+(2.5H_b-u_c)(1-A^{-u/Hb})\} \qquad (6.9)$$

Where u_c = uncorrected stack height (lesser of U_m or U_b) where $U_b = 1.95Q_h^{0.19}$ if $Q_h < 1MW$ or $1.7+0.25Q_h^{0.9}$ if $Q_h > 1MW$. $A = U_m/U_b$ (if there is no value or if $U_b > U_m$, then $A = 1$. H_b = building height.

- Stage 3 – other considerations, cross checks and adjustments:
  - Minimum discharge velocity and other flue exit conditions (>15m/sec);
  - Over riding minimum requirements for discharge stack heights (>3m above roof areas, gantries, etc.);
  - For discharge heights for plants with a wide operational range check stack height estimate for high and low discharge rates;
  - For multiple sources, nearby sources and combining emissions treat as a single stack if they are within 3 stack diameters of each other;
  - Difficult siting needs to be considered if tall buildings are nearby;
  - With highly abated emissions consider likely release rates during malfunction of abatement equipment, provision of emergency stacks, visibility of plumes, etc.

Figure 6.12 Plume rise

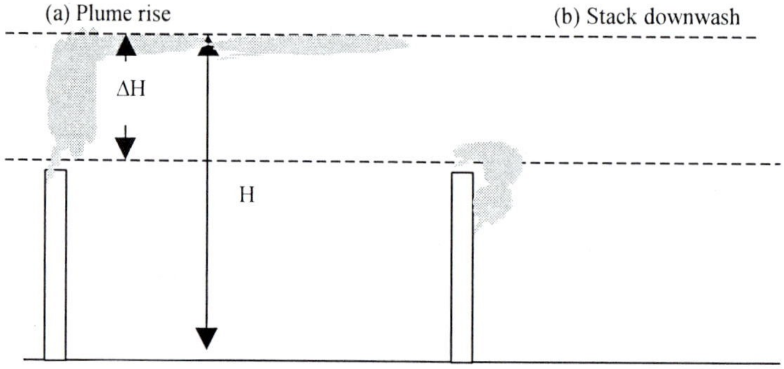

(a) Plume rise

(b) Stack downwash

ΔH

H

6.8 PLUME RISE

On being discharged from a stack gaseous or particulate emissions rapidly mix with the surrounding atmosphere, and the dilution increases as the plume is carried downwind of the emission's source (Fig. 6.12). Dispersion is most easily described by considering three separate phases of a plumes behaviour (Turner, 1994):

- Shortly after release to the atmosphere the plume may be affected by the aerodynamic disturbances created by the emitting stack, by adjacent buildings or by topographic irregularities;
- Because of its buoyancy and initial vertical momentum, the plume's rise will be relative to the mean motion of ambient air;
- The ultimate dilution of the plume will depend on the degree of turbulence in the atmosphere.

The initial behaviour of buoyant stack plumes has been widely investigated and the process has been reduced to mathematical approximations by a number of people. Moses and Strom (1961), having compared actual and calculated plume heights by means of six plume rise equations, reported that there was no one formula which was best in all respects. The formulae of Davidson-Bryant (1949), Bosanquet, Carey and Halton (1950), Holland (1953) and Bosanquet (1957) all gave generally satisfactory results in the test situations.

An important consideration in air dispersion modelling studies is to estimate the effective height of the emission source, H, at which the plume becomes essentially level. Rarely will this height correspond to the physical height of the stack. If the plume is caught in the turbulent wake of the stack or of buildings in the vicinity of the stack (Fig. 6.7), the emission will be mixed rapidly downward towards the ground. If the plume is emitted free of these turbulent zones, a number of emission factors and meteorological factors will influence its rise. The principal emission factors are the velocity of the emission at the top of the stack (momentum of the exiting gas), temperature of the emission at the top of the stack (stack gas buoyancy) and the diameter of the stack orifice. Another factor effecting plume rise is that of stack-tip downwash (Fig. 6.12b). This phenomena is due to regions of low pressure that form on the leeward side of the stack. The effect is that a cylindrical eddy with vertical axis forms in the low pressure region downwind of the stack and is shed from the stack and travels downwind in the prevailing wind. Immediately the low pressure region begins to be re-established and another eddy rotating in the opposite direction forms and is shed off into the flow. If the emitted effluent from the stack is not clear of these vortices the bottom part of the plume may be caught in the low pressure in the top of these eddies and will be lowered (Turner, 1994). The meteorological factors influencing plume rise are wind speed (u); temperature of the air (^{o}C or K); shear of the wind speed (m/sec) with height and atmospheric stability.

The equation of Holland (1953) was developed with experimental data from larger sources than those of Moses and Strom (1961). This equation frequently underestimates the effective height of emission; therefore its use often provides a slight 'safety' factor. Holland's equation (6.10) is given as:

$$\Delta H = [(vd)/u]\ [(1.5 + 2.68 \times 10^{-3}\ \rho\ ((T_s - T_a)/T_s)\ d)] \qquad (6.10)$$

Holland (1953) suggests that a value between 1.1 and 1.2 times ΔH should be used for unstable conditions and a value between 0.8 and 0.9 times ΔH should be used for stable conditions. Of the expressions available to estimate plume rise it is the formulae of Briggs (1969, 1975) that is the most complete (6.11). Selected proportions of Brigg's work have been incorporated into many diffusion modelling systems including those officially recommended by the USEPA. To calculate buoyant rise, the buoyancy flux (F) must first be calculated:

$$F = gvd^2\Delta T/(4T_s) \quad (6.11)$$

The formula (6.11) can then be used to determine plume rise under certain meteorological conditions (from Turner, 1994):

1. Final plume rise for unstable-neutral atmospheric conditions.

Where $F < 55$ $$\Delta H = 21.425\ F^{3/4}/u_h \quad (6.12)$$

Where $F > 55$ $$\Delta H = 38.71\ F^{3/5}/u_h \quad (6.13)$$

The ΔH calculated by either equation (6.12 or 6.13) is then compared to the unstable-neutral momentum rise (6.14). The highest of the momentum or the buoyancy value is used.

$$\Delta H = 3\ d\ v/u_h \quad (6.14)$$

2. Final plume rise for stable conditions.

To calculate the plume rise under stable atmospheric conditions requires the determination of the stability parameter(s):

$$s = (g\ d\vartheta/dz)/T \quad (6.15)$$

The change of potential temperature with height is related to the change of temperature with height by:

$$d\vartheta/dz = dT/dz + \Gamma \quad (6.16)$$

The final plume rise under stable conditions is determined by:

$$\Delta H = 2.6\ [(F/u_h\ s)]^{1/3} \quad (6.17)$$

The stable buoyancy rise for calm conditions is determined by:

$$\Delta H = 4\ F^{1/4}\ s^{-3/8} \quad (6.18)$$

The stable momentum is determined from the following expression (6.19). The lower value from equation 6.19 or 6.14 represents the final stable momentum rise. The ΔH calculated by either equation 6.17 or 6.18 is then compared to the stable momentum rise value. The highest of the momentum or the buoyancy value is used.

$$\Delta H = 1.5\ [(v^2 d^2 T_a)/(4 T_s u_h)]^{1/3}\ s^{-1/6} \qquad (6.19)$$

Providing the buoyant plume has a temperature greater than the surrounding atmosphere the plume will continue to rise. The period of time required to reduce the excess temperature is dependent upon the nature of the plume release and atmospheric conditions. A momentum plume will have an effect dissipated over a much smaller period of time and therefore it is assumed to have achieved its final rise very close to the source and gradual plume rise is seldom considered. For buoyant plumes, the gradual plume rise while the temperature dissipation is occurring is estimated for all atmospheric conditions (unstable, neutral or stable (Section 6.2.2)) with the following single Brigg's equation (Turner, 1994):

$$\Delta H = (1.60 F^{1/3}\ x^{2/3})/u_h \qquad (6.20)$$

6.9 GAUSSIAN MODELLING

One type of model widely used is the Gaussian, where the spread of a plume in the vertical and horizontal direction is assumed to occur by simple diffusion perpendicular to the direction of the mean wind (Fig. 6.13) (Turner, 1970). The equation (6.21) is the most common expression (Turner, 1994 (p 2-6, equation 2.1)). The concentration (χ) of a gas or aerosol (particulates less than 20 μm in diameter) at x_d, y_d, z_d from a continuous source with an effective emissions height, H, is given by:

$$\chi(x_d, y_d, z_d; H) = [(Q)/(2\pi\sigma_y\sigma_z u)]\ \exp\ [-(y_d^2)/(2\sigma_Y^2)]\{\exp\ [-((H-z_d)^2)/(2\sigma_Z^2)] + \exp\ [-((H+z_d)^2)/2\sigma_Z^2)]\} \qquad (6.21)$$

The following assumptions are made in the Gaussian model (Turner, 1970):

- The plume spread has a Gaussian distribution in both the horizontal and vertical direction with standard deviations of plume concentration distribution in the horizontal and vertical of σ_Y and σ_Z respectively;
- The mean wind speed affecting the plume is u;
- The uniform emission rate of pollutant is Q;
- Total reflection of the plume takes place at the earth's surface (i.e. there is no deposition or reaction at the surface).

To ensure that the Gaussian model is properly applied it is important to keep in mind some of the following general relationships:

- Downwind concentration at any location is directly proportional to the source emission rate;
- Downwind ground level concentration is generally inversely proportional to wind speed (Section 6.2.1). Plume rise also depends on wind speed in a complicated fashion that prevents a strict inverse proportionality;
- Dispersion coefficients in the vertical and horizontal directions increase as the downwind distance increases and the elevated plume centreline concentration declines with increasing distance.
- Ground level centreline concentrations increase, reach a maximum, and then decrease away from the stack;
- Dispersion parameters σ_y and σ_z increase with increasing atmospheric turbulence (instability). Thus unstable conditions decrease average downwind concentrations;
- Maximum ground level concentrations decrease as effective stack height increases. The distance from the stack at which the maximum concentration occurs increases with plume rise.

Equation 6.21 can be altered for specific situations. For receptor at ground level ($z_d = 0$) equation 6.21 reduces to (Turner, 1970):

$$\chi(x_d, y_d, 0; H) = [(Q)/(\pi\, \sigma_y\, \sigma_z\, u)]\ \exp[-{}^1/_2(y_d/\sigma_Y)^2]\ \exp[-{}^1/_2\ (H/\ \sigma_Z)^2] \qquad (6.22)$$

In order to make concentration estimates directly beneath the plume centreline ($y_d = 0$) at ground level ($z_d = 0$) equation 6.21 further reduces to:

$$\chi(x_d, 0, 0; H) = [(Q)/(\pi\sigma_y\sigma_z u)]\ \exp[-{}^1/_2\ (H/\sigma_Z)^2] \qquad (6.23)$$

In order to make concentration estimates along the plume centreline from a ground level release ($y_d = 0$, $z_d = 0$ and $H = 0$) equation 6.21 further reduces to:

$$\chi(x_d, 0, 0; 0) = Q/(\pi\sigma_y\sigma_z u) \qquad (6.24)$$

In practical applications it is often convenient and sufficient to determine only the principal properties of air pollution distribution. These are the maximum ground level concentration and its distance from the source. The 3 minute ground level concentrations are given by equation 6.25. Occurring at the distance where $\sigma_z = h/(\sqrt{2})$, which is about 15 stack heights (or effective stack heights) in neutral stability conditions (Parker, 1978).

$$C_{max} = [(2Q_s)/e\pi h^2 u_p)]\ (\sigma_y/\sigma_z) \qquad (6.25)$$

It has been found around large power stations that the ground level concentrations are in general less than or equal to the values calculated by the Gaussian formula detailed above. This appears to be owing to the fact that the vertical and crosswind spreads of the plume vary with distance from the source in

Figure 6.13 Co-ordinate system showing Gaussian distributions in the horizontal and vertical (Turner, 1994) (courtesy CRC Publishers)

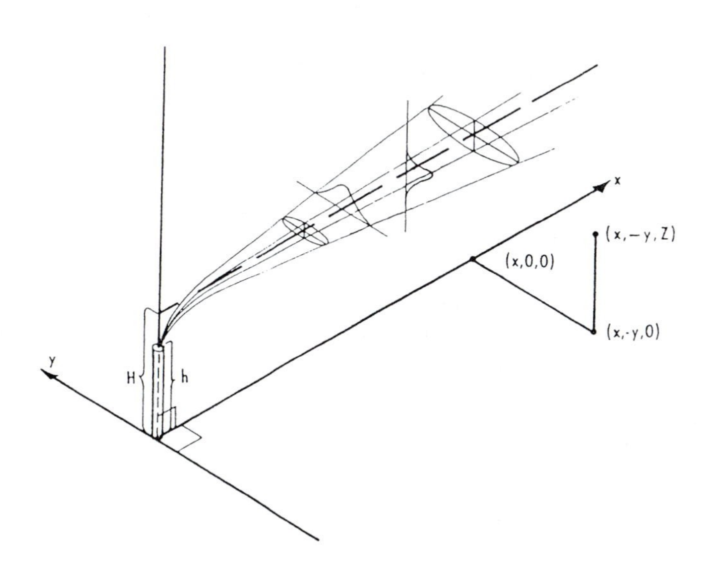

different ways (Parker, 1978). Moore (1973) has explained these differences by assuming that the size of the eddies transporting the material to the ground is generally smaller than the height of the plume. He derives the following expressions for predicting concentrations for neutral and stable conditions (equation 6.26) and unstable conditions (6.27):

$$C_m = (Q_s/H^2)\,[(A/H) + (C/u_p^2)] \times 10^8 \tag{6.26}$$

$$C_m = (Q_s/H^2)\,[(A/H)^2 + (C/u_p^2)^2]^{1/2} \times 10^8 \tag{6.27}$$

A and C are parameters (equations 6.26 and 6.27) whose values vary with vertical temperature gradient, but within stability groups are as follows:

	A	C
Unstable	1.65	0.58
Stable but no inversion	1.87	0.22
Stable but turbulent below and elevated inversion	1.35	0.20

The term containing A represents the effect of mechanical turbulence and the term containing C represents the effect of thermals on the turbulence in a stable environment. The maximum hourly value calculated on this basis is about one third of the 3 minute maximum value.

6.9.1 Extrapolating time average concentrations

Many models assume that the predicted ground level concentration is a 1 hour average. Beychok (1979) has derived criteria for the extrapolation of time average periods within the range of 10 minutes to 24 hours (Table 6.7). Conversion factors are given for 10 minute and 1 hour averages.

Table 6.7 Extrapolation of time average concentrations (within the range of 10 minutes to 24 hours) (Beychok, 1979)

Time period	*10 minutes*	*1 hour*
10 minutes	1.00	1.22
1 hour	0.82	1.00
3 hours	0.58	0.70
8 hours	0.33	0.40
24 hours	0.14	0.17

6.10 BOX MODELS

A simple atmospheric dispersion model, called a box model, can be used to calculate ground level concentrations of specific air pollutants of concern. A box model is based on the assumption that pollutants emitted to the atmosphere are uniformly mixed in a volume of air, or box of air (Canter, 1985, 1996). The most critical aspect of the usage of a box model is to establish the downwind, crosswind, and vertical dimensions of the box. In addition, the time period over which to consider pollutant emissions must also be established. A typical time period is 1 hour. The time and dimensional considerations are based on assuming steady state conditions, i.e., the emissions, wind speed and characteristics of air available for dilution do not vary over time (Ortolano, 1985). A box model is also based on the assumption that discharges mix completely and instantaneously with the air available for dilution and the released material is chemically stable and remains in the air. The box model can be used for single or multiple point, area or line sources of air pollutants, or combinations of source types (Ortolano, 1985; Canter, 1985):

$$C = QT/(xzy) \tag{6.28}$$

The box model results can be interpreted on a pollutant by pollutant basis, in relation to existing ambient air quality and the relevant AQS. It is important to compare the existing pollutant concentration plus the concentration from the proposed activity as calculated from the box model, in relation to the applicable standard. Greater significance should be attached to, and greater attention should be given to, those pollutants which do not meet or are near the allowable AQS. Timmis (1995) has demonstrated an application of the box model and treated an urban atmosphere as a ventilation box (Fig. 6.14).

Figure 6.14 An application of box modelling (Timmis, 1995)

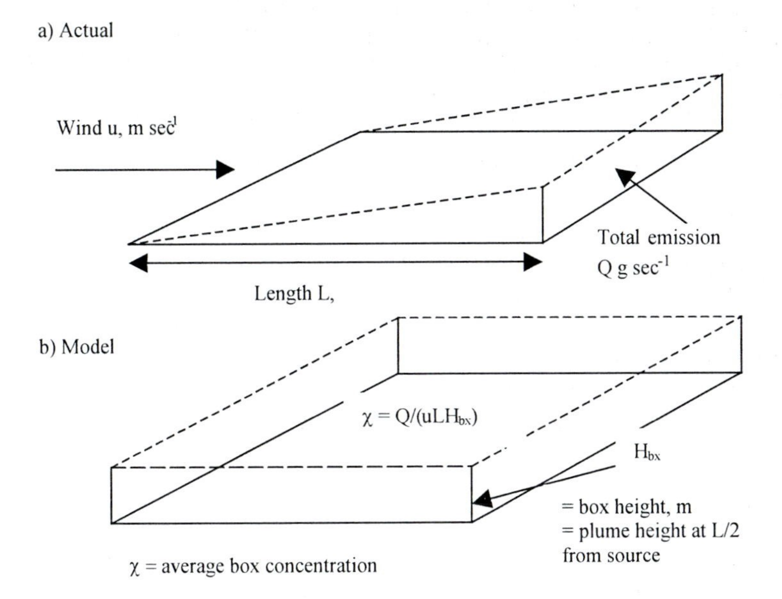

6.11 LINEAR MODELS

Most linear (e.g. highway) dispersion models employ Gaussian diffusion theory, though the formulation and choice of parameters tend to be specific to each model (McCrae *et al.*, 1988). A number of models have been developed to predict the impact of emissions from line sources, principally motor vehicles. These models include PREDCO (Hickman and Colwill, 1982), HIWAY (Petersen, 1980), CALINE (Benson, 1979) and SPAGLINK (Middleton *et al.*, 1979). A simple statistical model widely applied in the UK is the Design Manual for Roads and Bridges (DMRB) (DETR, 1999a). Another canyon model is the AEOLIUS suite of models (Buckland, 1998; Buckland and Middleton, 1999; Middleton, 1999). The suite of models consists of a screening version (AEOLIUS), a version to input emission data (AEOLIUSQ), a version that calculates hourly concentrations (AEOLIUSF) and a research version (AEOLIUSH). ADMS-Urban also caters for line source emissions. Case Study 11 (Chapter 10, Section 10.12) provides an example of the application of CALINE for a road way in Hong Kong. Case Study 14 (Section 10.15) provides an application of the UK DMRB.

6.12 PHYSICAL MODELS

Wind tunnel modelling requires the construction of a physical model (to scale) and the simulation of both the wind and emission under consideration (Fig. 6.15). If a wind tunnel experiment is carried out with care and the simulation is within certain

Figure 6.15 Wind tunnel model (photograph courtesy of BMT Fluid Dynamics, 1999)

constraints, then it is widely accepted that this is the most accurate and reliable method for making estimates of the dispersion of atmospheric emissions (Harvey and Obasaju, 1999). Within a wind tunnel, the intricacies of air flow and dispersion, the spectrum of eddies and in particular the effects of obstacles are mimicked, although usually at a smaller geometric scale.

6.13 TYPES OF AIR DISPERSION MODELS

There are many types of air dispersion models available (Table 6.8). However, advice on the most appropriate model to use is not so forthcoming. Models have been developed for a variety of pollutant types, time scales and operational scenarios. Short-term models are used to calculate concentrations of pollutants over a few minutes, hours or days and can be employed to predict worst case conditions (e.g. high pollution episodes). Long-term models are designed to predict seasonal or annual average concentrations, which may prove more useful in studying health effects and impacts on vegetation, materials and structures. The versatility of air dispersion modelling techniques (e.g. complex or simple terrain, rural or urban algorithms, etc) can be useful in assessing the impact of a development or process or where it should be sited. Examples of commonly used air dispersion models include ADMS (Carruthers, 1995) and Industrial Source Complex (ISC) (USEPA, 1987a). The USEPA is currently introducing a new air dispersion model, AERMOD. The growing UK preference has been towards the use of ADMS. The ADMS model has been developed by Cambridge Environmental Research Consultants Ltd (Carruthers, 1995).

The fundamental difference between ADMS and other older and more established models, such as R-91 and ISC, is that where as R-91 and ISC characterise the boundary layer structure in terms of a single parameter, the Pasquill-Gifford stability category, ADMS uses a parameterisation of the boundary

Table 6.8 Examples of computer-based air dispersion models (based on Canter, 1996, originally from Turner 1979; Environmental Analysis Co-operative, 1999)

Model	*Brief description*
R-91	Long-term and short-term predictions using Pasquill stability classes. DOS based.
AERMOD	Long-, short-term and percentile predictions. Recently developed. Built in contour plotting, multiple sources, GIS, deposition, complex terrain and building effects
ADMS	Long-term, short-term and percentile predictions. Volume, line, area and point sources. Built in contour plotting, multiple sources, GIS, deposition, complex terrain and building effects
CALINE 3	Line source model that can be used to predict CO concentrations near highways
HIWAY-ROADWAY	Two models which compute the hourly concentrations of non reactive pollutants downwind of roadways
INPUFF	Gaussian integrated puff model which addresses the accidental release of a substance over several minutes
ISCLT	Gaussian plume model for industrial source complexes
LONGZ-SHORTZ	Models the long-term and short-term receptor pollutant concentration from multiple stack, building, and area sources
MESOPUFF	Lagrangian model for the transport, diffusion and removal of air pollutants from multiple point and area sources at distances beyond 10-50 kms
MPTER	Multiple point source Gaussian model with optional terrain adjustments
MPTDS	Modification of MPTER that accounts for gravitational settling and/or deposition loss of a pollutant
PEM	Urban scale short-term average model for surface concentrations and deposition fluxes of two gaseous or particulate pollutants
PBM	Simple stationary single-cell urban area model with a variable height lid designed to provide volume-integrated hourly averages of O_3 and other photochemical smog pollutants for a single day of simulation
APRAC	Stanford Research Institute's urban CO model which computes hourly averages for any urban location
CDM	The Climatological Dispersion Model determines long-term quasi stable pollutant concentrations at any ground level receptor from point and area sources using a joint frequency distribution of wind direction, wind speed, and stability
VALLEY	A steady-state, univariate Gaussian plume dispersion algorithm designed to estimate either 24 hour or annual concentrations from point and area sources. Empirical dispersion coefficients are used and include adjustments for plume rise and limited mixing. Plume height is adjusted according to terrain elevations and stability classes
CRSTER	Estimates ground level highest and second-highest 1 hr, 3 hr and 24 hr concentrations from elevated stack emissions for an entire year. The algorithm is based on a modified form of the steady state Gaussian plume equation which uses empirical dispersion coefficients and includes adjustments for plume rise and limited mixing. Terrain adjustments are made as long as the surrounding terrain is physically lower than the lowest stack height input
PAL	This short-term Gaussian steady state algorithm estimates concentrations of stable pollutants from point, area, and line sources. Hourly concentrations are estimated and average concentrations from 1 to 24 hours can be obtained
PTMAX	Performs an analysis of the maximum short-term concentrations from a single point source as a function of stability and wind speed
PTDIS	Estimates short-term concentrations directly downwind of a point source at distances specified by the user. An option allows the calculation of isopleth half-widths for specific concentrations at each downwind distance
PTMTP	Estimates for a number of arbitrarily located receptor points at or above ground level, the concentration from a number of point sources. Downwind and crosswind distances are determined for each source receptor pair. Concentrations at a receptor from various sources are assumed additive.

layer structure based on the Monin-Obukhov length and the boundary layer height. The principal features of the model are:

- Concentration distributions are Gaussian in stable and neutral conditions, but the vertical distribution is non-Gaussian in convective conditions to take account of the skewed structure of the vertical component of the turbulence;
- Boundary layer structure is characterised by the height of the boundary layer and the Monin-Obukhov length, a length scale dependent on the friction velocity and heat flux at the ground surface;
- Plume spread depends on the local wind speed and turbulence and thus depends on plume height. This contrasts with Pasquill-Gifford methods where plume spread is independent of height;
- A meteorological pre-processor is included which calculates the required boundary layer parameters from a variety of input data: e.g. wind speed, day, time, cloud cover etc., surface heat flux and boundary layer height.

Case Studies 5, 7 and 10 (Chapter 10) provide applications of the ADMS model. Case Study 7 uses the ADMS-Urban model for a Local Authority LAQM study.

The ISC model is based on the Gaussian dispersion equation, which predicts short-term and long-term (annual) ground level concentrations. The ISC model is used to evaluate emissions from industrial sources. The principal features of the model are that it assesses the effects on aerodynamic building wakes and chimney downwash; effects of variations in terrain height; plume rise due to momentum and buoyancy as a function of downwind distance; dispersion of emissions from chimney, area, line and volume sources where line sources are simulated by multiple volume sources; physical separation of multiple sources; time-dependent exponential decay of pollutants and effects of gravitational settling and dry deposition.

A study (HMIP, 1995a) concluded that the ADMS model provides a more reliable prediction than the ISC short-term model for the location and magnitude of the maximum ground level concentration in convective conditions for elevated sources. For the locality these conditions occur relatively infrequently. A comparison of the predicted results for UK ADMS and ISC showed that predicted levels using UK ADMS may be as much as approximately 10 times greater than for the ISC model. The variance will depend upon the prevailing meteorological conditions. In some instances the study showed a greater variance. Recent studies have also compared the merits of ISC, R91, AERMOD and ADMS (Hall *et al.* 2000a; 2000b). Fig. 6.16 shows the dispersion of emissions around a building using the ADMS model.

Szepesi (1989) has considered the merits of various regulatory models; 183 from 27 different countries. No two models were the same and therefore each has its own merits. These merits should be considered prior to the selection of a model for an impact prediction exercise.

Figure 6.16 Dispersion of ground level concentrations ($\mu g/m^3$) around a building using ADMS (courtesy SGL Technic Ltd)

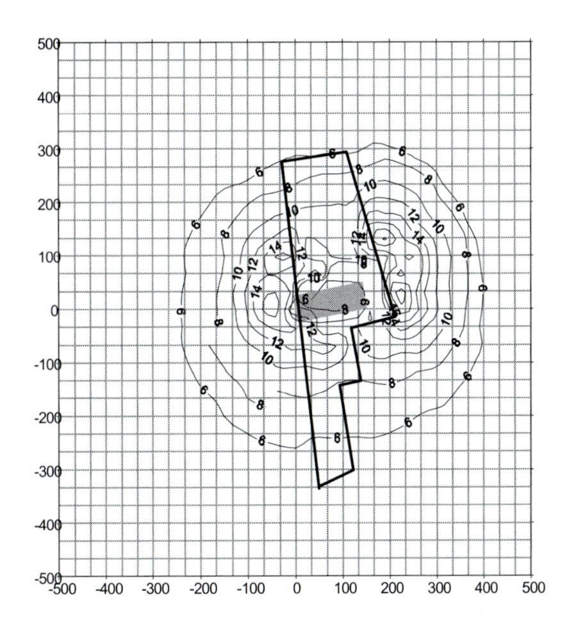

6.14 EXERCISE

An incinerator is proposed and planners wish to know the possible impact of the development on air quality. It is estimated that 72 g/sec of SO_2 would be emitted from a 45m stack with a diameter of 1.5m. The effluent gases will be emitted at a temperature of 395K with an exit velocity of 13 m/sec. The planners would like to know what the predicted ground level concentrations would be 500 m downwind of the stack on an overcast winter's morning with a surface wind speed off 6 m/sec. The atmospheric pressure is thought to be 970 mb and the ambient air temperature is 293K (Harrop and Nixon, 1999).

Using Holland's equation (6.10) the estimated plume rise is 8.1 m. Therefore the effective plume rise is the plume rise plus the stack height, which is 53.1 m. To determine the ground level concentration (χ) in the aforementioned meteorological conditions downwind of the stack requires an abridged version of equation (6.21):

$$\chi(x_d, 0, 0; H) = [(Q)/(\pi\sigma_y\sigma_z u)] \exp[-{}^1/_2 (H/\sigma_Z)^2] \tag{6.23}$$

The dispersion coefficients σ_y and σ_z may be determined from the graphs in Fig. 6.17 or by the following equations:

$$\sigma_y = ax_d^{0.903} \tag{6.29}$$

Figure 6.17 Dispersion coefficients σ_y and σ_z (Turner, 1994) (courtesy CRC Publishers)

(a) Pasquill-Gifford horizontal dispersion parameter σ_y as a function of Pasquill stability class and downwind distance from the source

(b) Pasquill-Gifford vertical dispersion parameter σ_z as a function of Pasquill stability class and downwind distance from the source

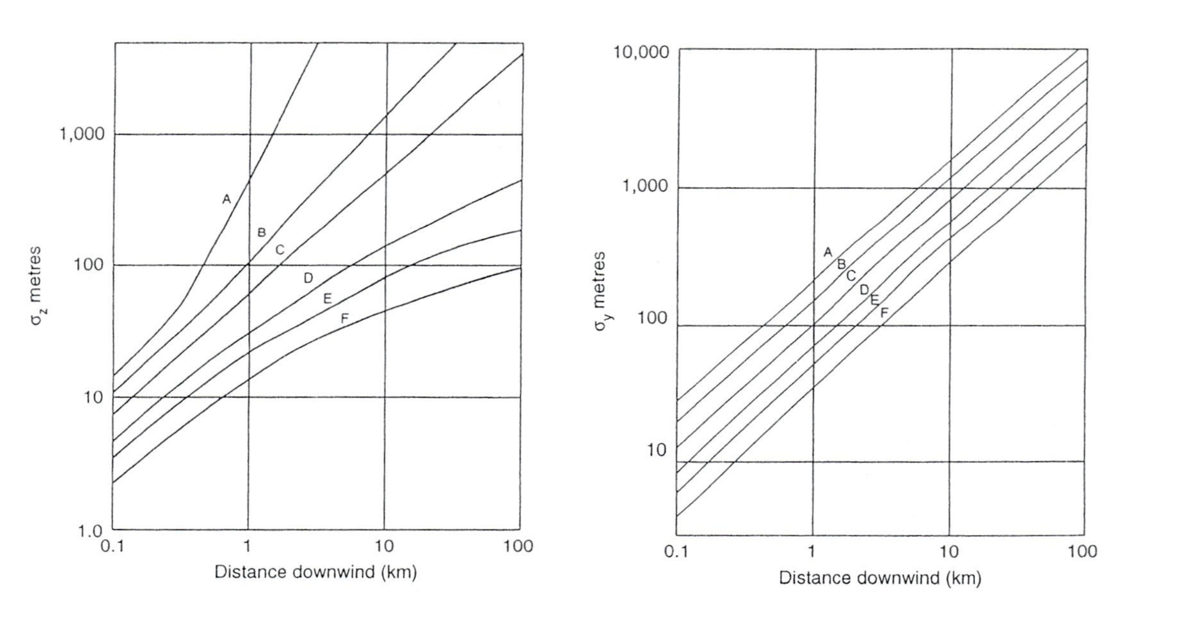

$$\sigma_z = bx_d^c \tag{6.30}$$

Where x_d is the downwind distance from the source, σ_y, σ_z and x_d are in metres. The parameters a, b and c can be gained using Tables 6.2 and 6.9 (Pasquill and Smith, 1982). From Table 6.2 it can be seen that overcast conditions with a wind speed of 6 m/sec, stability class D conditions (neutral) prevail.

From the information gained in Table 6.2 values a, b and c can be obtained from Table 6.9, where σ_y and σ_z become 35.6 and 17.9 m respectively (Fig. 6.17). Therefore the predicted short-term (approximately 10 minute) ground level concentration is approximately 74 μg/m^3. The predicted level may then be added to the known short-term background concentration and compared to the recommended ambient AQS (Chapter 7).

Table 6.9 Fitted constants for the Pasquill diffusion parameters (Pasquill and Smith, 1982)

	$x_d \leq x_{d1}$			$x_d > x_{d1}$ and $\leq x_{d2}$			$x_d > x_{d2}$	
	a	*b*	c	X_{d1}	*b*	c	X_{d2}	*b*	c
A	0.400	0.125	1.030	250	0.00883	1.510	500	0.00023	2.100
B	0.295	0.119	0.986	1000	0.05790	1.090	10000	0.0579	1.090
C	0.200	0.111	0.911	1000	0.11100	0.911	10000	0.111	0.911
D	0.130	0.105	0.827	1000	0.39200	0.636	10000	0.948	0.540
E	0.098	0.100	0.778	1000	0.37300	0.587	10000	2.85	0.366

CHAPTER SEVEN

Impact Significance and Legislation

7.1 INTRODUCTION

The principles of air pollution control legislation have evolved over many years. As nations have developed, so has their legislation and through it the setting of their own AQS and guidelines. Legislation may be either national or international. The former is variable due to the requirements and development of the specific country; this is shown by examples (Section 7.10). Protocols and conventions generally govern international air quality control. The first part of this chapter details the legislative criteria to determine impact significance. The second part details international and national air pollution control legislation.

7.2 IMPACT SIGNIFICANCE

Impact significance is an important aspect of AQM. The management of air pollution is important with regard to the preservation of human health and environmental protection. To assess the significance of pollution levels, criteria can be applied. These criteria, although in some instances arbitrary, allow an appreciation of importance to be set against ambient air quality. The criteria comprise guidelines and/or standards and are currently best scientific judgement. They are periodically revised since so much remains to be determined regarding the toxicity of air pollutants for humans. AQS and guidelines are concentrations that are considered to be acceptable in the light of what is known about the effects of each pollutant on human health. Although human health effects are a major consideration in establishing AQS, ecologically-based guidelines for preventing adverse effects on terrestrial vegetation are also considered. Unfortunately, a major shortcoming of currently accepted air quality criteria is that they generally only consider pollutants in isolation from each other rather than in combination. This shortcoming is because little is known about the synergistic effects of mixtures of pollutants and thus there is insufficient information presently available to enable the setting of such criteria.

AQS are criteria that have been legally accepted by an enforcing agency, whereas guidelines are generally accepted as recommended air quality criteria and may not be legally enforceable although they may be deemed good practice. The primary aim of air quality guidelines is to provide a basis for the protecting of public health from adverse effects of air pollution and for eliminating, or reducing to a minimum, those contaminants of air that are known or likely to be hazardous to human health and well-being (WHO, 1987). At its simplest, an AQS should be defined in terms of one or more concentrations and averaging times. In addition, other data should be added, including information on the form of exposure (e.g. outdoor), on monitoring which is relevant in assessing compliance with the standard, and on methods of data analysis, QA and QC. In some countries the standard is further qualified by defining an acceptable level of attainment or compliance. Levels of attainment may be defined in terms of the fundamental units that define the standard. For example, if the unit defined by the standard is the day, then a requirement for 99% compliance allows the standard to be exceeded by three days a year. Air quality criteria are in reality a benchmark. Criteria will invariably get more stringent as ambient air quality improves, therefore practitioners should aim to work not to the standard but below it. As air quality in regions improves the drive by interested parties to maintain the improvement will result in greater pressure to lower standards and guidelines.

An aim of AQS and guidelines is to provide a basis for protecting public health from adverse effects of air pollutants and eliminate, or reduce to a minimum, exposure to those pollutants that are known or likely to be hazardous to human health or well being. Many countries now have their own AQS, but for many they are limited in the scope of pollutants assessed. Therefore there is frequently a need to apply other national or international AQS or in the absence of suitable standards or guidelines, to derive them. For example, in the absence of standards applicable to the pollutants referred to in the EU Directives there are generally three sources of AQS which can be applied:

- Other nationally or internationally recognised/recommended standards and guidelines (e.g. USEPA);
- WHO air quality guidelines, which are intended to provide background and guidance to governments in making risk management decisions, particularly in setting AQS;
- Derived AQS.

In formulating AQS or guidelines it is sometimes appropriate that, for a criteria with a short averaging time, the objective for the pollutant in question should be expressed in terms of percentile compliance (see above). It is an approach that has been used in setting EU air quality limit and guide values and the UK AQS. For example, if the objective to be complied with is a 99.9th percentile, then 99.9% of measurements at each measuring point in the relevant period (e.g. 1 year) must be at or below the level specified. The use of percentiles accepts that for a small percentage of the time the threshold level given in the respective standard may be exceeded. Most criteria are presently set against an average temporal period. Long-term standards are generally months (i.e. 3 months) to 1 year. Short-term standards are based over periods of minutes, hours or days.

Probably the most frequently used reference guidelines are, other than specific national AQS, those of the WHO, the EU Directives, and the standards of the USEPA. The WHO and USEPA criteria have been set based on clinical, toxicological, and epidemiological evidence. Each organisation has determined their respective pollutant criteria in a systematic process. For example, the previous WHO guideline values of ambient particulate concentrations were established by determining concentrations with the lowest observed adverse effect (implicitly accepting the notion that a lower threshold exists under which no adverse human health effects can be detected). They were then adjusted by an arbitrary margin of safety factor to allow for uncertainties in extrapolation from animals to humans and from small groups of humans to larger populations. Standards determined by the

USEPA also reflect the technological feasibility of attainment. USEPA consider the adverse effect of a pollutant concentration as 'any effect resulting in functional impairment and/or pathological lesions that may affect the performance of the whole organism or which contributed to a reduced ability to respond to an additional challenge' (USEPA, 1980). The EU guidelines have been determined by consultation and legislative decision-making processes that took into account the environmental conditions and the economic and social development of the various regions and acknowledged a phased approach to compliance. A potential trade off was also recognised by the guidelines for the combined effects of SO_2 and particulate matter.

An example of a periodic review of AQS is that currently being undertaken by New Zealand. The Environment Ministry has begun a review of their 1994 ambient air quality guidelines. The guidelines contain advice on how air quality should be managed under the effects-based principles of the Resource Management Act 1991 and guideline values for eight common air pollutants, namely CO, PM_{10}, SO_2, NO_2, H_2S, F, O_3 and Pb. The Ministry began the review by commissioning experts to prepare reports on the latest international research on the effects of air pollutants. The Ministry will use these reports to develop new guideline values and guidance on how to use guideline values in AQM. Experts are currently preparing a number of draft technical reports and the Ministry is developing its draft proposals for new guidelines based on these reports. A group of around 50 air quality experts from councils, consultants, non-government organisations and industry will review all the draft reports. In 2000 the Ministry held meetings with the reviewers and Maori to discuss the draft proposals. The information from these meetings will be used in developing proposed new guidelines for public release and comment.

7.2.1 EU air quality standards

EU AQS have existed for the following pollutants:

- SO_2 and suspended particulates (80/779/EEC);
- NO_2 (85/203/EEC);
- O_3 (92/72/EC);
- Lead-in-air (82/884/EEC).

These Directives detail limit values, which should not be exceeded, and guide values, proposed limits to be attained, for each of the pollutants (Table 7.1). In addition, the EU Directives set out procedures and techniques for the monitoring of pollutant ambient concentrations. The EC Air Quality Framework & Daughter Directives (96/62/EC) identifies 12 pollutants for which limit and target values will be set in subsequent Daughter Directives. The main aim of the Directive is to protect human health and the environment through the following measures:

- Fixing limit values and alert thresholds for airborne pollutants;
- Specifying common methodologies to assess ambient air quality in EU Member States;

- Obtaining adequate information on ambient air quality and making this information publicly available;
- Maintaining ambient air quality where it is good and improving it where necessary.

Table 7.1 EU air quality directives (μg/m^3)

Pollutant	*Standard*
SO_2 and smoke (80/779/EEC)	Annual 98th percentile limit value • 350 for smoke <150 • 250 for smoke >150 Annual limit value - SO_2 is 120 for smoke <40 Annual guide value - 40 to 60 24 hour guide value - 100 to 150
Suspended particulates (smoke) (80/779/EEC)	Annual 98th percentile limit value - 250 Annual limit value - 80 Annual guide value - 40 to 60
Lead-in-air (82/884/EEC)	Limit value annual mean - 2. No guide value
O_3 (92/73/EEC)	8 hour non-overlapping standard - 110 for human health 1 hour public information threshold and public warning threshold - 180 and 360 respectively
NO_2 (85/203/EEC)	98th percentile limit value - 200 98th percentile guide value - 135 50th percentile guide value - 50

Table 7.2 Proposed EC air quality standards (μg/m^3) EC Directive (COM 98/386)[a]

Pollutant	*Averaging time*	*Level*	*Objective date*	*Eppcy*
SO_2	1 hr	350	1.1.2005	24
	24 hr	125	1.1.2005	3
NO_2	1 hr	200	1.1.2010	18
	Annual	40	1.1.2010	
Lead	Annual	0.5	1.1.2005	
PM_{10}[b]	24 hr	50	1.1.2005	35
	Annual	40	1.1.2005	
CO	8 hr	10000	EC proposal (2.12.98)[c]	

a) Directive came in to force 22.4.01
b) The Directive includes indicative values for PM_{10} for 2010 which are subject to review. Annual mean of 20 μg/m^3 and a 24 hr mean of 50 μg/m^3 not to be exceeded more than 7 times per year
c) Limit value with a date of compliance by 1.1.2005
Eppcy: Exceedences permitted per calendar year

The first Daughter Directive covers SO_2, NO_2, Pb and PM_{10} (Table 7.2) and provides limit values for the protection of human health, as well as annual limit values for SO_2 and NO_2 for the protection of vegetation. The requirements of the Daughter Directives will supersede earlier directives covering the same pollutants.

Table 7.3 WHO air quality guidelines (μg/m^3) (WHO, 1987; 2000b)

Pollutant	*Averaging time*	*Level*
Pb	Annual	0.5
NO_2	1 hr	200
	Annual	40
O_3	8 hr	120
SO_2	10 min	500
	24 hr	125
	Annual	50
CO (mg/m^3)	15 min	100
	30 min	60
	1 hr	30
	8 hr	10
Benzene	Unit risk/lifetime	6×10^{-6} (μg/m^3)$^{-1}$
Carbon disulphide	24 hr	100
As	unit risk/lifetime	1.5×10^{-3} (μg/m^3)$^{-1}$
Cd	Annual	5×10^{-3}
Hg	Annual	1.0
1,3-butadiene	No guideline	
1,2-dichloroethane	24 hr	700
Dichloromethane	24 hr	3000
HCHO	30 min	100
H_2S	24 hr	150
PAH (benzo-a-pyrene)	Unit risk/lifetime	8.7×10^{-2} (ng/m^3)$^{-1}$
Styrene	1 week	260
Tetrachloroethylene	Annual	250
Toluene	1 week	260
Trichloroethylene	Unit risk/lifetime	4.3×10^{-7} (μg/m^3)$^{-1}$
Cd	Annual	5 ng/m^3
Cr^{VI}	Unit risk/lifetime	4×10^{-2} (μg/m^3)$^{-1}$
Mn	Annual	150
Ni	Unit risk/lifetime	3.8×10^{-4} (μg/m^3)$^{-1}$
Acrylonitrile	Unit risk/lifetime	2×10^{-5} (μg/m^3)$^{-1}$
Vinyl chloride	Unit risk/lifetime	1×10^{-6} (μg/m^3)$^{-1}$
V	24 hr	1
Platinum[a]		
F[a]		
PM_{10}[a]		
PCB[a], PCDD[a], PCDF[a]		

a) No guideline value because there is no evident threshold for effects on morbidity and mortality

7.2.2 World Health Organisation air quality guidelines

The WHO provides air quality guidelines for more than twenty pollutants. The guidelines are intended to provide background information and guidance to national or international authorities in making risk assessment and risk management decisions (Table 7.3). In providing pollutant levels below which exposure, for lifetime or for a given period of time, does not constitute a significant public health risk, the guidelines form the basis of setting (inter)-national standards or limit values for air pollutants.

For air pollutants which are considered to be genotoxic carcinogens, the WHO have used mathematical models to calculate a unit risk factor. These calculations are based on information from animal or from epidemiological studies. The unit risk factor is generally expressed as the additional cancer risk in a given population that inhales this compound at a concentration of 1 $\mu g/m^3$ during its entire lifetime. For example, for benzene the unit risk factor is $6x10^{-6}$, which means that in a population of one million persons, 6 additional cases of cancer may be expected following a continuous inhalation of 1 $\mu g/m^3$ during its lifetime, for PAH the unit risk tells you that 87 persons in a population of 1 million will die from cancer following a lifetime exposure to 1 ng/m^3. It should be noted, however, that these theoretical risk estimates contain large safety factors as well as a number of uncertainties.

Table 7.4 Summary of relative risk estimates of particulates on human health (WHO, 2000b)

Endpoint	*Relative risk for $PM_{2.5}$ (95% CL)*		*Relative risk for PM_{10} (95% CL)*	
Bronchodilator			0.0305	(1.0201-1.0410)
Cough			1.0356	(1.0197-1.0518)
LRS			1.0324	(1.0185-1.0464)
PEF change			-0.13%	(-0.17% - -0.9%)
Respiratory hospital emissions			1.0080	(1.0048-1.0112)
Mortality	1.015	(1.011-1.019)	1.0074	(1.0062-1.0086)

Table 7.5 Summary of the relative effects of particulates on human health (WHO, 2000b)

Endpoint	*Relative risk for $PM_{2.5}$ (95% CL)*		*Relative risk for PM_{10} (95% CL)*	
Morbidity	1.14	(1.04,1.24)	1.10	(1.03, 1.18)
Mortality	1.07	(1.04,1.11)	na	
Bronchitis	1.34	(1.04,1.11)	1.29	(0.96, 1.83)
Percentage change in FEV1, children	-1.9%	(-3.1%, -0.6%)	-1.2%	(-2.7%, -0.1%)
Percentage change in FEV1, adults			-1.00	(not available)

The WHO (2000b) provide a summary of relative risk estimates for bronchodilator use, cough and lower respiratory system (LRS) reporting, peak expiratory flow (PEF) changes and respiratory hospital emissions and daily mortality associated with a 10 μg/m^3 increase in the concentration of PM_{10} or $PM_{2.5}$ (Table 7.4). In addition the WHO provide risk estimates for the effects of long-term exposure to particulate matter on morbidity and mortality associated with a 10 μg/m^3 increase in the concentration of PM_{10} or $PM_{2.5}$ (Table 7.5).

Table 7.6a Summary AQS for selected countries for principal air pollutants (μg/m^3) (Murley 1993; 1995; Harrop, 1999)

Country	*CO^a*		*NO_2*		*Particulates*		*O_3*	*SO_2*		
	1hr	*8hr*	*1hr*	*Annual*	*24hr*	*Annual*	*1hr*	*1hr*	*24hr*	*Annual*
Australia	-	10	320	-	-	90**	240	-	-	60
Brazil[b]	40	10	320	100	240** 150* 150^{+}	80** 50* 60^{+}	160	-	365	80
Finland	30	10	300	-	150**	60**	-	500	200	40
Japan	<u>20</u> <u>10</u> daily average		-	-	100^{**c}	-	*60*	*100*	*40*	-
Kuwait	<u>35</u>	<u>10</u>[d]	-	-	350	90	*80*	*170*	*60*	*30*
New Zealand	30	10	300	-	120*	40*	150 <u>100</u>	350	125	50
Norway	25	10	100	-	70*	-		400[e]	90	-
Peru	-	20	-	-	350	150	<u>200</u>	-	*300*	*60*
South Korea	<u>25</u>	<u>9</u>	*150*	*50*	150**	80**	<u>60</u>	-	*150*	*50*
Sweden[f]	-	6	110	50[g]	-	-	-	200	100	50[g]
Taiwan	<u>35</u>	<u>9</u>	*250*	*50*	250** 125*	130** 65*	*120* *[60]*	*250*	*100*	*30*

a) mg/m^3
b) Primary standards
c) 1 hr level of 200**
d) 24 hr level of <u>8</u>
e) 15 min
f) Target
g) 6 months
<u>ppm</u>
ppb
* = *PM_{10}*
<u>*120*</u> = *8 hr average for O_3*
[120] = *8 hr average for O_3 ppb*
\+ = *Smoke (S/m^3)*
** = *total suspended particulates (TSP)*

Table 7.6b AQS for selected countries for principal air pollutants (mg/m^3) (World Bank, 2000)

Country	*SO_2*			*NO_x*			*Particulates*			*CO*	
	Annual	*24 hr*[g]	*Daily*	*Annual*	*24 hr*[g]	*Daily*	*Annual*	*24 hr*[g]	*Daily*	*Annual*	*8 hr*
Germany	0.14	-	0.40	0.1	-	0.3	0.1[a] 0.2[b]	-	0.2[a] 0.4[b]		
India	-	0.03-0.12	-	0.03-0.12[c]	-	0.0925	0.1-0.5[d]	-	-	1-5[e]	
Indonesia	-	-	0.26	-	-	0.093	-	-	0.26		22.6
Philippines	-	0.85[f]	0.37	-	0.19[h]	-	-	0.25[f]	0.15		
Poland	0.032	-	0.2	0.05	-	0.15	0.05	-	0.12		
Thailand	0.10	-	0.30	-	0.32[h]	-	0.10	-	0.33		
World Bank	0.10	0.5[i]	1.0[h]	0.1	-	0.5	0.10	0.50	-		

a) <10 um
b) >10 um
c) 0.03 mg/m^3 for "sensitive" areas, 0.08 mg/m^3 for "residential and mixed use" areas
d) 0.1 mg/m^3 for "sensitive" areas, 0.2 mg/m^3 for "residential" and "rural" areas, and 0.5 mg/m^3 for "industrial and mixed use" areas
e) 1 mg/m^3 for "sensitive" areas, 2 mg/m^3 for "residential" and "rural" areas, and 5 mg/m^3 for "industrial and mixed use" areas
f) 1 hr average
g) Maximum
h) Inside
i) Outside

Table 7.7 UK AQS for human health and vegetation and ecosystems (μg/m^3) (DETR, 2000e)

Pollutant	*AQS*
SO_2	15 min mean of 266 not to be exceeded more than 35 times per year (tpy) by 31.12.05; 1 hr mean of 350 not to be exceeded more than 24tpy by 31.12.04 and 24 hr mean of 125 not to be exceeded more than 3tpy by 31.12.04
Benzene	Annual mean of 16.25 by 31.12.03
1,3-butadiene	Annual mean of 2.25 by 31.12.03
CO	Running 8 hr mean of 11600 by 31.12.03
Pb	Annual mean of 0.5 by 31.12.04 and 0.25 by 31.12.08
NO_2	1 hr mean of 200 by 31.12.05 not to be exceeded more than 18tpy and annual mean of 40 by 31.12.05
O_3	8 hr mean of 100 not to be exceeded more than 10tpy by 31.12.05
PM_{10}	Annual mean of 40 by 31.12.04 and 24 hr mean of 50 by 31.12.04
Vegetation and ecosystems	
NO_2	Annual mean for the protection of vegetation of 30 by 31.12.00
SO_2	Annual mean for the protection of vegetation 20 by 31.12.00 and 20 winter average (1.10–31.3) by 31.12.00

7.2.3 National air quality standards

Table 7.6 (a and b) details the AQS for selected countries (Harrop, 1999; World Bank, 2000). Tables 7.7, 7.8 and 7.9 specifically detail the air quality criteria for the UK, US and the People's Republic of China for comparative purposes. Tables 7.6 to 7.9 show a wide range of formats for the presentation and reporting of AQS as well as a difference in actual concentrations for specific pollutants. With regard to carcinogens, countries have to choose their acceptable level of risk for each hazard, balancing risks and benefits, and establishing the degree of urgency of public health problems among sub populations inadvertently exposed to carcinogens. For example, in the UK a threshold value has been derived, as opposed to a unit risk factor (WHO air quality guidelines Section 7.2.2), where the standard for benzene is presently 5 ppb (16.3 $\mu g/m^3$) (Table 7.7).

In the UK the National Air Quality Strategy (NAQS) (DETR, 2000e), sets out the AQS for eight pollutants which are considered presently harmful to human health and flora. The NAQS set out a national framework to improve ambient air quality and ensure that it will not cause significant harm to human health and the environment (Case Study 7, Chapter 10, Section 10.8). The adopted AQS for the eight pollutants include those covered by the EU Directives (Tables 7.1 and 7.2). Specific objectives were included in the Strategy detailing the degree of compliance for each pollutant to be reached by 2005 (DETR, 2000e; SEPA, 2000).

In the US the CAA requires the USEPA to set AQS for pollutants considered harmful to public health and the environment (Table 7.8). The CAA established two types of AQS. Primary standards set limits to protect public health, including the health of sensitive populations such as asthmatics, children and the elderly. Secondary standards set limits to protect public welfare, including protection against decreased visibility, damage to animals, crops, vegetation and buildings. The USEPA Office of Air Quality Planning and Standards (OAQPS) has set AQS for six principal pollutants, which are called criteria pollutants.

In the US, State and Federal governments have set different health standards for pollutants, specifying levels beyond which the air is unhealthful. For example, California's State standards for air pollutants are more stringent than the Federal government's. It is up to each individual State to determine if they want to set more stringent standards. Standards are set to provide an adequate margin of safety in the protection of public health. Under the Federal CAA, the USEPA must base standards solely on health considerations and not economics or technology. A comparison of the Federal standards to those for California is:

- O_3 (1 hour average) - Federal standard should not exceed 0.12 ppm more than one day per year. The State standard should not equal or exceed 0.09 ppm;
- CO - Federal standard should not exceed 35 ppm for 1 hour average or 9.4 ppm for an 8 hour average, whereas the State standard should not exceed either 20 ppm for 1 hour average or 9 ppm for an 8 hour average;
- PM_{10} – The Federal standard is 150 $\mu g/m^3$ for a 24 hour average (arithmetic mean); where as the State standard is 50 $\mu g/m^3$ for a 24 hour average;
- NO_2 - Federal standard should not exceed 0.52 ppm for a 1 hour average, whereas the State standard is 0.25 for a 1 hour average.

Table 7.8 US AQS ($\mu g/m^3$) (USEPA, 2000)

Pollutant	*Averaging period*	*Standard value* [a]	*Standard type*
CO	8 hr	10000	Primary
	1 hr	40000	Primary
NO_2	Annual	100	Primary & secondary
O_3	1 hr[b]	235	Primary & secondary
	8 hr	157	Primary & secondary
Pb	Quarterly	1.5	Primary & secondary
PM_{10}	Annual	50	Primary & secondary
	24 hr	150	Primary & secondary
$PM_{2.5}$	Annual	15	Primary & secondary
	24 hr	65	Primary & secondary
SO_2	Annual	80	Primary
	24 hr	365	Primary
	3 hr	1300	Secondary

a) Value is an approximately equivalent concentration

b) The O_3 1 hr standard applies only to areas that were designated non-attainment when the O_3 8 hr standard was adopted in July 1997. This provision allowed a smooth, legal and practical transition to the 8 hr standard

Table 7.9 People's Republic of China AQS (mg/m^3) (State Environmental Protection Administration, 1998)

Pollutant	*Averaging period*	*Class 1*[a]	*Class 2*[a]	*Class 3*[a]
SO_2	Annual	0.02	0.06	0.10
	24 hr	0.05	0.15	0.25
	1 hr	0.15	0.5	0.70
TSP	Annual	0.08	0.20	0.30
	24 hr	0.12	0.30	0.50
PM_{10}	Annual	0.04	0.10	0.15
	24 hr	0.05	0.15	0.25
NO_x	Annual	0.05	0.05	0.10
	24 hr	0.10	0.10	0.15
	1 hr	0.15	0.15	0.30
NO_2	Annual	0.04	0.04	0.08
	24 hr	0.08	0.08	0.12
	1 hr	0.12	0.12	0.24
Pb ($\mu g/m^3$)	3 month		1.50/1.00	
BaP ($\mu g/m^3$)	24 hr		0.01	
F urban areas ($\mu g/m^3$)	24 hr and 1 hr		7[a] and 20[a]	
F agricultural areas	Monthly		1.8[b] (3.0)[c]	
($\mu g/(dm^2$ day))	Growing season		1.2[b] (2.0)[c]	

a) Applicable to the urban area, b) Livestock areas, mixed farming and silkworm areas and c) Arable farming areas

Note: From October 1999, a new standard for SO_2 and NO_x was introduced. The new standard for SO_2 was changed form 0.25 mg/m^3 to 0.70 mg/m^3 and for NO_x from 0.15mg/m^3 to 0.30 mg/m^3.

In China (People's Republic of China, State Environmental Protection Agency, 1996) the AQS stipulate the classification of the air quality against different land use types (Table 7.9) e.g.:

- Class 1 - natural conservation areas, scenic areas, historic sites and regions requiring special protection;
- Class 2 - residential areas, areas of mixed activity (e.g. commercial and traffic, residential, cultural, industrial and rural);
- Class 3 - special industrial areas.

Separate AQS exist for Hong Kong and these are given in Table 7.10.

Table 7.10 Hong Kong AQO ($\mu g/m^3$)[a, g]

Pollutant	*1 hr*[b]	*8 hr*[c]	*24 hr*[c]	*3 months*[d]	*1 year*[d]
SO_2	800		350		80
TSP			260		80
RSP[e]			180		55
NO_2	300		150		80
CO	30000	10000			
Photochemical oxidants (as O_3[f])	240				
Pb				1.5	

a) *Measured at 298K (25°C) and 101.325 kPa (one atmosphere)*
b) *Not to be exceeded more than three times per year*
c) *Not to be exceeded more than once per year*
d) *Arithmetic means*
e) *RSP means suspended particles in air with a nominal aerodynamic diameter of 10 μm or smaller*
f) *Photochemical oxidants are determined by measurement of O_3 only*
g) *API corresponding air quality objective levels:*
   - *25 – air pollution is low – at half the annual AQO, or at a quarter of the 1 hr or 24 hr AQO level*
   - *50 – air pollution is medium – at the annual AQO, or at half the 1 hr or 24 hr AQO level*
   - *100 – air pollution is high- at the 1 hr or 24 hr AQO level*
   - *200 – air pollution is very high – at 2 times the 1 hr or 24 hr AQO level*
   - *500 – air pollution is severe – at 3 to 12 times the 1 hr or 24 hr AQO*

7.2.4 Derived air quality standards

Although many countries have their own AQS, these can be limited in the scope of pollutants assessed. There is therefore often a need to apply other national or international AQS or in the absence of suitable adopted standards, to derive them. As an alternative to statutory AQS and guidelines, a safety factor of 100 has previously been applied to the workplace occupational exposure limit (OEL) values (Hong Kong Labour Department, 1995; HSE, 2000; US OSHA, 2000) to extend its use to the general public. Procedures for setting standards for air quality in the workplace are well established and there is considerable agreement on the levels to be used for most substances in the workplace between countries, though there are significant discrepancies for some substances.

There are two main types of workplace standards: short-term exposure limits (STEL), which are intended to protect exposed individuals against acute effects of a brief exposure, and long-term exposure limits (LTEL). It should be emphasized that both short-term and long-term exposure limits are averaged figures for the levels in air over a defined time period (e.g. 15 minutes and 8 hours). In addition to the STEL and LTEL standards the OEL outline the maximum exposure limit (MEL). For a substance to be assigned a MEL, exposure to the substance has, or is liable to have, serious health implications for workers; or socioeconomic factors indicate that although the substance meets the criteria for an occupational exposure standard (OES), a numerically higher value is necessary if the controls associated with certain uses are to be regarded as reasonably practicable (HSE, 2000). A derived AQS of taking the occupational air quality criteria and dividing by 100 is derived from:

- The duration of exposure per week could be as much as 168 hours (7 days x 24 hours) rather than 40 hours (5 days x 8 hours). Moreover exposure might extend to 52 weeks in any one year as opposed to an average working year of 44 weeks. Applying these assumptions, the minimum safety factor of 4.96 (168/40 x 52/44) is derived. On the basis that in principle there may be no recovery period between exposure sessions, a safety factor of 10 would be more applicable.
- A complete spectrum of the population may be exposed rather than just 'healthy' workers aged between 16 and 65. A safety factor of 10 should be introduced to allow for possible variability in human response to individual chemicals in other spheres of toxicological safety calculations.

Therefore, multiplying the two safety factors of 10 results in an overall safety factor of 100 (Environment Agency, 1997). For substances classified as requiring a MEL the long-term criteria is derived by dividing the MEL by 500. The 8 hour OEL is used to derive long-term environment assessment levels (EAL). It is recommended that the short-term OEL STEL be divided by 10 and 50 if using the MEL STEL. The EAL are given in Table 7.11 (Environment Agency, 1997). It should be remembered that derived guidelines should only be used in the absence of appropriate national and international standards. The Environment Agency (1997) provides a comprehensive list of derived air quality criteria.

Table 7.11 EAL derived air quality standards

	Long-term EAL (annual average)	*Short-term EAL (1 hr average)*
OES 8 hr time weighted average	OES/100	-
MEL 8 hr time weighted average	MEL/500	-
OES STEL 15 minute average[a]	-	OES STEL/10
MEL STEL 15 minute average[a]	-	MEL STEL/50

a) For those substances for which a STEL is not listed a figure of 3 times the 8 hr time weighted average may be used (Environment Agency, 1997).

7.3 AIR POLLUTION INDICES

The Pollutant Standards Index (PSI) was developed by the USEPA to provide information about daily levels of air pollution (USEPA, 2000). The index provides the USEPA with a uniform system of measuring pollution levels for those air pollutants regulated under the US CAA. Following measurement the PSI figures are reported in all metropolitan areas with populations exceeding 200,000. Index figures enable the public to determine whether air pollution levels are for example 'good', 'moderate', 'unhealthful' or 'worse'. In addition, the PSI can be used as a public information tool to advise the public about the general health effects associated with air pollution, and to describe whatever precautionary steps may be needed if air pollution levels rise into the unhealthy range. The USEPA uses the PSI to measure five major pollutants for which it has established AQS (e.g. PM, SO_2, CO, NO_2, and O_3) (Table 7.8). The PSI is now widely used by other countries such as Australia, Taiwan, Singapore and Hong Kong (Table 7.10).

The PSI converts the measured pollutant concentration to a number on a scale of 0 to 500. The most important number on the scale is 100, since this corresponds to the standard established under the US CAA. For example, a 0.14 ppm (365 $\mu g/m^3$) reading for SO_2 or a 0.12 ppm (235 $\mu g/m^3$) reading for O_3 (Table 7.8) would translate to a PSI level of 100. A PSI level in excess of 100 means that a pollutant is in the 'unhealthful' range; a PSI level or below 100 means that a pollutant reading is in the satisfactory range (Table 7.12). The intervals on the PSI scale relate to the potential health effects of the daily concentrations of each of the five pollutants. Each value has built into it a margin of safety that, based on current knowledge, protects highly susceptible members of the public. USEPA determines the index number and then reports the highest of the five figures for each major metropolitan area, and identifies which pollutant corresponds to the figure that is reported. On days when two or more pollutants exceed the standard (that is, have PSI values greater than 100), the pollutant with the highest index level is reported, but information on any other pollutants above 100 may also be reported.

Table 7.12 USEPA PSI (USEPA, 2000)

	PSI					
	Good	*Moderate*	*Unhealthful*	*Very unhealthful*	*Hazardous*	
Index value	0-50	5-100	101-200	201-300	301-400	401-500
Pollutant	*Concentration ranges (ppm)*					
CO	0-4.5	4.5-9	9-15	15-30	30-40	40-50
NO_2					1.2-1.6	1.6-2
O_3 1-hr	0-0.06	0.06-0.12	0.12-0.2	0.2-0.4	0.4-0.5	0.5-0.6
PM_{10} [a]	0-50	50-150	150-350	350-420	420-500	500-600
SO_2	0-0.03	0.03-0.14	0.14-0.3	0.3-0.6	0.6-0.8	0.8-1.0

a) $\mu g/m^3$

The USEPA (2000) has identified the following actions that should be followed based on certain PSI levels. A PSI level above 100 may trigger preventive action by State or local officials, depending upon the level of the pollution concentration. This could include health advisories for citizens or susceptible individuals to limit certain activities and place potential restrictions on industrial activities. The 200 level is likely to trigger an 'Alert' stage. Activities that might be restricted by local governments, depending on the nature of the problem, include incinerator use, and open burning of leaves or refuse. A level of 300 on the PSI will probably trigger 'Warning', which is likely to prohibit the use of incinerators, severely curtail power plant operations, cut back operations at specified manufacturing facilities, and require the public to limit driving by using car pools and public transportation. A PSI level of 400 or above would constitute an 'Emergency', and would require a cessation of most industrial and commercial activity, plus a prohibition of almost all private use of motor vehicles. If air pollution were to reach such extremely high levels, death could occur in some sick and elderly people, and even healthy people would likely necessitate restrictions on normal activity. Before determining which stage is to be called, officials examine both current pollutant concentrations and prevailing and predicted meteorological conditions (USEPA, 2000).

The PSI places maximum emphasis on acute health effects occurring over very short time periods, 24 hours or less, rather than chronic effects. However, what the PSI cannot be used as is the sole method for ranking the relative healthfulness of different cities. A variety of other factors would have to be considered including the number of people actually exposed, transportation patterns, industrial composition and the representativeness of the monitoring sites would also need to be taken into account in developing an accurate ranking of metropolitan areas. In addition, the PSI does not specifically take into account the damage caused by air pollution, however, there is likely to be a correlation between increased PSI levels and increased damage. The PSI in addition does not take into account the possible adverse effects associated with combinations of pollutants (synergism). In June 2000 the USEPA updated the index and renamed it the Air Quality Index (AQI). The PSI and AQI are similar, they both:

- Focus on health risks of brief exposure to pollutants;
- Involve air pollutants regulated by the CAA (i.e. criteria pollutants);
- Use the same method to calculate index values;
- Use an index value of 100 to represent pollutant concentration at the level of the AQS set by the USEPA.

In addition the AQI includes:

- A new health risk category, 'unhealthy for sensitive groups';
- Two additional pollutants (e.g. O_3 averaged over 8 hours, and $PM_{2.5}$);
- A specific colour associated with each of the health risk categories.

Table 7.13 provides the revised health categories for PSI and AQI, the range of index values and pollutant concentrations for each category. Fig. 7.1 shows the usage of pollution indexes for air quality in California, US. Malaysia has adopted a similar PSI system called the API (Table 7.14). The Malaysian API was modified to suit Malaysian guidelines developed by their DoE. The API provides a uniform system of measuring air pollution levels. The measured levels of five pollutants (PM_{10}, CO, SO_2, NO_2, and O_3) are converted to an index from 0 to 500. Again, the most important number on this index is 100, since it corresponds to the level at which health effects can occur. The API used in Malaysia is the same as that used in Singapore. Fig. 7.2 provides an application of the API in the People's Republic of China for principal cities. The figure shows the lowest and highest index value for June–July 2000 (National Environmental Monitoring Centre of the People's Republic of China, 2000, data from China Daily, 2000). Fig. 7.3 shows the frequency of occurrence of the API categories for the City of Shenyang, China.

Table 7.13 USEPA PSI (USEPA, 2000)

	AQI						
	Good	*Moderate*	*Unhealthy for sensitive groups*	*Unhealthy*	*Very Unhealthy*	*Hazardous*	
Index	0-50	51-100	101-150	151-200	201-300	301-400	401-500
	Pollutant concentration ranges (ppm)						
CO	0-4.5	4.5-9	9-12	12-15	15-30	30-40	40-50
NO_2						1.2-1.6	1.6-2.0
O_3[a] 1 hr					200-400	400-500	500-600
O_3[a] 8 hr	0-60	60-80	80-100	100-120	120-370		
$PM_{2.5}$[b]	0-15	15-40	40-65	65-150	150-250	250-350	350-500
PM_{10}[b]	0-50	50-150	150-250	250-350	350-420	420-500	500-600
SO_2[a]	0-30	30-140	140-220	220-300	300-600	600-800	800-1000

a) ppb
b) μg/m³

Table 7.14 Malaysian API

API	*Classification*	*API*	*Classification*
0-50	Good	101-200	Unhealthy
51-100	Moderate	201-300	Very Unhealthy
101-200	Unhealthy	301-500	Hazardous

In the UK, the DETR has established an Air Pollution Public Information System. The DETR's System is based upon three air pollution thresholds (Standard, Information and Alert) and four bands (Low, Moderate, High and Very High Pollution (Table 7.15):

- Low (below the Standard): effects are unlikely to be noticed even by individuals who know they are sensitive to air pollutants;
- Moderate (between the Standard and Information Threshold): mild effects, unlikely to require action, may be noticed by sensitive individuals;

- High (between the Information and Alert Thresholds): significant effects may be noticed by sensitive individuals and action to avoid or reduce these effects may be needed (e.g. reducing exposure by spending less time in polluted areas outdoors);
- Very High (above the Alert Threshold): the effects of sensitive individuals exposed to high levels of pollution may worsen.

Table 7.15 UK DETR air quality bands

		Standard threshold	*Information threshold*	*Alert threshold*
	Low	*Moderate*	*High*	*Very high*
SO_2 (ppb) 15 min average	<100	100-199	200-399	>400
O_3 (ppb)	<50[a]	50-89[b]	90-179 [b]	>180 [b]
CO (ppm)[a]	<10	10-14	15-19	>20
NO_2 (ppb) 1 hr average	<150	150-299	300-399	>400
PM_{10} ($\mu g/m^3$) 24 hr running average	<50	50-74	75-99	>100

a) 8 hr running average
b) 1 hr running average

Figure 7.1 An application of API for air quality in California (Kings County), 1999 (USEPA, 2000) (courtesy USEPA internet site www.epa.gov)

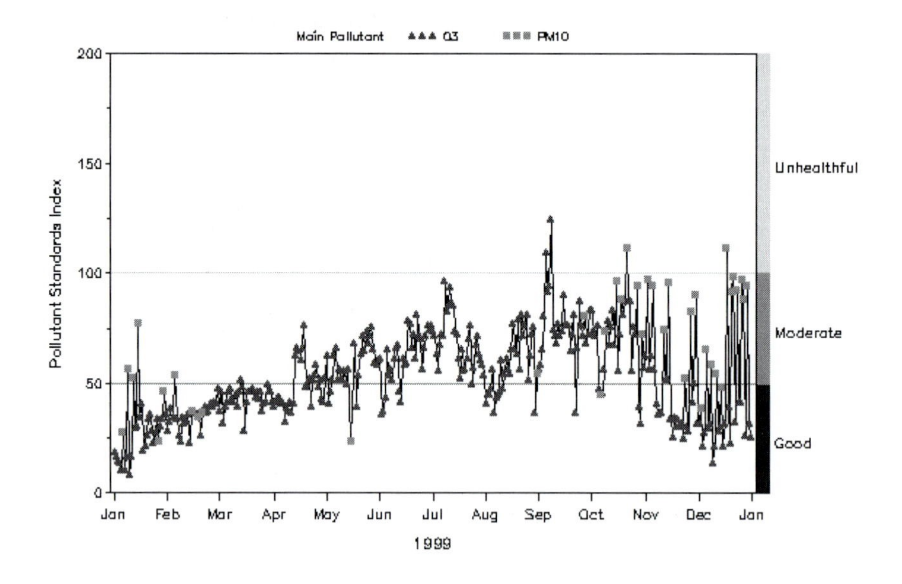

Figure 7.2 Air pollution index in major Chinese cities (June - July 2000) (Data from the China Daily, 2000)

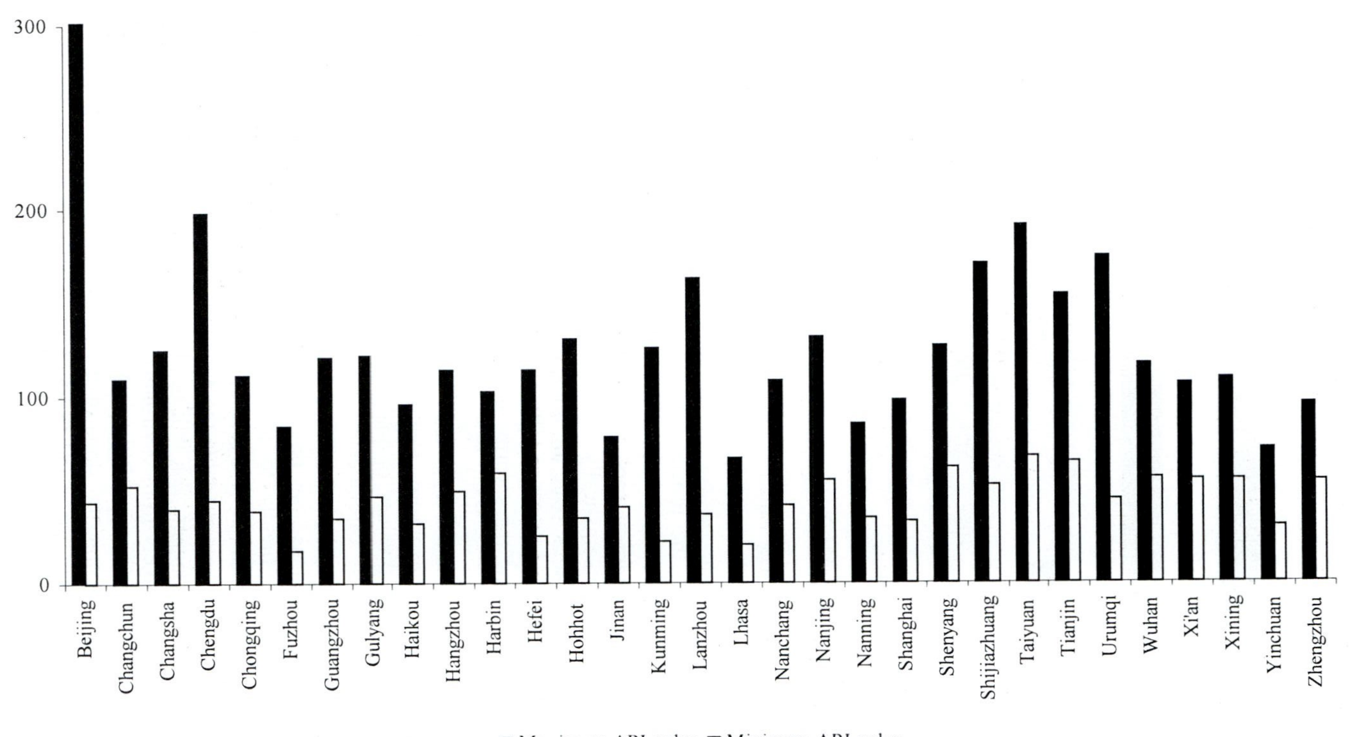

Figure 7.3 Frequency of occurrence of the Chinese API for the City of Shenyang, China 1999 (Data from Sustainable Shenyang Project Office (1998))

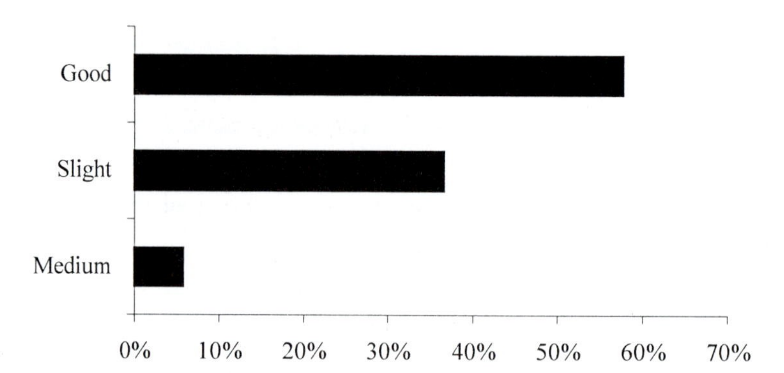

7.4 RISK ASSESSMENT

A human health assessment criterion now includes the concept of risk. Risk combines the probability of an adverse event (a hazard) occurring, with an analysis of the severity of the subsequent consequences. Risk assessment therefore can be simply defined as the process of assimilating and analysing all the available scientific information associated with a hazard or set of hazards.

As a management tool, quantitative risk assessment (QRA) attempts to express risks mathematically by modelling exposures to source(s) present, aggregating exposures over all relevant exposure routes and key sources, and expressing estimated risks for ASR. Health risk assessment may be defined as the means of evaluating the severity and likelihood of harm to human health or the environment occurring from exposure to a potentially hazardous substance or activity (Cohrssen and Corvello, 1989). Risk is the probability of harm occurring under the specific exposure circumstances. It may be expressed in a quantitative form with numerical values, or qualitatively (as high, low, insignificant, etc).

For potential carcinogens, risks are estimated as probabilities. The carcinogenic potency factor, which is the upper 95 per cent confidence limit on the probability of response per unit intake of a chemical over a lifetime, converts estimated intakes directly to incremental risk. If the exposure assessment is conservative, the resultant risk predicted is an upper bound estimate. Consequently, predicted risk may overestimate the actual risk at a site. However, this method is used so that carcinogenic risk will not be underestimated. Because relatively low intakes are most likely from environmental exposures, it can be assumed that the dose-response relationship will be in the linear portion of the dose-response curve. Under this assumption, the slope of the dose-response curve is equivalent to the carcinogenic potency factor and risk will be directly related to intake at low levels of exposure. The key stages for assessing the risks from environmental exposures for toxic air pollutants (TAP) are (Harrop, 1999; Harrop and Nixon, 1999):

- Stage 1: hazard identification;
- Stage 2: exposure assessment;
- Stage 3: dose-response assessment;
- Stage 4: risk characterisation.

Risk assessment attempts to express risks by modelling exposures to relevant source(s) present, thereby aggregating exposures over all relevant routes (e.g. inhalation) and sources, expressing risk estimates for individual receptors or groups of receptors. The above four stages refer to the identification of pollutants posing health effects; the assessment of the amount of exposure (intake) received by sensitive receptor(s); the assessment of the relationship between exposure and the toxic response; and finally, the characterisation of exposure intakes through their translation into risk estimates (USEPA, 1989). In Stage 4, risks are characterised within the context of background exposures and air quality criteria (Tier I Assessment) and then possible risks are compared with existing risk acceptability criteria (Tier II Assessment). Exposure pathways can be varied and include:

- Inhalation of air;
- Ingestion of food;
- Ingestion of drinking water;
- Dermal (skin) contact with soil;
- Intentional ingestion of soil;
- Dermal contact with water.

Frequently dermal contact and intentional ingestion of soil (e.g. pica) are screened out as being a significant exposure pathway due to the infrequent nature of the events and to the very low absorption factors for these routes of exposure. In addition it is unlikely that this pathway would provide a significant and plausible dose. Studies (Environment Agency, 1999) have found that the mean incremental dose from ingestion of soil is at least one order of magnitude lower than by other pathways. Other studies (Pasternach, 1989) have also found that soil dermal absorption and ingestion are at least one order of magnitude less efficient than lung absorption. Similar arguments have been proposed with respect to the elimination of aquatic pathways from consideration with regard to swimming, fishing and other recreational activities as they are sporadic and unlikely to lead to significant exposures or uptake of any human contamination. For example, exposure via drinking water would require its contamination to the point of consumption, which is often a remote possibility, also making this pathway insignificant in terms of total potential uptake. Inevitably, therefore, the focus of attention on pathway exposure focuses on either inhalation or ingestion of food. Frequently, however, a rapid assessment of risk of exposure can be gained by reviewing the inhalation exposure route. The equations used for the calculation of unit carcinogenic risk for inhalation (e.g. inhalation) are as follows (USEPA, 1986):

Risk = cancer potency factor x inhalation exposure (7.1)

Where:

Inhalation exposure =
(total dose) ÷ (body weight (kg) x lifetime (days)) (7.2)

Where:

Total dose (mg) = contaminant concentration (mg/m^3) x inhalation rate (m^3/day) x duration of exposure (days) (7.3)

It is customary in risk analysis to assume a worst case scenario, contaminants are at the maximum limit permitted by the emission source and are continuous over 365 days per annum for 70 years, the average lifetime of a hypothetical exposed individual. Further, the individual is assumed to have spent an entire lifetime at the point of maximum impact exposure, the emissions from the source and absorbing 100% of the inhaled material. The total excess cancer risk for a number of contaminants may be gained from the following equation:

Total Excess Cancer Risk = risk 1 + risk 2 + risk 3 + ... risk n (7.4)

Where:

Risk 1 = individual excess cancer risk from a lifetime exposure from the first substance;
Risk n = individual risk of additional substances.

The Royal Commission on Environmental Pollution (RCEP) (1993) states that there is a consensus among regulatory authorities in the UK and US that annual incremental risks of death greater than 1 in 10,000 (10^{-4}) are too high to be acceptable and that a risk of 1 in 1,000,000 represents a reasonable upper bound beyond which measures to achieve a further reduction in the risk would not be justified in terms of the benefit gained. The UK HSE have presented criteria designed to advise planning authorities on the development of land in the vicinity of 'major hazard' installations. The HSE propose that housing developments providing for more than about 75 people should be subjected to a calculated individual risk of less than 1 in a million per year. The same value is recommended by the RCEP. The Netherlands Framework for Risk Management (Murley, 1993) recommended that the maximum acceptance level should be 10^{-6} per year with a target for negligible risk of 10^{-8} per year. The Hong Kong Environmental Protection Department (HKEPD), in 1999, has previously considered health risk assessment criteria for carcinogens. These are given in Table 7.16. Chapter 10 (Section 10.7) provides a case study on the application of QRA for an inhalation exposure route. Risks predicted by this methodology are additional to the general risks that an individual may be exposed to during their every day life. The UK HSE have published the lifetime risk factors associated with daily life (Table 7.17).

For non-carcinogens health criteria are based on reference doses (RfDs). The RfD is an estimate of the daily exposure that presents a low risk of adverse effects during a lifetime of exposure. The purpose of the RfD is to provide a benchmark against which other doses might be considered. A dose that is less than the RfD is unlikely to be of concern. Conversely, doses which are greater than the RfD may indicate inadequate margins of exposure safety from the pollutant. The non-carcinogenic effect of a pollutant can therefore be determined in terms of the Hazard Quotient (HQ), which is calculated as the predicted intake of a pollutant divided by the RfD. If the HQ were less than unity (1.0) the exposure would be considered not to be an adverse non-carcinogenic effect. The Hazard Index (HI) is the summation of the HQs and assumes that there is no synergistic or antagonistic health effects (Environmental Resource Management Ltd, 2000b).

Table 7.16 HKEPD preliminary health risk guidelines

Reference levels (each chemical)
Unacceptable - To bring the risk from exposure to individual TAP to below 10^{-6} per year (lifetime as 7×10^{-5})
As Low As Reasonably Practicable - For those TAPs with risk level between 10^{-6} per year and 10^{-8} per year (lifetime between $7 \times 10^{-5} - 7 \times 10^{-7}$), adopt the best available control technology to ensure emissions of individual TAP are as low as reasonably practicable
Acceptable - Ultimate goal would be to bring the risk from exposure to individual TAP to below the negligible level of 10^{-8} per year (lifetime as 7×10^{-7})
The total risk from exposure to all TAPs should not exceed a level of 10^{-5} per year (lifetime risk of 7×10^{-4})

TAP = toxic air pollutant

Table 7.17 Involuntary fatal risks (Environmental Resource Management Ltd, 2000b)

Activity	*Lifetime fatal risk*		
Road accidents	1 in 143	or	7.0×10^{-3}
Accidents in the private home	1 in 154	or	6.5×10^{-3}
Fire	1 in 952	or	1.1×10^{-3}
Excessive cold	1 in 1786	or	5.6×10^{-4}
Drowning	1 in 2381	or	4.2×10^{-4}
Gas incidents	1 in 7937	or	1.3×10^{-4}
Lightning	1 in 143000	or	7.0×10^{-6}

7.5 NUISANCE

For many pollutants the use of an AQS or guideline may be appropriate in the assessment process. Particular problems, however, may arise in the assessment of the significance of nuisance complaints, for example dust or odours. The inevitable subjective nature of nuisance complaints makes them difficult to assess and quantify.

7.5.1 Particulates

In many countries AQS for particulates focus on concentrations in air (Tables 7.6 to 7.9), and are therefore of partial value when considering particulates (dust) nuisance issues. Developments such as quarries, open cast sites, cement works, landfill sites, etc may create sufficient particulate emissions to cause soiling of surfaces and hence a nuisance. It is appropriate, therefore, to consider air quality criteria to quantify the level of nuisance. Unfortunately, there are problems with the measurement of dust, in establishing reliable criteria and setting credible limits for it. Human reaction to overall deposition of dust can relate to the rate of deposition surfaces. Significant nuisance is likely when the dust coverage of surfaces is visible in contrast with adjacent clean areas, especially when it happens regularly. Severe dust nuisance occurs when the dust is perceptible without a clean reference surface (Harrop, 1999).

Dust is considered to be any solid matter occurring from on-site workings. Dust pollution, as for all air pollution borne on the wind, may vary rapidly. Typically peaks may be 2-5 times the monthly average (Bates *et al.*, 1990). The main potential effects of dust are (DoE Minerals Division, 1995a):

- Visual - dust plume, reduced visibility, coating and soiling of surfaces leading to annoyance, loss of amenity, the need to clean surfaces;
- Physical and/or chemical contamination and corrosion of artefacts;
- Coating of vegetation and contamination of soils leading to changes in growth rates of vegetation and possibly reduced value of agricultural products;
- Health effects due to inhalation, e.g. asthma, or irritation of the eyes.

A criteria previously used for assessing nuisance that may be set against a deposition rate is the value of 200 mg/m^2/day averaged over a month (DoE, 1992). The criteria are relatively unsophisticated and should be treated with caution. Nevertheless, the level is an indication of potential nuisance. British Coal Opencast has previously used this figure (British Coal Opencast Northern, 1993). However, recent guidance has expanded upon these criteria and Table 7.18 summarises a range of criteria cited by the UK DoE (DoE, 1995a).

Table 7.18 Dust control guidelines and standards (mg/m^2/day) (DoE, 1995a)

Standard and origin	*Description of guideline/standard*	*Value*
Nuisance standard, Washington State US	Residential area	187[a]
Nuisance standard, West Germany (TA Luft Vol. 27.2 1986)	Possible nuisance	350[a]
	Very likely nuisance	650[a]
Nuisance standard, West Australia	Loss of amenity first perceived	133[a]
	Unacceptable reduction in air quality	333[a]
UK OEL (applies to all workplaces)	Total inhalable dust health standard	10 (8 hr average)

a) monthly mean

Some countries like Malaysia and Israel have introduced dust nuisance criteria into their own AQS. In Malaysia they have set a dust deposition level of 133 mg/m^2/day. Whilst in Israel they have a total dust settling loading of 20 ton/km^2 over a 30 day period and a soiling index of 2 CoH/10^3 1 ft for 30 minutes and 1 CoH/10^3 1.ft for 2 hours (Murley, 1993; 1995).

7.5.2 Odour

Odour annoyance is a common problem, and is frequently difficult to resolve satisfactorily. An odour is simply what can be smelt. It can be pleasant or offensive. In short an odour is a sensation (Artis, 1984). Moncrieff (1967) suggests that any further search for a definition would be axiomatic. Other sources have defined an odour as the product of the activation of the sense of smell, an olfactory experience (Artis, 1984).

Harrop (1999) has previously summarised odour nuisance concerns. Odorous emissions generally consist of complex mixtures of compounds, which are odorous both individually and collectively. There is no simple relationship between the odorous characteristics of the individual compounds and the characteristics of the mixture. Furthermore, most compounds are odorous when present at low concentrations. However, whilst considerable progress has been made in identifying and quantifying the constituents of odorous gas streams, the human nose still remains, for the most part, the only reliable guide to the presence and the strength of an odour. The process by which odour molecules are detected by the olfactory system is a complex one, and even now it is still not fully understood (Booker and Moorcroft, 1988). If a substance is to be odorous then it must, to some extent, be volatile. This means that it must release molecules into the air which are then capable of being absorbed onto the surface of the olfactory cells in the human nose. Common examples are many organic solvents which have a degree of volatility at room temperature and which, therefore, exhibit characteristic odours. Most gases and vapours, being readily absorbed, are odorous, although light gases are exceptions. The threshold of detection of an odorant will depend on both the chemical nature of the molecules and on the physiology and psychology of the human receptor. The response to compounds, which stimulate the olfactory sense, will also vary with time and concentration. Regular exposure to an odour will lead to a level of odour fatigue and poor odour recognition. The physiological and psychological responses of individuals to odours vary between individuals. There are a number of non-olfactoral parameters to consider, for example environment and personality, and they serve to illustrate the subjective nature of any attempt to quantify the nuisance. Extensive and useful odour threshold level lists are provided by van Gemert and Nettenbreijer (1977), WSL (1980) and AEA Technology (1994). Table 7.19 lists the odour thresholds of selected principal air pollutants.

An alternative to the use of an odour threshold value is an approach adopted by the State Environmental Protection Agency of the People's Republic of China (State Environmental Protection Administration, 1993) where acceptable odour levels are set for the boundary of the emitting source. Criteria vary for whether the process is existing or new (Table 7.20). Emission standards (kg/hr) are also set for odorous pollutants depending upon the height of the released emission above ground level.

Table 7.19 Odour threshold levels (mg/m^3) (AEA Technology, 1994; Environment Agency, 1997)

Pollutant	*Threshold*	*Pollutant*	*Threshold*
Acetic acid	0.043	Ethanol	0.28
Acetic anhydride	0.0013	Ethyl acetate	2.41
Acetone	13.9	Formaldehyde	0.49
Acrylic acid	0.0013	2-Furaldehyde	0.25
Acrylonitrile	4.65	Hydrogen sulphide	0.00076
Ammonia	0.1-11.6	1-Methoxypropan-2-ol	0.0122
1,3 Butadiene	0.45-1.1	Methyl methacrylate	0.38
Benzene	1.5-108	4-Methylpentan-2-one	0.54
Butan-1-ol	0.09	2-Methylpropan-2-ol	71
Butan-2-ol	3.3	NO_2	0.02-2.0
Butan-2-one	0.87	O_3	0.05-0.1
2-Butoxyethanol	0.0051	1-Nitropropane	28.2
Butyl acetate	0.047	Propan-2-ol	1.185
Carbon disulphide	0.02-0.1	Styrene	0.16
Carbon tetrachloride	280	SO_2	0.00009
Cyclohexane	315	Tetrachloroethylene	4.68-8.0
Cyclohexanone	0.083	Toluene	0.47-0.79
1,2 Dichloromethane	0.02-5.7	Trichloroethylene	6.5-34.9
Dichloromethane	3.42	Trimethylamine	0.0026
1,4-Dioxane	306	Xylene	0.078

Odours are often expressed in terms of odour units (ou) or dilutions to threshold (Chapter 5). If an odour were attributed to one compound present in the air, then the odour unit would reflect the threshold concentration for that compound. For example, the odour threshold for formaldehyde is approximately 1 ppm. Air with formaldehyde of 10 ppm would contain 10 ou (Environment Agency, 1997). The concept of ou is useful when the sampled air is likely to contain more than one odorous compound and their relative contribution to the odour is unknown. In such cases, the threshold of detection or recognition can be defined for the sample as a whole. Valentin and North (1980), Woodfield and Hall (1994) and Environmental Analysis Co-operative (1999) provide further information on the quantification of odours.

For the avoidance of community annoyance the aim should be that all emissions from a source are free from offensive odour (AEA Technology, 1994; Scottish Office, 1994) outside the process boundary, as perceived by the regulatory authority. Other regulatory bodies (e.g. The Draft Minnesota Odour Rule #7029) (Glasgow Scientific Services, 1998) have required odour emissions from a waste water treatment works not to create a community nuisance as determined by a system of recording and checking public complaints. In the US and Australia the 'no nuisance' approach is still used in terms of 'no justified complaints at the fence' (Schulz and van Harreveld, 1996). Whereas in Northern European countries the trend is towards a quantitative odour policy, thus acknowledging that some annoyance may exist and aims to reduce the annoyance to a level deemed acceptable by society (Schulz and van Harreveld, 1996).

Table 7.20 Boundary standard values of odour pollutants (mg/m^3) (People's Republic of China, State Environmental Protection Administration, 1993)

Pollutant	*Class 1*	*Class 2*		*Class 3*	
		BER[a]	*Existing*	*BER*[a]	*Existing*
Ammonia	1.0	1.5	2.0	4.0	5.0
Trimethylamine	0.05	0.08	0.15	0.45	0.80
Hydrogen sulphide	0.03	0.06	0.10	0.32	0.60
Thio-alcohol	0.004	0.007	0.010	0.020	0.035
Dimethyl sulphide	0.03	0.07	0.15	0.55	1.10
Dimethyl disulphide	0.03	0.06	0.13	0.42	0.71
Carbon disulphide	2.0	3.0	5.0	8.0	10
Styrene	3.0	5.0	7.0	14.0	19
Odour concentration[b]	10	20	30	60	70

Class 1 - natural conservation areas, scenic areas, historic sites and regions requiring special protection.
Class 2 - residential areas, areas of mixed activity (e.g. commercial and traffic, residential, cultural, industrial and rural).
Class 3 - special industrial areas.
a) Newly built, extended or rebuilt (BER)
b) Dimensionless

Although odour nuisance can not be regarded as an adverse health effect in a strict sense, it still affects the quality of life (WHO, 1979b). Therefore for this reason the WHO (1987) have used odour threshold levels for pollutants where relevant and used them as a basis for separate guideline values. The WHO (1987) have considered the following aspects and respective levels in the evaluation of sensory effects:

- Intensity - where the detection threshold level is defined as the lower limit of the perceived intensity range (by convention the lowest concentration that can be detected in 50% of the cases in which it is present) (Chapter 5);
- Quality - where the recognition threshold level is defined as the lowest concentration at which the sensory effect, e.g. odour, can be recognised correctly in 50% of the cases;
- Acceptability and annoyance - where the nuisance threshold level is defined as the concentration at which not more than a small proportion of the population (less than 5%) experiences annoyances for a small part of the time (less than 2%); since annoyance will be influenced by a number of psychological and socioeconomic factors, a nuisance threshold level cannot be defined on the basis of concentration alone.

Table 7.21 shows the WHO recommended odour criteria (WHO, 1987).

Table 7.21 WHO odour threshold levels (μg/m^3) (WHO, 1987), rationale and guideline values based on sensory effects or annoyance reactions, using an averaging time of 30 minutes

Pollutant	*Detection threshold*	*Recognition threshold*	*Guideline value*
Carbon disulphide[a]			20
Styrene	0.2-2.0	0.6-6.0	7
Tetrachloroethylene	70	210-280	70
Toluene	8000	24,000-32,000	8000
Hydrogen sulphide	1000	10,000	1000

a) *If carbon disulphide is used as the index substance for viscose emissions, odour perception is not to be expected when carbon disulphide peak concentration is kept below 1/10 of its odour threshold value, i.e. below 20 μg/m^3*

There are a number of published sources of odour detection and recognition thresholds (Table 7.19) (Valentin and North, 1980; WHO, 1987; AEA Technology, 1994). The reliability of published data varies considerably and it is generally preferable to refer to several sources and compare the variability (Environment Agency, 1997). The variability in thresholds is generally due to the measurement techniques applied and also to the sensitivity of the human nose to odours.

The odour emission rate (E) from a source is given by the following expression where D_o is the dilution factor which is a measure of the strength of the odour and v_v is the gas flow rate (m^3/sec) (WSL, 1980) (see equation 5.3):

$$E = D_o v_v \tag{7.5}$$

The maximum distance from the source at which complaints are likely to occur has been estimated by using equation 7.6 (Bailey and Bedborough, 1979; Keddie, 1982), where d_{max} is the maximum distance from the source within which complaints are likely to occur:

$$d_{max} = (2.2E)0.6 \tag{7.6}$$

Table 7.22 shows and application of the equation 7.6 to define the maximum distance from a sewage treatment works at which odour complaints are likely to occur. Table 7.23 shows typical odour emission levels for different component parts of a sewage works. Fig. 7.4 shows predicted ou levels around a sewage treatment works.

A common approach to odour nuisance assessment has been to set criteria of 5 or 10 ou at or beyond the site boundary as a 98th percentile of 1 hour concentrations. In other words, this is a level of 5 or 10 ou that cannot be exceeded for more than 2% of the time. The main problem with this is that it allows high levels of odour for anything up to 175 hours per year (2% of a year) which is sufficient to cause a significant nuisance. An emission level that occurred for less than 2% of the time would give a 98th percentile value of zero but could have a significant impact during times when it did occur. Similarly, an area which is downwind of a source of odour for less than 2% of the year would give a 98th percentile value of zero but could be experiencing high odour concentrations whenever there was a change in wind direction. Percentiles are often also difficult to explain to the general public and can be misinterpreted, however this approach has been used in other studies (Valentin and North, 1980; Hall and Kukadia, 1994; CIWEM, 1997).

Table 7.22 The odour emission rate and d_{max} for the operational phase of a sewage treatment works (Territory Development Department South East New Territories Development Office, Hong Kong, 1992)

Source	*Dilution factor*	*Emission rate (m^3/sec)*	*D_{max} (m)*	*Range of uncertainty*
• Phase 1				25
Wet well		171	50	to
Dry well		125		100
• Interim phase			1460	735
Wet well		171		to
Dry well		125		2900
Lime assisted settlement tank	455	5300		
Sludge holding tank	6300	79,960		
• Phase 2 - Stage 1			2600	1300
Wet well		171		to
Dry well		125		5200
Primary settlement tank	455	31,800		
Aeration tank	1200	114,240		
Final settlement tank	460	30,600		
Sludge holding tank	3650	46,320		
• Stage 2			3600	1800
Wet well		171		to
Dry well		125		7200
Primary settlement tank	445	53,000		
Aeration tank	1200	190,400		
Final settlement tank	460	45,900		
Sludge holding tank	3650	92,640		

A simpler odour assessment approach (Simms *et al.*, 2000) is to consider directly the likely frequency of occurrence of levels of impact of 5,10, 20 and 50 ou. Five ou is considered to be the minimum level for detection in the open environment where inhabitants are subject to a wide range of natural odours. 10 and 20 ou are levels that if occurring frequently enough may be sufficiently detectable to constitute a nuisance, while investigation of the exceedence of 50 ou gives information about the extent of high levels of impact. This is far easier to interpret and visualise in relation to perceived odour nuisance, which is a highly subjective matter. Odour detection duration is important and it is clear that the nose can detect elevated concentrations of an odorous material lasting only a few seconds. What is not clear is the point at which intermittent detection should be considered a nuisance.

Table 7.23 Summary of typical source odour emission rates (ou/sec) from a small sewage treatment works (Water Research Centre, 1999)

Source	*Odour emission rate*
Flow channels (inlet works to secondary treatment)	510
Grit separation	5136
Screens	894
Storm tank	11,344
Secondary treatment aeration	17,611
Secondary treatment settlement tanks	1069
Unthickened sludge storage	114,278

Figure 7.4 Dispersion of ou levels around a sewage treatment plant (base map reproduced by kind permission of Ordnance Survey © Crown Copyright NC/01/24240)

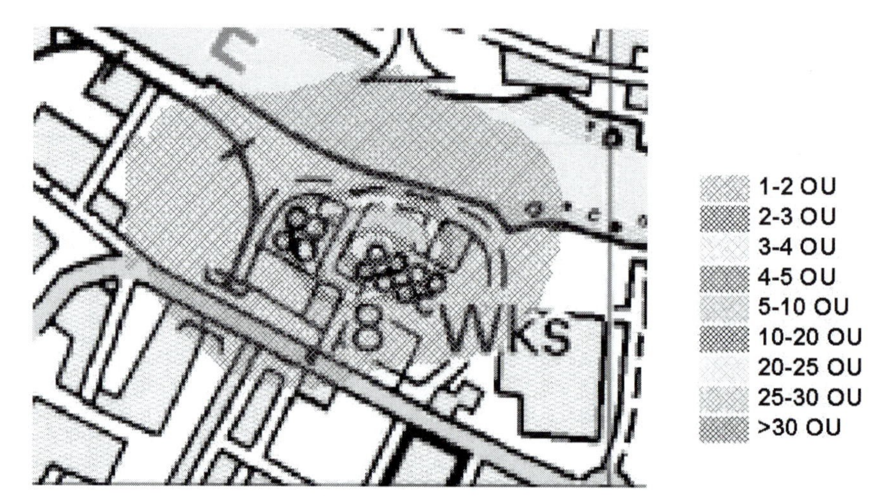

7.5.3 Visibility

Visibility can be a source of complaint by individuals. In the absence of reliable data researchers (Table 7.24) have derived a methodology for quantifying the concentration of particles causing haze. As long as no reliable information on the current air pollution level is available, it is possible to approximate the PM_{10} concentration using visibility and relative humidity. The method is based on an increase in particle concentration and humidity in ambient air reduces visibility.

Table 7.24 Correlation PM_{10} concentration ($\mu g/m^3$) with visibility (km) and relative humidity (%) (USEPA, 2000 citing work from Hoyningen-Huene)

km	*Relative humidity*						
	50%	*60%*	*70%*	*80%*	*85%*	*90%*	*95%*
10	48	48	49	36	29	21	13
5	95	96	97	71	58	42	25
4	119	120	121	89	72	52	32
3	158	159	161	119	95	70	25
2.5	190	192	194	142	114	85	50
2	238	239	242	178	143	105	62
1.5	317	320	323	238	191	140	83
1	475	479	484	356	286	211	124
0.7	679	684	691	509	409	301	177
0.5	950	959	968	712	572	420	248
0.4	1180	1198	1210	890	716	526	311
0.3	1584	1597	1614	1186	954	701	414
0.2	2375	2395	2420	1780	1430	1052	622
0.1	4751	4791	4838	3559	2861	2103	1244

Note: This table reflects only statistical values, which can differ considerably from actual PM_{10} concentrations. However, this method is sufficient for rough estimates.

Haze can be characterised using two measures, either the API or the Visibility Index (VI). Some practitioners use visibility as a basic indicator of air pollution. While these two measures are interrelated, the relationship may not be a simple direct one. The haze phenomenon is due to impairment of visibility as a result of the scattering and absorption of light by particles and gases.

Visibility is usually measured as the furthest distance from which an individual can see an object. However, the same amount of pollution is known to have different effects on visibility and it is useful to know that changes in visual range are not proportional to an individual's perception. A 5 km change in visual range can either be very apparent or imperceptible depending on the amount of pollution before the change.

7.6 FLORA

Available air quality criteria for the assessment of flora are provided as either critical levels or loads. A critical level is the concentration in the atmosphere above which direct adverse effects on receptors, such as plants, ecosystems or materials may occur (DoE, 1994a). Frequently the principal pollutants of concern are SO_2, NO_x and O_3. Critical loads are a quantitative estimate of exposure to one or more pollutants below which significant harmful effects on sensitive elements of the environment do not occur (SEPA, 2000).

Critical loads and levels are conveniently represented on maps and this allows the combined effects of physical, chemical and ecological, geological and hydrological factors on sensitivity to pollutant inputs to be quantified.

Table 7.25 UNECE critical loads (kg N/ha /yr) for terrestrial ecosystems (Grennfelt and Thornelof, 1992)

Receptor	*Critical load*	*Indication*
Lowland dry heath	15-20	Transition of heather to grass
Lowland wet heath	17-22	Transition of heather to grass
Species-rich lowland heath grassland	7-20	Decline of sensitive species
Calcareous species-rich grassland	14-25	Increase in tall grasses, decline in diversity
Acidic managed coniferous forest	15-20	Change in ground flora and fruit bodies
Acidic managed deciduous forest	<15-20	Change in ground flora
Arctic and alpine heath	5-15	Decline in lichens, mosses and evergreen dwarf shrubs; increase in grasses and herbs

Table 7.26 Critical levels for vegetation (μg/m^3 (ppb))

Pollutant	*Source*	*Time*	*Level*
NO_2	WHO[a]	Annual	30
	UNECE	Annual	29 (15)
	EU	Annual	30[b]
SO_2	WHO[c]	Annual	10-30[d]
	UNECE	Annual	20 (7.5)
		24 hr	70 (26)
	EU	Annual	20[b]
O_3	WHO	3 month (crops)	3[e]
		6 month (forests)	10[e]
		3 month (semi-natural)	3[e]
			0.2-0.5[e]
	UNECE	5 days visible injury	65
	EU	24 hr running mean	200
		1 hr	

a) Critical load 5 – 35 kgN/ha/yr
b) Proposed
c) Critical load 250 – 1500 eq/ha/yr
d) Lichens annual mean 10 μg/m^3, forest ecosystems annual and winter mean 20 μg/m^3, natural vegetation annual and winter mean 20 μg/m^3 and agricultural crops annual and winter mean 30 μg/m^3
e) AOT40 in ppm-h. (AOT40 = Accumulated exposure over a threshold of 40 ppb)

Once a critical loads or levels map is available for a particular receptor pollutant combination, comparison with current deposition loads or exposure levels for that pollutant provides the spatial distribution of areas of exceedence, which may also be mapped (DoE, 1994a). Critical loads are most frequently calculated for sulphur, acidifying nitrogen and eutrophying nitrogen. The UNECE has developed various approaches for estimating the critical loads for nitrogen and sulphur (Grennfelt and Thornelof, 1992). Until recently, identification of critical loads was based on empirical approaches. The initial approach (Level 0) was based on mineralogy and was set to prevent soil acidification. The succeeding approach (Level 1) used sensitive biological indicators (usually fine roots of trees) which were used to gauge the health of the plant. This was used to set a chemical critical load above which the plant would be likely to show an adverse response. Finally, the latest approach (Level 2) uses a mass balance which allows consideration of acidity produced within the soil-plant system as a result of the net uptake of base cations and the potential neutralising effect of atmospheric inputs of calcium and magnesium. Some critical loads have been defined in terms of ranges for a variety of ecological receptors; these are shown in Table 7.25.

The EU, WHO and UNECE have each proposed air quality guidelines (critical levels) for the protection of vegetation. Table 7.26 summarises these critical levels for vegetation (WHO, 1987; Association of London Authorities, London Boroughs Association and The South East Institute of Public Health, 1995). The International Union of Forest Research Organisations (IUFRO) (WHO, 1987) recommends a SO_2 annual level of 25 $\mu g/m^3$ as an acceptable level for the protection of trees in areas where growth is poor owing to other environmental stresses. Table 7.27 details examples of national air quality criteria for flora (SEPA, 2000; State Environmental Protection Administration, 1998).

7.7 INDOOR AIR POLLUTION

The WHO (2000a) provides air quality criteria for radon, man made vitrous fibre (MMVF) and environmental tobacco smoke (ETS) (Table 7.28). Quantitative risk estimates have been established for radon and for some MMVF, the refractory ceramic fibres (RCF). No guideline value was recommended for the other MMVF. For ETS no guideline value was established, because it has been found to be carcinogenic and to cause other serious health effects (cardiovascular effects) at typical environmental exposure levels. Within centrally air conditioned offices, indoor air quality is becoming a more prominent issue and people are now more aware of sick building syndrome (SBS) (Chapter 2, Section 2.6). The HKEPD are preparing new legislation on indoor AQS (Murley, 1995). Whilst in Singapore indoor air quality criteria already have been promulgated (Table 7.29).

7.8 POLLUTION EPISODES

In addition to national AQS some countries have also introduced air quality criteria for acute pollution episodes. For example, Brazil has set criteria for three levels of acute exposure to air pollution (Table 7.30) ranging from caution to emergency.

Table 7.27 National critical levels for vegetation (μg/m^3) (DETR, 2000e; People's Republic of China, State Environmental Protection Agency, 1996; Murley, 1995)

Pollutant	*Country*	*Time*	*Level*
NO_2	UK[a]	Annual	30
SO_2	UK[a]	Annual	20[b]
SO_2	Finland	Annual	25[c]
NH_4		Annual	8

People's Republic of China

Pollutant	*Sensitivity*	*Average level during growing period*[d]	*Daily average concentration*[e]	*Any once*[f]
SO_2[g]	Sensitive	0.05	0.15	0.50
	Semi-sensitive	0.08	0.25	0.70
	Resistant	0.12	0.30	0.80
Fluoride[h]	Sensitive	1.00	5.00	
	Semi-sensitive	2.00	10.00	
	Resistant	4.50	15.00	

a) Objectives apply in areas which are more than 20 km from a conurbation of more than 250,000 people and greater than 5 km from industrial sources regulated under Part 1 of the EPA 1990, motorways and built up areas of more than 5000 people

b) 20 μg/m^3 (8 ppb) winter average (October to March)

c) Outside towns and bigger villages (15 μg/m^3 under consideration). To avoid acidification effects the sulphur deposition level should be under 0.5 gS/m^2 (under 0.3 gS/m^2 under discussion)

d) 'Average concentration during growing period' limit value to the daily average concentration in any growing period

e) 'Daily average concentration' is the limit value to the average concentration in any day

f) 'Any once' is the limit value to any sampling

g) mg/m^3

h) μg/(dm^2.d)

Table 7.28 WHO air quality criteria for indoor air pollutants (WHO, 1987; 2000a)

Pollutant	*Guide value*	*Averaging time*	
Asbestos	500 F*/m^3 (0.0005 F/ml)	Lifetime	10^{-6}–10^{-5a} 10^{-5}–10^{-4b}
MMVF (RCF)	1x10^{-6} (fibre/litre)$^{-1}$	Lifetime	
Radon	3-6x10^{-5}/Bq/m^3	Lifetime	
ETS	No guide line		

a) Lung cancer in a population where 30% are smokers

b) Mesothelioma

F* = Fibres measured by optical methods

ETS = Environmental tobacco smoke

MMVF = Man made vitrous fibres

RCF = refactory ceramic fibres

Table 7.29 Singapore indoor guideline maximum concentrations for specific indoor air contaminants (Singapore Government, 2000)

Parameter	*Averaging time*	*Limit for acceptable air quality*
CO_2	8 hrs	1800 mg/m^3
CO	8 hrs	10 mg/m^3
HCHO	8 hrs	120 μg/m^3
O_3	8 hrs	100 μg/m^3
SPM[a]		150 μg/m^3
VOC		3 ppm
Total bacteria counts		500 cfu/m^3
Total fungal counts		500 cfu/m^3

a) Respirable particles with aerodynamic diameters less than 10 μm sampled with a size selective device (commonly used devices; cyclones and impactors) having a median cut size of 4 μm
SPM = Suspended particulate matter

Table 7.30 Criteria for acute air pollution episodes[a] in Brazil (Murley, 1995) (μg/m^3)

Pollutant	*Levels*		
	Caution	*Alert*	*Emergency*
SO_2 (24 hr)	800	1600	2100
TSP (24 hr)	375	625	875
SO_2 x TSP (24 hr)	65,000	261,000	393,000
Inhalable particles (24 hr)	250	420	500
Smoke (24 hr)	250	420	500
CO (8 hr)	15	30	40
O_3 (1 hr)	400	800	1000
NO_2 (1 hr)	1130	2260	3000

a) National Environment Council (CONAMA) Resolution Number 03/90 of June 1990 criteria for acute pollution episodes

7.9 AIR POLLUTION LEGISLATION

Countries are subject to international responsibilities and rights in air pollution management and control through membership of international institutions (e.g. EU and the UNECE). This is because air pollution does not recognise or confine itself to administrative boundaries. Therefore many air pollution concerns are recognised as being global or transboundary problems requiring international solutions. The introduction of international air pollution protocols and conventions has greatly strengthened air quality control. It is in this arena that the UN organisations and its agencies (e.g. UNEP) have played an important role in providing a forum for discussion and co-operation on transboundary air pollution.

7.9.1 Greenhouse gases

A UN conference ('The Earth Summit') in Brazil in 1992, under the auspices of the United Nations Conference on Environment and Development (UNCED), identified the importance of sustainable development. The principal aim of the conference was to halt and reverse the effects of environmental degradation while promoting sustainable and environmentally sound development in all countries.

Of particular concern to the conference was the issue of global warming caused by greenhouse gases (Chapter 3, Section 3.8.3). The conference therefore endorsed the Framework Convention on the Atmosphere (Climate Treaty) requiring developed countries to reduce greenhouse gas emissions to 1990 levels by the year 2000, as well as provide assistance to developing countries to achieve reductions. The Convention came into force in 1994. The Framework's other obligations included compiling emission inventories, producing and publishing programmes of measures to limit emissions and promoting research and environmental awareness about climate change. In December 1997, in Kyoto Japan, signatories to the Climate Treaty agreed to make legally binding cuts in the emissions of six greenhouse gases (CO_2, CH_4, N_2O, HFCs, PFCs and SF6). Signatories of developed countries agreed to reduce their emissions (between 2008 and 2012) by an average of 5.2% below 1990 levels (for CO_2, CH_4, N_2O and any year between 1990–1995 for HFCs, PFCs and SF6). The emission reduction targets varied between countries. The average reduction for EU countries was 8%, US 7% and Japan 6%. The UK agreed to reduce their emissions by 12.5%. The agreed protocol allowed for emissions trading between countries as a means of meeting targets. The first emissions trading agreements were implemented in the UK in February 2001.

7.9.2 Ozone layer

Another example of the international co-operation on the control of air pollution has been the control of emissions of pollutants damaging the stratospheric O_3 layer. A series of meetings held under the auspices of the UNEP culminated in 1987 with the signing of an Agreement on Substances that Deplete the Ozone Layer - the Montreal Protocol. The protocol committed industrialised nations to reducing consumption of CFCs by 50% by 1999 and to freeze production of halons in 1992. The protocol was subsequently reviewed in 1990 and 1992 and strengthened in the light of further scientific evidence. Current controls under the Montreal Protocol and subsequent amendments include (NSCA, 2000a):

- CFC consumption reductions based on 1986 consumption levels and a 75% reduction by 1 January 1994 and a phasing out by 1 January 1996. Less developed countries are required to phase out CFC use by 2010.
- Halons consumption reductions based on 1986 levels and a phasing out by 1 January 1994. Less developed countries are required to phase out halons by 2010.
- Carbon tetrachloride consumption reductions based on 1989 levels and 85% reduction by 1 January 1994 and a phasing out by 1 January 1996.

- 1,1,1-trichloroethane consumption reductions based on 1989 levels and a 50% reduction by 1 January 1994 and a phasing out by 1 January 1996.
- Methyl bromide consumption reductions for developed countries frozen at 1991 production and consumption levels from 1 January 1995. A 25% reduction by 1999 followed by a 50% reduction by 2001, 70% by 2003 and a phasing out by 2005, subject to exemption for critical agricultural uses. Less developed countries are to reduce consumption by 20% by 2005 based on an average of 1995-98 consumption. There is to be a phasing out of the substance by 2015.
- HCFCs consumption reductions for developed countries frozen at 2.8% of CFC and HCFC consumption in 1989 from 1 January 1996 levels; with further reductions of 35% by 2004; 65% by 2007; 90% by 2010 and 99.5% by 2013. There should be a phasing out by 2020, subject to limited exemptions for existing equipment to 2030. Less developed countries are to freeze their consumption of HCFCs in 2016 and phase them out entirely by 2040.
- Hydrobromofluorocarbons consumption phased out from 1 January 1996.

Signatories of the Montreal Protocol, meeting in Beijing in 1999, agreed further reductions in the production and consumption of certain O_3-depleting substances. These reductions included:

- Freezing on the production of HCFC from 2004 in developed countries and 2016 in developing countries.
- Annual reports to be made to the UNEP on methyl bromide use as a quarantine and pre-shipment fumigant, with applications for such use to be limited to 21 days before export. An advisory panel is to report on the feasibility of alternatives in 2003.
- Production and consumption of bromochloromethane to be banned from 2002.

From January 2000 signatories to the Montreal Protocol agreed that imports and exports of new and recycled CFC, halons, HCFCs and methyl bromide will require licensing to curb the illegal trade in their usage. EU Member States implemented the Montreal Protocol through the EC Regulation (3322/88). Under the current regulation (3093/94) the production and consumption of CFCs and carbon tetrachloride was phased out from 1 January 1995, a year ahead of the Montreal Protocol controls. HCFC consumption will be phased out by 2015 with interim cuts starting in 2004 and production and consumption of methyl bromide was cut by 25% from January 1998. Some production is allowed after the phase out date for any essential uses, which have been agreed by Parties to the Montreal Protocol, and for export to developing countries. The Regulation also includes use controls on HCFCs. The intention of these is to ensure that HCFCs are only used where they or CFCs have been used previously and where there is no acceptable alternative. Limits are also placed on the amount of O_3-depleting substances which each of the producers in the EU can place on the EC market. In addition, quotas are set for the amounts which can be imported from outside the Community, and import licensing monitors these. An annual list of critical uses in developed countries will be published by the UNEP from 2004.

At a national level regulators are also applying further controls on the emissions of O_3-depleting substances from industrial processes. For example, in Scotland SEPA are able to control emissions through the UK EPA1990. The same also applies to the Environment Agency (EA) in England and Wales. The new UK IPPC Regulations will increase the range of processes from which UK regulators, like SEPA, can control emissions. Regulators also play their part in furthering understanding of the process, extent and implications of O_3 depletion. This includes the extent of banks of O_3-depleting chemicals, and the possible environmental implications of compounds developed as alternatives to CFCs and HCFCs. Many manufacturers now use alternatives to CFCs and wherever possible industry is using products that do not contain O_3-depleting chemicals. Consumers are also being encouraged to switch to CFC-free products. As refrigerators which do not contain CFCs become available, consumers are being encouraged to replace their old units.

7.9.3 International conventions

The CLRTAP was adopted in 1979, and came into force in 1983 under the auspices of the UNECE. The Convention requires countries to endeavour to limit and, as far as possible, gradually reduce and prevent air pollution, including long-range transboundary pollution. This should be achieved through the use of the Best Available Technology (BAT) that is economically feasible. The convention arose as a result of concern that long-range transport of certain pollutants (mainly SO_2 and NO_x) was having an adverse effect on the environment. The convention also deals with the long-range transport of nitrogen and chlorine compounds, PAH, heavy metals and particles of various sizes. The convention has subsequently been followed by a series of protocols that have laid down more specific commitments. These protocols include (NSCA, 2000a):

- The protocol on Long-term Financing of Co-operative Programme for Monitoring and Evaluating Long-range Transmission of Air Pollutants in Europe (EMEP) was adopted in 1984 and came into force in 1988. The protocol committed parties to mandatory annual contributions to the EMEP budget approved by the Conventions Executive Body.
- The protocol of the Reduction of Sulphur Emissions or their Transboundary Fluxes by at least 30% (Helsinki Protocol or '30% club') was adopted in 1985 and came into force in 1987. The protocol committed parties to a 30% cut in total national emissions by 1993 based on 1980 levels. Although some countries did not become a party to the protocol they still achieved comparable emission reductions. For example, the UK achieved a reduction of 37% by the end of 1993. The EU did not ratify the protocol.

  The protocol has subsequently been renegotiated using data based on critical load assessments. Countries are required to reduce by 2000 their sulphur emissions to meet a UNECE-wide target of 60% of the gap between sulphur emissions and the critical load. Under CLRTAP there was a commitment that subsequent reductions in SO_2 and NO_x would be considered against 'best available scientific and technical developments'. In addition, the reductions would be considered against internationally accepted critical loads. Under CLRTAP, EMEP prepare critical maps for Europe. Reductions for individual countries have been based on their contribution to acid deposition. For example, for the UK to meet the 60% reduction level they have agreed to reduce their SO_2 emissions by 50% by 2000, 70% by 2005 and 80% by 2010 based on 1980 levels. The new protocol was signed in 1994 and came into force in 1998.

- The protocol concerning the Control of Emissions of Nitrogen Oxides or their Transboundary Fluxes (Sofia Protocol) was adopted in 1988 and came into force in 1991. The protocol committed parties to bring NO_x emissions back to their 1987 levels by 1994. Agreement has now been reached for a new multi-effects protocol covering NO_x, SO_2, VOCs and NH_3.
- The protocol concerning the Control of Emissions of VOCs or their Transboundary Fluxes was adopted in 1991 and came into force in 1997. The protocol committed most parties to secure a 30% reduction in VOC emissions from 1988 levels by 1999. Other requirements, two years after the protocol came into force, included the setting of national and international emissions standards to new sources of VOCs and to promote the use of products that have a low solvent content. Further controls included the application of BAT to existing stationary sources in areas of O_3 exceedence and implementation of technologies to reduce VOC emissions from petrol distribution and motor vehicle refuelling and reduce the volatility of petrol.

Other protocols include one requiring Cd, Pb and Hg emissions to be reduced below their 1990 levels (or an agreed year between 1985 and 1995). The protocol was signed in 1998 in Aarthus, Denmark. The protocol aimed to cut emissions from new and existing industrial sources and waste incinerators by setting stricter emission limits based on BAT. In addition, in 1998 a protocol was promulgated to phase out production and use of certain organic substances (e.g. pesticides, PCB, dioxins and furans and PAH) as well as requirements to eliminate discharges, emissions and losses, and to ensure safe disposal methods. A multi-pollutants protocol was signed in Gottenburg in 1999 and aims to achieve reductions in SO_2, NO_x, VOCs and NH_3 emissions. Parties to the protocol have signed up to a national emissions ceiling which requires them to reduce total emissions of each pollutant to their respective ceiling level or below by 2010. A 1984 protocol ratified by the EU in 1986 provides funding for EMEP comprising over 100 monitoring stations in over 30 countries (Chapter 5, Table 5.4).

7.9.4 European Union

The need to reduce emissions from all sources is a major thrust of the EU's policy to improve air quality. The more recent global threats on air quality have also been an important factor in shaping EU policy. Indeed in 1998 the Commission published a discussion paper 'Clean Air for Europe' which proposed integrating all the EU's air pollution control programmes into a single strategy. EU policy on pollution control has been clearly defined by a collection of instruments which, in one form or another, have been concerned with, approved or adopted by different

Member States. The declaration of the Council of the European Communities laid the framework for the first Programme of Action on the environment and called for a reduction in pollution and nuisances. This has been followed by further Action Programmes on the environment, and more than 100 specified measures (Directives, Decisions, Resolutions and Agreements) (Table 7.31) have been adopted to fill out in more detail various aspects of the general policy. The EU Directives provide a framework and define objectives or requirements, but leaves the Member States themselves to take action needed to achieve the desired results. Thus the terms of individual directives are often enacted under existing statutes by means of regulations. Some very important initiatives have been taken by the EU in relation to air pollution control through 'production standards' to reduce emissions from motor vehicles and limit the Pb and sulphur content of fuels (Table 7.32). These set levels for pollutants or nuisances, which are not to be exceeded in the composition or the emission of a product; or specify properties or characteristics of design of a product; or are concerned with the way in which products are used. Other initiatives also include emission standards for industrial processes.

7.9.4.1 Industrial emissions

EU legislation on the control of industrial emissions is based on their strategy to combat acidification, control of LCP, control of solvent emissions and the control of major hazards.

The control of emissions contributing to acid deposition (NO_x, SO_2, NH_3, VOCs, etc.) has been a focal point of the EU 5EAP in that there should be no exceedences of critical loads (Chapter 7, Section 7.6) for acidification anywhere in the EU. However, it has been estimated that 9 million hectares of the EU will still exceed the critical load in 2010. In an effort to cut this deficit by 50% the EU has published a Draft Communication on Community Strategy to Combat Acidification (COM(97)88). The principal elements of the Communication are the setting of national emission ceilings for four critical atmospheric pollutants for each Member State (proposed Directive COM(99)125). The limits are consistent with the attainment of the 50% gap closure target for acidification and the agreed air quality objective for tropospheric O_3 (NSCA, 2000a).

EU agreement was reached on LCP (Directive (88/609/EEC)) in 1988 committing Member States to specific reductions in SO_2 and NO_x from large fossil fuel burning plant >50MWe (e.g. power stations). An amendment (94/66) to the Directive extended the cover to SO_2 emissions for new solid fuel plants between 50–100 MWe and set an emission limit of 2000 mg/m^3. The EU published proposals (COM(98)415), in 1998, for amending the 1988 Directive to take account of technical developments since the adoption of the original Directive. The proposals set emission limits for NO_x and SO_2 based on BAT for new LCP.

EU Directive (99/13) was adopted on the limitation of emissions of VOCs due to the use of organic solvents in certain activities and installations. The Directive is due to come into force in 2001 and will mainly apply to paint coating and pharmaceutical industries.

Following an explosion at a chemical factory in Seveso Italy (Chapter 2) in 1976 the EU adopted a Directive on the Major Accident Hazards of Certain

Table 7.31 Selected European air pollution control directives (NSCA, 2000a)

Number	*Title*	*OJ No.*
Industrial pollution		
84/360	'Framework' Directive on combating EEC air pollution from industrial plants	L188
88/609/EEC	Directive limiting emissions of certain pollutants into the air from LCP	L336
96/61/EC	Council Directive concerning IPPC	L257
9913/EC	Council Directive on limitation of emissions of VOCs due to the use of organic solvents in certain activities and installations	L85
Air quality		
96/62/EEC	Directive on AQA and management	L296
99/30/EC	Council Directive relating to limit values for SO_2, NO_2, NO_x, PM and Pb	L163
97/101/EC	Decision (superseding 82/459), establishing a reciprocal exchange of information and data from networks and individual stations measuring ambient air quality within Member States	L35
80/779/EEC	Directive on air quality limit values and guide values for SO_2 and suspended particulates	L229
	Resolution relating to transboundary air pollution by SO_2 and suspended particulates	C222
81/462/EEC	Decision on the conclusion of the CLRTAP	L171
86/277/EEC	Council Decision concluding a protocol to the 1979 Geneva CLRTAP concerning the funding of a long-term programme of co-operation for the constant monitoring and evaluation of the long distance atmospheric transfer of pollutants in Europe (EMEP)	L181
93/361	Council Decision on the accession of the community to the protocol on the CLRTAP on the control of emissions of NO_x or their transboundary fluxes	L149
77/312/EEC	Directive on biological screening of the population for Pb	L105
82/884/EEC	Directive on limit value for Pb in the air	L378
85/203/EEC	Directive on air quality standards for NO_2	L087
87/217/EEC	Directive on prevention and reduction of environmental pollution by asbestos	L085
89/427/EEC	Directive modifying 80/779 on limit values and guide values of air quality for SO_2 and suspended particulates	
92/72/EEC	Directive on air pollution by O_3	L297
Incinerators		
89/369/EEC	Directive on air pollution from new municipal waste incinerators	L163
89/429/EEC	Directive on air pollution from existing municipal waste incinerators	L203
Chlorofluorocarbons		
	Resolution on CFCs in the environment	C133
87/412	Decision on the signing of a protocol of the Vienna Convention for the protection of the O_3 layer, relating to the control of CFCs	
88/540/EEC	Decision implementing the Vienna Convention on protection of the O_3 layer and the Montreal Protocol on substances which deplete the O_3 layer	L297
594/91/EEC	Regulation on substances that deplete the O_3 layer (amendment Directive 3952/92)	L67
3093/94/EC	Regulation on substances that deplete the O_3 layer	L333
Energy efficiency		
93/76/EEC	Directive to limit carbon dioxide emissions by improving energy efficiency (SAVE)	L237

Table 7.32 Selected European air pollution control directives for motor vehicles (NSCA, 2000b)

Number	*Title*	*Date*
70/220/EC	Directive relating measures to be taken against air pollution by gases from positive ignition engines of motor vehicles (amended 74/290/EEC, 77/102/EEC, 78/665/EEC, 83/351/EEC, 88/76) (The Luxembourg Agreement)	6.4.70
89/458/EEC	Amended Directive on gaseous emissions from private motor cars with a capacity of less than 1400cc. (amendment Directives 89/491/EEC, 91/441/EEC, 94/12/EC, 98/69/EC, 93/59/EC, 96/69)	3.8.89
72/306/EEC	Directive on measures to be taken against emissions of pollutants from diesel engines for use in motor vehicles	20.8.72
88/436	Amended Directive limiting gaseous pollutants from diesel vehicles up to 3.5 tonnes	6.8.88
88/77	Directive on emission of gaseous pollutants from diesel lorries and buses (amended 91/542, 96/1)	9.2.88
77/537/EEC	Directive relating to measures to be taken against the emission of pollutants from diesel engines for use in wheeled agricultural forestry tractors	29.8.77
97/68	Directive concerning emissions from non-road machinery	27.2.98
75/716/EEC	Directive relating to the approximation of the laws of Member States on the sulphur content of certain liquid fuels (amends Directive 87/219/EEC)	27.11.75
93/12	Directive relating to the sulphur content of certain liquid fuels	27.3.93
98/70/EC	Council Directive relating to quality of petrol and diesel fuels	28.12.98
99/32/EC	Council Directive relating to sulphur content of certain liquid fuels	11.5.99
78/611/EEC	Directive concerning the lead content of petrol	22.7.78
85/210/EEC	Directive on the approximation of Member State legislation on lead content of petrol, and the introduction of lead free petrol (amendment Directive 85/210/EEC)	3.4.85
92/6/EEC	Directive on the installation and use of speed limitation devices for certain categories for motor vehicles	2.3.92
92/55/EEC	Directive on vehicle emission testing	10.8.92
93/116/EEC	Directive relating to fuel consumption of motor vehicles	30.12.93
94/63/EC	Council Directive on the control of VOCs resulting from the storage of petrol and its distribution from terminals to service stations	31.12.94

Industrial Activities (82/501/EEC) (commonly known as the 'Seveso Directive'). The Directive was amended in 1987 and 1988 and in 1999 it was repealed, as this was the date by which EU Member States had to have implemented its replacement Directive on the Control of Major Accident Hazards involving dangerous substances (96/82/EC).

7.9.4.2 Transport

Since 1970 the EU has adopted a number of air pollution control Directives (Table 7.32) aimed at reducing emissions from all classes of motor vehicles. These control measures are basically product related. Product-related legislation relates to the fuel type and its constituents. Other control measures relate directly to the mass exhaust emissions depending on the type and size of motor vehicle.

7.9.5 World Health Organisation

The WHO provides recommended air quality guidelines (Section 7.2.2). These are used by countries for setting their own national AQS. Indeed the WHO guidelines were used for setting the EU standards and AQS in other countries.

7.10 NATIONAL AIR POLLUTION CONTROL REGIMES

As well as international directives and strategies, individual national governments have introduced their own air pollution control legislation. The following section details selected examples of national control regimes.

7.10.1 UK air pollution control regime

In England, Wales and Scotland the EA and SEPA are responsible for the regulation of industrial processes under Part 1 of the EPA 1990. The EPA Act introduced a system of prior authorisation for polluting industries through two systems of pollution control. Major processes (Part A) are controlled under Integrated Pollution Control (IPC) (Table 7.33) and the smaller and potentially less polluting, although more common processes (Part B), are controlled under Local Air Pollution Control (LAPC).

The two systems are based on the concept of Best Available Techniques Not Entailing Excessive Costs (BATNEEC). Under BATNEEC, industry is required to take all steps (subject to a test of excessive costs) to prevent or, where that is not practicable, to minimise release of certain substances (prescribed in regulations) into any environmental medium (air, water, land); and to render harmless both any such substances which are released and any other substances which may cause harm if released into any environmental medium. BATNEEC is usually expressed as emission limits for the prescribed substances released by the process. For IPC processes there is a requirement to apply the principles of Best Practicable Environmental Option (BPEO). This option is used when emissions are released to more than one environmental media and ensures that there is least impact to the environment as a whole. Before IPC was established, the releases of polluting substances from industry were regulated separately according to the environmental medium into which they were released. After the Fifth RCEP it was proposed that one body should regulate the release of prescribed substances. This integrated approach enabled an overall assessment to be made on the impact of the releases into the environment as a whole.

Under IPC, industrial processes are regulated according to individual site-specific judgements of what is BATNEEC and, in the case of the processes likely to give rise to emissions to more than one medium, what is the BPEO. The BPEO procedure establishes, for a given set of objectives, the option that provides the most benefit or least damage to the environment as a whole, at acceptable cost, in both the long- and short-term. The site specific nature of BATNEEC judgements means that appropriate account can be taken of local factors, such as air dispersion conditions at the site, peculiarities of site configuration and particularly sensitive

Table 7.33 Summary of the principal components of IPC

	Summary
Purpose	• Concerned with the release of polluting substances to air, land and water
Objectives	• To use the BATNEEC in order to prevent or minimise the release of prescribed substances and to render harmless any such substances which are released • To ensure consideration of releases from industrial processes to all media in the context of the effect on the environment as a whole

Principal features	• Operators of the potentially most polluting processes ('prescribed processes' which are specified in the amended Environmental Protection (Prescribed Processes and Substances) Regulations 1991/472) have to apply for prior authorisation from the regulatory agency to operate the process
	• IPC requires operators to consider the total impact of all releases to air, water and land when making an application
	• The operator has to advertise that an application has been made. Details of this are held in a public register (obtainable at Local Authority and/or regulatory offices) which is available for the public and statutory consultees to inspect
	• The EPA 1990 only allows for exclusion of the application from the public register on grounds of commercial confidentiality or National Security Commercial confidentiality is allowed only when the operator can prove that release of the information would reduce a commercial advantage or produce an unreasonably commercial disadvantage
	• After careful consideration, the regulatory agency may either grant an authorisation or reject the application. In granting an authorisation the regulatory agency must include conditions to ensure that:
	  • BATNEEC are used
	  • If a process involves releases to more than one environmental medium, the operator must use the BPEO
	• Compliance with any direction given by the Secretary of State for the Environment to implement EC or international obligations, or any statutory environmental quality standards or objectives, or other statutory limits, plans or other requirements
	• Operators have to monitor their emissions and report them to the regulatory agency on a yearly basis
	• If the regulatory agency believes that the operator is breaching the conditions of their authorisation the agency has enforcement options available and can serve the following notices:
	  • Revocation: when the regulator has reason to believe that withdrawing an authorisation is necessary
	  • Enforcement: if an operator is not adhering to the conditions set out by the authorisation, the regulator specifies the steps to take and time limit for the problem to be remedied
	  • Prohibition: if the regulator believes that operation of a process under authorisation involves imminent risk of serious pollution, the authorisation will be suspended until the risk is removed
	• The Secretary of State has the authority to direct the regulatory agency on authorisation and conditions to be included in authorisation
	• Operators can appeal: if the regulatory agency refuse to issue or vary an authorisation; against the conditions of the authorisation; against all the different types of notice the regulatory agency may serve against an authorisation

local environmental receptors. The requirement for a BPEO judgement in the case of most IPC processes means that air quality may not be the sole criterion (e.g.

aqueous discharges, waste), unlike for LAPC, according to which decisions about permit conditions are taken. The primary purpose of IPC is to facilitate consideration of the effect of industrial installations on the environment as a whole, of which air quality is frequently a major consideration. Under LAPC a range of small and potentially less polluting industrial processes are regulated for emissions to air only, including mobile plant (i.e. those designed to move or be moved on roads). Processes releasing substances in 'trivial' amounts (except where the release results in an offensive smell outside the premises) are exempt from the EPA 1990, as are engines for propelling most forms of transport (e.g. aircraft, road vehicles).

From October 1999 all EU Member States are required to incorporate the requirements of EC Directive 96/61 on IPPC into their national legislation. All new industrial installations must comply with IPPC and existing installations which come under IPPC are allowed a transitional period until 2007. There are a number of important differences between IPPC and IPC. Under the Directive there is a requirement that certain industrial installations prevent or reduce pollution from their operations using BAT. IPPC applies to installations whereas IPC applies to the process. An installation is a stationary technical unit where one or more activities are carried out, along with any other directly associated activities that have a technical connection which could have an effect on emissions and pollution. This now means that associated activities (e.g. fuel storage, etc.) will have to apply BAT, rather than just the process. IPC only takes into account immediate effects on land, air and water.

In addition to these media, IPPC will require consideration of the condition of the site when the installation closes; energy efficiency; waste reduction; noise, vibration and heat; consumption of raw materials and accident prevention. IPPC will in the future provide regulatory agencies with greater powers to control and regulate industry. The implementation of the IPPC Directive will result in certain Part B processes requiring authorisation (or permitting) under IPPC (e.g. animal rendering plants) with different industry sectors being phased in between 2001-2008 (SEPA, 2000). Table 7.34 summarises how IPPC will help UK regulatory authorities to control pollution.

Table 7.34 Summary of the attributes of IPPC (SEPA, 2000)

Attributes
• Many more installations will be regulated under IPPC than currently come under IPC
• IPPC takes into account more environmental impacts than IPC, including noise, accident prevention and energy efficiency
• IPPC applies to installations, whereas IPC applies only to processes
• IPPC does not exempt installations that only give rise to trivial emissions
• IPPC allows EU Member States to make general binding rules instead of site-specific conditions in authorisations
• IPPC allows for information exchange between EU Member States and within industry

7.10.2 US air pollution control regime

In 1881 the cities of Chicago and Cincinnati passed the first air pollution statutes in the US in an effort to control smoke and soot, primarily from furnaces and locomotives. By the early 1900s county governments began to pass their own pollution control laws. The next significant development was in 1952 when Oregon became the first state to legislatively control air pollution. Other States soon followed, enacting their own air pollution statutes generally aimed at controlling smoke and particulates. The Federal government intervened in 1955 with the passage of the Air Pollution Control Act that limited their involvement to providing financial assistance for the State's air pollution control research and training efforts. A greater synopsis of the US air pollution control regime is given in USEPA (2000) and summarised here. The shift towards greater Federal government involvement began with the passage of the 1963 CAA. The CAA was a comprehensive Federal law that regulated air emissions from area, stationary, and mobile sources. The Act also provided permanent Federal support for air pollution research, continued and increased Federal assistance to States for developing their own air pollution control agencies. The CAA also provided a mechanism through which Federal government could assist States with transboundary air pollution problems. The CAA was amended in 1965 to set the first Federal emission standards for motor vehicles.

In 1970 the Federal government created the USEPA and charged it with the responsibility of setting National Ambient Air Quality Standards to protect public health and the environment. The setting of maximum pollutant standards was coupled with directing the States to develop State Implementation Plans (SIPs) applicable to appropriate industrial sources in the State. States have to develop SIPs that explain how each State will perform its responsibilities under the CAA. A SIP is a collection of the regulations a State will use to clean up areas of concern. The States must involve the public, through hearings and opportunities to comment, in the development of each SIP. USEPA must approve each SIP, and if a SIP is not acceptable, USEPA can take over enforcing the CAA in that State. The USEPA assists the States by providing scientific research, expert studies, engineering designs and money to support clean air programs. The Act was amended in 1977 primarily to set new goals (dates) for achieving attainment of AQS since many areas of the country had failed to meet the initial deadlines. The USEPA was also given the authority to develop national emission standards for motor vehicles and set emission performance standards (known as New Source Performance Standards (NSPS)) for all new sources of all the common air pollutants. Under the CAA, the only major responsibilities that the States retained was that of determining how to control existing sources.

In 1977 the US Congress also made additional modifications to the CAA laying the groundwork for more significant changes to occur with the passage of the CAA Amendments. The 1990 amendments to the CAA in large part were intended to meet non addressed or insufficiently addressed problems such as acid rain, ground level O_3, stratospheric O_3 depletion and air pollutants.

Although the 1990 CAA is a Federal law covering the entire country, the individual States do much of the work to carry out the Act. For example, a State air pollution agency holds a hearing on a permit application by a company or fines a company for violating air pollution limits. Under this law, USEPA sets limits on how much of a pollutant can be in the air anywhere in the US thus ensuring that all Americans have the same basic health and environmental protection. The law allows individual States to have stronger pollution controls, although States are not allowed to have lesser pollution controls than those set for the whole country. The Act recognises the need for States to take the lead in carrying out the CAA, as pollution control problems often require special understanding of local industries, geography, housing patterns, etc. The 1990 CAA also provides for interstate commissions on air pollution control, which are to develop regional strategies for improving air quality. The 1990 CAA includes other provisions to reduce interstate air pollution. The Act also covers transboundary pollution between the US, Mexico and Canada.

A principal component of the 1990 CAA was its permit programme for larger source emissions. Under the new programme, permits are issued by States or, when a State fails to carry out the CAA satisfactorily, by the USEPA. The permit includes details on which pollutants are being released, how much may be released, and what kinds of steps the company is taking to reduce pollution, including plans to monitor emissions. The permit system is especially useful for businesses covered by more than one part of the law, since information about all of a source's air pollution are now in one place. The permit system simplifies and clarifies the obligations of businesses for air pollution control. Permits are publicly available. The CAA also gave new enforcement powers to USEPA. Previously it used to be very difficult for USEPA to prosecute a company for breaching the Act. USEPA had to go to court for even the most minor of offences. The Act now enables USEPA to fine offenders. Other parts of the law increase penalties for breaking the Act and bring the CAA's enforcement powers in line with other environmental laws. The Act sets deadlines for USEPA, States, local governments and businesses to control air pollution. The deadlines in the Act were designed to be more realistic than previous versions of the Act. The CAA has many features designed to control air pollution as efficiently and inexpensively as possible, letting companies make choices on the best way to reach pollution abatement targets. These new flexible programmes are called market or market-based approaches (e.g. the acid rain control programme offers companies choices as to how they reach their pollution reduction goals and includes pollution allowances that can be traded, bought and sold).

7.10.3 Japanese air pollution control regime

In Japan the Air Pollution Control Law (Law No. 97 of 1968 amended by Law No. 32 of 1996) details general provisions of the act; regulations pertaining to soot and smoke emissions; particulate regulation; promotion of measures concerning hazardous air pollutants; maximum permissible limits of exhaust of motor vehicles; monitoring air pollution levels; compensation for harm; miscellaneous provisions and penal provisions. The Offensive Odour Control Law (Law No. 91 of 1971 amended by Law No. 71 of 1995) details general provisions and regulations; promotion of offensive odour control measures; miscellaneous provisions and penal provisions.

7.10.4 Singaporean air pollution control regime

In Singapore the CAA was enacted in 1971 and it's Clean Air (Standards) Regulations 1972 control air quality. The control of vehicular emissions is under the Road Traffic Act and its subsidiary legislation, the Road Traffic (Motor Vehicles, Construction and Use) Rules.

7.10.5 People's Republic of China legislative system

The National People's Congress is the highest legislative body of China. It is responsible for drawing up laws, supervising their execution, and examining and approving environmental legislation. The State Council promulgates necessary regulatory requirements. To co-ordinate environmental issues the State Council has founded an Environmental Protection Commission to co-ordinate natural resource environmental protection and management and policy issues across all government departments. Many of the Commission's functions have been delegated to its executive body – the State Environmental Protection Bureau. A similar structure also exists at a provincial government level to enable national law enforcement and the promulgation of additional regulations specific to local environmental protection. The structure is repeated again at a municipality level. For example, in the province of Liaoning environmental protection is the remit of the Liaoning Provincial Environmental Protection Bureau (LPEPB) and in the city of Shenyang it is the Shenyang Environmental Protection Bureau (SEPB). The principal purpose of SEPB is to implement and enforce environmental protection laws and regulations.

7.11 PUBLIC AWARENESS

Public access to information is an important element of AQM. Access to information will facilitate:

- Public awareness;
- Information and knowledge transfer;
- Training.

Information availability is variable and is largely dependent on the political ideology of the country. The level of disclosure ranges from little/restrictive to full availability. For example, the US Freedom of Information Act allows individuals to glean a large level of information on air quality data, emission sources, etc. whilst the level of information in some countries is more restrictive in its availability. The proliferation of the internet has further aided the availability of information. Public perception and awareness of air pollution has heightened in recent years. Pollution incidents have further increased public interest. Media interest and a growing appreciation of environmental protection have galvanised public attitudes for stricter legislation and greater enforcement of legislation. Many countries monitor public attitudes towards law enforcement and register the level of complaints received as an indication of regulatory performance. The carrying out of regular inspections (Chapter 9) of industrial processes and rapid response to complaints by the regulator provides confidence to the general public on the regulator's performance. For example, SEPA, in Scotland, has introduced an Operator Performance Assessment (OPA) scheme. The scheme provides a systematic methodology for judging whether an operator has complied with the terms and conditions of their authorisation/permit. In 1998/99 approximately half of all the IPC processes were assessed, 90% of which were found to be satisfactory. More than half of all LAPC processes were assessed, 74% of which were found to be satisfactory (SEPA, 2000). Nevertheless, the agency still received 1986 complaints concerning IPC and LAPC processes in 1998/99, of which 82% were responded to within 24 hours.

Frequently most complaints relate to either dust or odour, as these problems are more easily discernible to the public. For example, in 1998 the Singapore Government (2000) received 348 complaints on air pollution (Table 7.35), of which 163 were verifiable incidences. The principal complaint was dust or odours resulting from poor maintenance, improper operation and/or overloading of pollution abatement equipment (Singapore Government, 2000).

To further the dissemination of information and increase public awareness a 1996 framework Directive on ambient air quality assessment and management (96/62/EC), (Official Journal (OJ) of the EU, 1994), sets amongst its general aims the requirement to make available to the public information on ambient air quality. The EU has also embraced the need for public access to environmental information through the Directive on Public Access to Environmental Information (90/313/EEC), which was implemented in 1992. The Directive allows members of the public to have access to all information about the environment held by government and public authorities in their own or other EU Member States (unless exempt on the grounds of commercial confidentiality or national security). The UNECE Convention on Access to Information, Public Participation in Decision Making and Access to Justice in Environmental Matters was adopted in 1998. It covers access to information, public participation and access to justice. The Convention aims to make it easier for people to take action through the courts or other public authorities if they feel that national environmental law is being contravened. In addition, the Convention details projects where public participation would be obligatory.

Table 7.35 Complaints and incidences of air pollution in Singapore (Singapore Government, 2000)

Type of air pollution	*Number of complaints*		*Number of incidences*	
	1997	*1998*	*1997*	*1998*
Odour	148	88	71	33
Fumes/dust	146	112	82	65
Smoke/soot	43	25	19	13
Noise	98	81	42	29
Others	35	42	13	23
Total	470	348	227	163

Public participation is also enshrined in the US CAA. Throughout the Act, the public is given opportunities to take part in determining how the law will be carried out. For instance, the public can take part in hearings on the State and local plans for controlling air pollution. They can also sue the government or a company to get action when USEPA or their State has failed to enforce the Act. The public can request action by the State or USEPA against offenders. Also the reports required by the Act are publicly available. A great deal of information is collected on pollutant emission releases and air quality monitoring. The 1990 CAA requires the USEPA to set up clearing houses to collate and provide technical information (Chapter 5). Typically, these clearing houses serve the public as well as State and other air pollution control agencies. Similar data is also made available by other agencies (e.g. UK).

Another example of public participation at a local level is employed by the South Coast Air Quality Management District (AQMD) of the US. The District is a four county regional government agency responsible for ensuring good air quality throughout the South Coast Air Basin through the regulation of mobile and stationary emission sources. Members of the general public are encouraged to dial 1-800-CUT-SMOG to report their observations of excessive dust, odours, smoke, or other air contaminants to the AQMD. An air quality complaint is defined as any injury, detriment, nuisance, or annoyance occurring as a result of air contaminants or other materials, including (but not limited to) smoke, dust, or odours. When an individual calls to report a 'smoking vehicle', they are requested to provide the vehicle's license plate number, vehicle model or type, and when (date and time) and where (city and street) the vehicle was observed. The AQMD checks this information against California Department of Motor Vehicles records to determine the name and address of the vehicle's owner. The owner is then sent an advisory letter, which describes when and where the vehicle was seen and recommends vehicle maintenance to help reduce emissions. The letter also asks the registered owner to return a postage-paid form telling AQMD what was done to reduce visible emissions from the vehicle.

CHAPTER EIGHT

Mitigation, Control and Management

8.1 INTRODUCTION

Air pollution control is fundamental to AQM. Pollution control is a hierarchy phenomenon, where pollution prevention is the primary control and abatement is the last result, should prevention not be possible. Prevention is always better that cure. The hierarchy for control is the corner-stone of European legislation and consists of:

- Prevention techniques;
- Technical control and operational practicalities;
- Abatement equipment and recycling within the process;
- Abatement equipment with release to atmosphere.

Prevention techniques are the use of alternative processes/techniques to minimise pollution control. Technical control and operational practices are the use of alternative control techniques or the use of different operational practices to minimise or reduce emissions. Abatement equipment and recycling is the process of using the emission from one process and recycling the emission or using it for another process. The final scenario is the minimising of emissions by abatement control for final release to atmosphere. This chapter provides a general introduction to air pollution control and management techniques.

8.2 WHAT IS MITIGATION AND CONTROL?

To mitigate is to 'make or become milder, less severe'. To control is to 'constrain or regulate'. Mitigation and control measures refer to the project activity design or operation features that can be used to minimise the magnitude of an air quality impact. This is achieved through a number of mechanisms. It can also be broadly implemented through a series of policies and strategies. These may include (UNECE, 1995):

- Regulatory provisions: legislative and regulatory frameworks; AQS; target loads or deposition standards; fuel quality standards; emission standards; licensing of potentially polluting activities; product-oriented requirements and labelling; etc;
- Economic instrument: emission charges and taxes; product charges, taxes, tax differentiation, including fuel taxes; user and administrative charges; emission trading; subsidies and other forms of financial assistance; etc;
- Measures related to emission control: technology requirements in legislation and regulations; control technology requirements for stationary sources; control technology requirements for mobile sources; the availability of

unleaded fuel; the role of technology-related research and development; monitoring and assessment of air pollution effects; etc.

The effectiveness of certain types of mitigation and control measures are determined by their level of application. Direct emission source control is often the most effective at the project level where site-specific issues can be mitigated against. Strategic level control will be most effective at mitigating cumulative and indirect impacts.

8.3 CONTROL OF FUGITIVE EMISSIONS

Many complaints about a process or activity relate to fugitive emissions. These are principally either odour or particulate (dust) related. Good housekeeping procedures by operators can reduce complaints and abatement costs considerably (Case Study 9, Section 10.10). The adoption of environment, and health and safety management systems help to ensure that management and employees take ownership and responsibility for reporting and controlling fugitive emissions.

Environmental Management Systems (EMS) have experienced an unprecedented surge of interest in recent years; whether in formal schemes, such as ISO 14001 or the EU's Eco-Management and Audit Scheme (EMAS), or informal self-certified schemes, the uptake of EMS has been almost exponential (ENDS, 1999). Despite this, the linkage with other environmental management tools, such as pollution control, has been slow to develop.

The use of EMS in business and industry process is growing in importance as organisations seek to manage proactively the environmental consequences of their activities and improve their environmental performance. The Business Charter for Sustainable Development (International Chamber of Commerce, 1991) includes as one of its 16 principles of environmental management one calling for an environmental assessment to be performed 'before starting a new activity... and before decommissioning a facility or leaving a site'. Such a proactive stance will help to identify potential concerns relating to AQM. Many companies seeking to demonstrate actively improvements in environmental performance have, or are in the process of developing and implementing, an EMS to the requirements of the international standard ISO 14001 (International Organisation for Standardisation, 1996). One of the key tasks in the early stages of developing an EMS is the identification of environmental impacts and an evaluation of their significance.

8.3.1 Fugitive dust control

Suitable measures for fugitive dust control tend to be dependent upon good housekeeping issues. It is therefore essential that appropriate EMS are applied to assure dust control measures are carried out and are effective. Previous studies have applied the following preventative measures to control emissions:

- Work being carried out in such a manner that avoidable dust is not generated;
- Use of screens or other methods to prevent generation of dust;

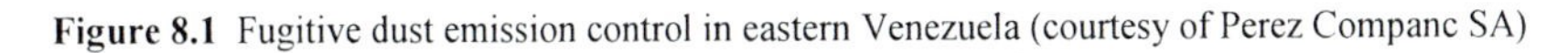

Figure 8.1 Fugitive dust emission control in eastern Venezuela (courtesy of Perez Companc SA)

- The use of dust sheets and/or tarpaulins for off-site material handling activities;
- Spraying or covering of materials, including soil/overburden material, from which dust may be generated when being transported to or from the site;
- Watering or wetting by agent sprays material which has the potential to create dust prior to being loaded into or unloaded from vehicles (Fig. 8.1);
- Fitting of side and tailboards to any vehicle with an open load carrying area used for moving materials. Materials having the potential to create dust should not be loaded to a level higher than the side and tail boards;
- Regular removal of dust on hard surfaced routes and road edges within the site. Access roads to the site should be kept entirely free of dust, mud or other wastes and in good repair to avoid the occurrence of dust clouds (Fig. 8.2);
- Water spraying of stored materials;
- Design and control of plant equipment and site vehicles to minimise dust release;
- Effective tyre and wheel cleaning equipment to be installed and adequately maintained at all times;
- On-site speed restrictions for site traffic.

Operational practices that are likely to cause the generation of dust should be continuously monitored and all findings, including prevailing meteorological conditions, recorded. In addition there may be a need to deploy effective methods for suppressing undue amounts of dust, including an adequate supply of water for

Figure 8.2 Fugitive dust emissions from an unswept road in northern Peru (courtesy of Cordah Limited)

spraying equipment. Case Study 9 (Chapter 10, Section 10.10) details the dust control measures used to abate emissions from an open cast coal site (OCCS).

In addition to the above mitigation measures for quarrying or OCCS all long-term storage stockpiles closest to the sensitive receptors should be seeded to grass and adequately maintained. The provision and maintenance of a windsock will also benefit in the management of the site to help assessing operational activities in conjunction with wind speed and direction and ASR.

Dust control and mitigation measures should be rigidly applied to minimise dust impacts during all excavation, and restoration activities. A complaints response system should also be established with a contact name and telephone number. This would enable the public to call if a dust incident occurs and for appropriate action to be taken on-site to reduce dust emissions. In the event of a particular activity on the site seeming likely to create a specific dust nuisance in adverse weather conditions, the site management could stop operation and only allow it to be resumed when satisfactory control has been achieved.

8.3.2 Fugitive gaseous control

The principal issue of fugitive gaseous emission generally relates to odour control. Section 8.4.4 details odour abatement technology for controlled emission sources. With regard to fugitive emissions principal sources are:

- Handling and transfer of process materials;
- Handling and transfer of waste materials;
- Produced materials;
- The process itself.

Fugitive emission control can be catered for primarily by good housekeeping practices. These include the covering and careful storage of waste and odour materials associated with the process; slight negative pressure in the process building to cause the movement of air into the building; improved integrity of the building/process area to prevent the egress of odorous emissions; sealing of process equipment to prevent fugitive emissions. Tables 8.1 and 8.2 detail an approach to receiving and investigating an odour complaint.

Table 8.1 Checklist for recording odour complaints

Complainants details (e.g. name and address):	______________
Was the complainant accompanied?	Y or N
Was the odour witnessed?	Y or N
Date of complaint:	______________
Time of complaint:	______________
Description of the odour:	______________
Location of odour:	______________
What was the duration of the odour?	______________
Has regulatory authority been informed of the complaint?	Y or N
Has the Local Authority Environmental Health Department been informed of the complaint?	Y or N

8.3.3 Visual inspections

Continued vigilance of a process including its operation, management and pollution control and abatement systems will help to minimise complaints and exceedence of discharge consent (permit) conditions. All processes should be visually inspected on a routine basis (e.g. daily) to ensure satisfactory operation and compliance. Observations should be recorded and reported to the appropriate personnel. Simple visual inspections should note and include the following:

- Odours and their source;
- Dark smoke;
- Malfunctioning and abnormal process operation;
- Pollution control and abatement equipment performance;
- Operator concerns;
- Discoloration of surfaces and vegetation in vicinity of process;
- Meteorological conditions (e.g. wind speed, direction, etc).

Table 8.2 Actions to be undertaken by the operator following receipt of an odour complaint

- Report receipt of odour complaint to:
  - Site Manager (Name);
  - Operating company (Name);
  - Regulatory authority.
- Visit the location of the odour immediately providing the complainant has stated the odour is presently occurring:
  - Identify the nature of the odour;
  - Identify the spatial extent of the odour;
  - Identify the prevailing wind direction;
  - Identify the duration of the odour;
  - Identify other activities in the locality that might generate odours (e.g. agricultural practices);
- Walk the site boundary of the process to identify whether other odours are discernible;
- On-site following receipt of the complaint the following actions should be undertaken:
  - Note wind direction and speed at the process location if possible;
  - Note process operations on site and their location;
  - Identify nature of odours on-site and compare them to off-site observations;
  - Ascertain whether operators on-site have detected a discernible odour.
- If an odour is detected and the process is responsible, undertake the following actions:
  - Liaise with the Site Manager on the appropriate actions to be taken to rectify the problem (these actions may include):
  - Liaise with regulatory authority on the appropriate action taken
- If an odour is detected, and the process is not responsible, report the findings of the investigations to the following:
  - Site Manager (Name);
  - Operating company (Name);
  - Regulatory authority.
- Record investigation findings on the complainant record sheet and advise the complainant of the situation.

The method for visual assessment of smoke emissions by comparison of the darkness of the smoke with the standard shades of grey on a chart placed in a suitable position is called the Ringelmann Chart. Ringelmann Charts are published by the BS Ringelmann Chart (British Standards Institute 2742) and the US Bureau of Mines. The chart consists of a cardboard sheet on which are printed five squares, four of which are cross hatched by 20 horizontal and 20 vertical lines so that in use the cross hatched black lines merge into the white background and produce for each shade, apparently a uniform grey. The number of shades, Ringelmann numbers, range from 0 to 4, each shade increasing by comparison with the previous number by 20% obscuration so that:

- Ringelman 1 = 20% obscuration

- Ringelman 2 = 40% obscuration
- Ringelman 3 = 60% obscuration
- Ringelman 4 = 80% obscuration

Dark smoke is smoke that is as dark or darker than shade 2 on the chart. Black smoke is smoke that is as dark or darker than shade 4 on the chart. Another assessment of the optical density of process emissions within the flue is by reference to the Bacharach Scale (a scale of ten shades, from white to black).

8.4 TECHNIQUES FOR THE CONTROL OF GASEOUS EMISSIONS

8.4.1 General abatement techniques

There are many different types of pollution control equipment, each have their own merits and applications when used for the removal of differing effluent emissions. The following section details the principal techniques used for the control of gaseous and odorous emissions from point and area sources.

8.4.1.1 Source separation

An effective method for the control of some acid gases (SO_2, HCl, etc) is source separation. Separation of source components such as plastics will reduce emission levels of HCl due to the reduction in chlorine. The use of low base sulphur fuels will also reduce SO_2 emissions from combustion processes.

8.4.1.2 Absorption

Gas absorbers (or scrubbers as they are frequently referred to) are a widely used method for pollution control. Scrubbers can be designed to operate so that they will remove both particulate and gaseous emissions. Absorption refers to the selective transfer of material from a gas to a contacting liquid (Cooper and Alley, 1986). The basic principle involved in the process is the preferential solubility of a gaseous component in the liquid. Generally the scrubbing medium is water. Vapours and gases are absorbed from a gas stream into chemical solutions. The liquid phase is then generally recirculated with a small volume of liquid being added and bled off. Examples of the application of scrubbing include (Cooper and Alley, 1986):

- Removal and recovery of NH_3 in fertiliser manufacture;
- Removal of HF in glass manufacture;
- Control of SO_2 from combustion flow gases;
- Recovery of water soluble solvents (e.g. acetone, methyl alcohol);
- The control of odours from animal rendering plants.

A principal factor controlling the absorption process relates to the solubility (or chemical reaction) of the gas to be removed in the liquid used for scrubbing. The method of gaining intimate contact between the gas and liquid streams heightens the absorption process. To ensure sufficient retention time to facilitate effective scrubbing packing is placed in the scrubbing tower (Fig. 8.3). Water can be used to remove gases of high solubility. However, in some situations caustic soda or salt solutions may be used because they react chemically with less soluble gaseous pollutants (Palmer, 1974). Essentially acid or alkaline gases are absorbed into pH control solutions, whilst other odours are absorbed into appropriate chemical solutions according to their composition. Oxidants are frequently used for aldehyde and ketone control for example (Wright and Woodfield, 1994). The requirements of satisfactory packing material of a scrubbing unit are a high wetted area per unit volume, minimal weight, sufficient chemical resistance; low liquid hold up, low pressure drop and cost effectiveness. Typical packing material includes Berl saddle, Intalox saddle, Raschig ring, Lessing ring, Pall ring and Tellerette (Copper and Alley, 1986). Table 8.3 details the characteristics of common absorption units (Parker, 1978).

Table 8.3 Characteristics of absorption units (Parker, 1978)

Type	*Typical flow arrangement*	*Superficial gas velocity(m/sec)*
Packed towers Berl saddles, Raschig rings, wood grid, etc.	Countercurrent	0.5-1.5
Plate towers Bubble caps, perforated plates, impingement plates	Overall countercurrent Plate crosscurrent	0.3-1.5
Hydraulic dispersion Spray, cyclone, venturi washer	Cross or cocurrent	1.5-12
Mechanical dispersion Agitated tanks, Felds washer	Crosscurrent	0.3-1
Fluidised bed principle Turbulent bed of hollow sphere, glass marbles	Cross and countercurrent	2.5-7

The common basic types of absorption system include (Wright and Woodfield, 1994; Leeds University, 2000):

- Moving bed absorbers consist of mobile packing (i.e. plastic or glass spheres) where gas and scrubbing liquid can be mixed. The system consists of support grids on which the packing material is placed, inlets and outlets for gas and scrubbing liquid and a mist eliminator. The system provides a large surface area and produces intimate mixing of the gas and liquid for good mass transfer, due to agitation and turbulence in the moving bed.
- Packed bed absorbers consist of an outer shell containing a bed of packing material on support grids, liquid distributors, gas and liquid inlets and outlets and a mist eliminator. Liquid is distributed continuously over the packing material forming a film, which provides a large surface area for gas/liquid contact. The gas stream flows through the packed bed and is subsequently

Figure 8.3 Simple gas absorption unit (scrubber)

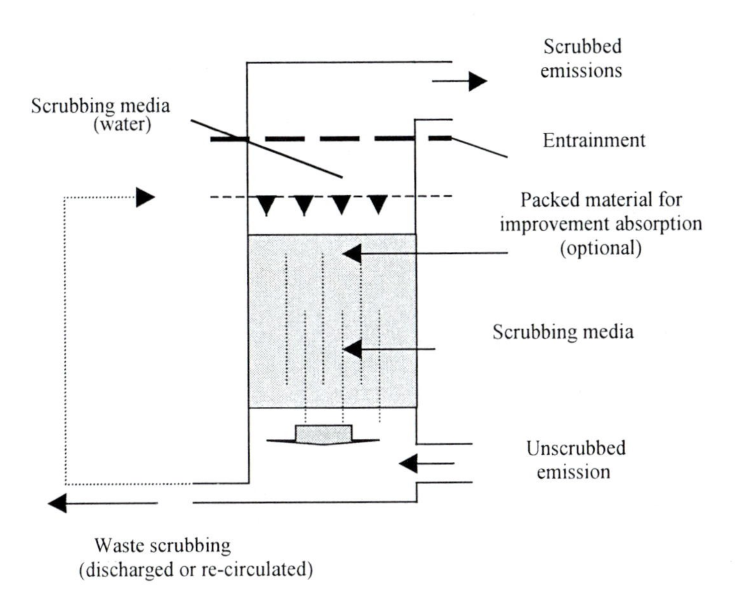

scrubbed. The liquid and gas flows may be countercurrent, cocurrent or cross current.

- Fibrous packing absorbers consist of a chamber with gas inlet and outlet, and containing mats of fibrous packing material (e.g. glass, plastic and steel) which are sprayed (continuously or intermittently) with liquid. The system may be designed for horizontal or vertical gas flow.
- Plate absorbers consist of a vertical tower with several horizontal perforated trays stacked in it. Baffles are situated a short distance above the apertures in the plates. The scrubbing liquid enters the top of the chamber and flows along each of the trays successively. The gas stream enters the bottom of the chamber and flows upwards passing through the perforated plates. The velocity of the gas is sufficient to prevent liquid seeping through the perforations.
- Spray absorbers are when liquid enters the top of the chamber via spray nozzles. The gas stream passes up the chamber. Preformed spray scrubbers can be designed for vertical or horizontal gas flow, as well as countercurrent, cocurrent and cross current flow. Liquid droplets produced by the spray nozzles falls down the chamber through the rising gas stream. The liquid/gas contact at the droplet surface enables mass transfer to occur (Fig. 8.3).

Table 8.4 details the comparative advantages of the commonly used different types of absorption systems.

Table 8.4 Comparison of the advantages and disadvantages of common absorption equipment (Wright and Woodfield, 1994)

Technique	*Advantage*	*Disadvantage*
Packed bed	Highly efficient; Low maintenance; Relatively low pressure drop.	Relatively large size; Likely to be expensive compared to spray scrubber.
Fibrous packing	Capable of treating greater volumes of gas than conventional packed tower.	Prone to blockage by particulates or growth of biomass within fibrous matting.
Moving bed	Larger surface area of absorption; Suitable for particulate and trace gas removal.	Greater maintenance requirement.
Plate	Absorption efficiency increase by increasing number of plates; Plate absorbers can operate at high liquid flows compared to packed towers; Tolerate fluctuations in gas flow and temperature; Absorption efficiency increases as pressure drop across a plate increases.	Scaling on plate absorbers; Not suitable for use with slurries.
Spray tower	Inexpensive; Simple to operate and maintain.	Less effective than packed bed absorbers; Improved efficiency by adding consecutive stages to system; Droplet diameter and liquid distribution critical to system efficiency.

8.4.1.3 Adsorption

Adsorption is a heterogeneous process in which gas molecules are retained on a solid surface. Some adsorbents can preferentially, and reversibly, adsorb specific chemical species thus removing them from flue gas streams (Wright and Woodfield, 1994). Adsorption may be either physical or chemical. In the case of the former, gaseous molecules are held into the adsorbent surface by van der Waals' forces. In chemical adsorption chemical bonds are formed between the gas molecule and the surface. In order for adsorption to occur there must be a mass transfer of molecules from the bulk of the gas to the gas-solid interface. There must also be diffusion through the media's pores until the molecule is finally adsorbed on to the internal surfaces. The principal types of adsorber system include (Wright and Woodfield, 1994):

- Disposable cartridge systems (fixed bed) for simple processes with intermittent flows and low concentration of contaminant (e.g. VOCs);
- Regenerating fixed beds where two units are required for continuous use, one in use, one being regenerated;
- Fluidised bed adsorbers that are suitable for large quantities of gas and can tolerate higher flue gas temperatures because cooling tubes can be incorporated. They can also tolerate some particulate material. The gas passes through a suspension of absorbent, or in a continuous moving bed adsorber where the adsorbent falls by gravity through the rising stream of gas;
- Continuous moving bed.

An important consideration in the abatement process is the choice of the adsorbent, which is dependent upon the nature of the flue gas composition. For the removal of organic vapours, activated carbon is frequently used. Table 8.5 shows the relative adsorptivity of certain molecules on activated carbon, however, the capacity of the absorbent is ultimately dependent upon a number of factors. These include (Wright and Woodfield, 1994):

Table 8.5 Relative adsorptivity of molecules on activated carbon (Wright and Woodfield, 1994)

High capacity adsorbing 20-50% of own weight		*Satisfactory capacity adsorbs 10–25% of own weight*	
Acetic acid	Hospital odours	Acetone	Chlorine
Alcohol	Household smells	Acrolein	Diesel fumes
Benzene	Isopropyl alcohol	Anaesthetics	H_2S
Body odours	Masking agents	Animal odours	Solvents
Butyl alcohol	Mercaptans	*Not highly adsorbent, may be sufficient under certain conditions*	
Butyric acid	Ozone		
Cancer odour	Perfumes, cosmetics		
Caprylic acid	Perspiration	Acetaldehyde	Butane
Carbon tetrachloride	Phenol	Amines	Formaldehyde
Chloropicnin	Pyridine	Ammonia	Propane
Cigarette smoke	Ripening fruits	*Adsorption not satisfactory*	
Cresol	Smog	CO_2	
Disinfectants	Stuffiness	CO	
Ethyl acetate	Toluene	Ethylene	
Essential oils	Turpentine		
Gasoline			

- Gas molecules reaching the internal pores of the adsorbent;
- Concentrations of the absorbate around the adsorbent;
- Surface area of the material;
- Volume of pores;

- Characteristics of the absorbate molecules, their weight, electrical polarity and shape;
- Characteristics of the adsorbent surface.

8.4.1.4 Biological treatment

The principle of the technique is that certain compounds can be metabolised by naturally occurring micro-organisms and fungi (Table 8.6 details the types of bio-filters). The technique is commonly used for the abatement of odours and when effectively applied has an efficiency of >99%. Basically, the process consists of a moist layer of material (e.g. peat, heather, compost, wood bark, soil, etc.) supporting a diverse microbial population together with a system for supplying gaseous emission uniformly to the filter. Wright and Woodfield (1994) state that the residence time of the gas stream in the bio-filter should be a minimum of 30 seconds, however, for strong malodorous compounds the time should be a minimum of 50 seconds. If the emission is contaminated with particulates or fatty acids a pre-filtering system (e.g. scrubber) may be required to prevent deterioration of the bio-filter.

Soil bio-filters have a more varied biological population in a less controlled environment than for bio filters containing peat. They therefore require a longer residence time for the emission. In a soil bio filter the residence time is typically in the range 3–5 minutes (Wright and Woodfield, 1994). Another form of control using biotechnology is the use of bio-scrubbers. An emission is passed through a flow of water containing a population of microbes suitable for oxidising the gas emission. The bio-scrubber contains a packing material on which the microbes adhere to form a biological bed. Trickle bed filters are a hybrid between a conventional bio-filter and a bio-scrubber. The gas stream is passed through a packed bed (e.g. heather, peat or an inert substance) which supports a microbial film. Water is sprayed on to the top of the bed and is continuously recirculated. Table 8.6 details the advantages and disadvantages of their use. All filters require a sizeable land take.

8.4.1.5 Thermal processing

Thermal oxidation and catalytic incineration abatement techniques are highly effective. However, both techniques are constrained by capital cost, operational costs, the nature of emissions, availability of fuel and land take. In thermal oxidation emitted gases are oxidised by heating them in the presence of oxygen or air in a furnace. Complete combustion will essentially result in the emission of CO_2, H_2O and NO_x. The efficiency, however, of the process is dependent principally on temperature, turbulence and time (the 3t). It is therefore necessary for an incinerator to contain three stages: the burner, mixer and afterburner to ensure that the 3t requirements are met:

- The burner is where fuel is burnt with combustion air (clean or part contaminated) to produce a flame at a high temperature;
- The mixer is where the contaminated air residue is mixed with hot gas from the flame to bring it to uniform temperature. Mixing of the gas can be achieved by

either natural diffusion between turbulent streams, impingement of gas streams at an angle or changes in direction of flow, round corners or past baffles;
- The afterburner is where gases are held until oxidation has been completed. The afterburner should ensure that there is a high and even temperature across the chamber, a long residence time, a sufficiently high oxygen concentration for complete oxidation to occur, plug flow with little back mixing in the axial direction and maximum turbulence in the mixing stage.

Table 8.6 Merits and disadvantages of gas abatement technologies (Wright, 1994)

Method	*Advantage*	*Disadvantage*
Absorption	Useful for removal of odours of known composition >90% efficient with multi-stage scrubbing system	Needs multi-stage scrubber if acidic and basic compounds are to be removed Plate scrubbers can have scaling problems
Adsorption	>99% efficiency for certain chemical species Partially useful for low concentration odours	Odour break through can occur in high concentration systems, temperature and humidity applications Efficiency deteriorates through time due to solvent saturation. Regeneration to sufficient quality is often not cost effective Fatty droplets or abrasive particulate streams can degrade adsorber performance Some pre-treatment may be required (e.g. dust filters)
Incineration	Very efficient, often >99%; Almost universally applicable as most odours can be oxidised to non-odorous products at high temperatures Can be a cost effective option for small scale odour abatement	Usually cost is a limiting factor especially if large amounts of air are involved or if very high temperatures are required Compounds containing sulphur and C_l atoms give SO_2 and HCl/Cl_2 which then require further treatment Dust and soot cause abrasion of machine parts, and tend to accumulate, so attrition can be high
Biological	Soil filters can be >99% efficient Peat heather bio-filters can be 95% efficient and cheaper than soil filters	Soil bio-filters require large areas of land and have higher running costs than peat/heather bio-filters Need regular irrigation (especially in dry weather) Microbiological populations must be kept viable so water content, pH, nutrient level and temperature must be controlled Adaptation period of micro-organisms is slower to fluctuations in many industrial processes making bio-techniques unsuitable for some processes Monitoring can be difficult

Other issues that may require consideration in the use of incineration is the pre-treatment of the waste stream. Such pre-treatments may include reduction in moisture content, removal of solid or liquid contaminants, preheating to reduce fuel requirements. Another consideration may be the need for heat recovery to reduce running costs. Similar equipment used for the incineration of gaseous waste streams include kilns, boilers and silica bed incinerators. Catalytic incineration is a technique similar to that of incineration except the oxidation reaction takes place on a catalyst surface (e.g. platinum, palladium, rhodium, copper chromite, and oxides of copper, chromium, manganese, nickel and cobalt supported on a base metal) instead of in free air. The equipment consists of a preheat section, a catalyst bed, one or more burners, a control panel, safety equipment and a heat recovery system. Table 8.6 details the merits of abatement processes.

8.4.2 Control of oxides of nitrogen

Fuel NO_x, formed by the reaction between oxygen and organic nitrogen in the fuel, and thermal NO_x, formed by the reaction between nitrogen and oxygen in the air used for combustion are the two primary sources of NO_x requiring control. Source separation of the material entering the combustion process may be a method for the partial control or removal of organic nitrogen. Whereas thermal NO_x control can be accomplished through either combustion control or by flue gas treatment, combustion control methods include either flue gas recirculation, where gases are recycled back into the combustion process, or low excess air operation and staging of combustion, where the input air is controlled to the combustion process by dividing it into primary and secondary flows. Therefore, part of the combustion process operates on partially starved air and the remainder of the combustion process operates on excess air reducing the amount of thermal NO_x formed in the combustion process.

The two principal flue gas treatment techniques include selective catalytic reduction (SCR) and selective non-catalytic reduction (SNCR). In SCR ammonia is injected into the flue gas, followed by a passage of air over a catalyst bed (i.e. base metals such as copper, iron, chromium, nickel, molybdenum, cobalt, etc.) where the following reaction occurs in the temperature range 530°F to 800°F (Tchobanoglous *et al.*, 1993).

$$NO + NH_3 + 1/4O_2 \rightarrow N_2 + 3/2\ H_2O$$

The SNCR treatment employs ammonia injection, but no catalyst is involved. Gaseous ammonia is directly injected into the combustion process, in a temperature range 1300 to 2200°F the subsequent reaction occurs (Tchobanoglous *et al.*, 1993):

$$NO + NH_3 + O_2 + H_2O + H_2 \rightarrow N_2 + H_2O$$

8.4.3 Control of sulphur dioxide

Two basic approaches are available to control SO_2 emissions and include:

- Removal of sulphur from fuel before it is burnt;
- Remove of SO_2 from exhaust gas emissions.

Table 8.7 summarises the principal options for controlling sulphur emissions. The two basic methods of classifying flue gas desulphurisation systems are throwaway or regenerative and wet or dry. A process is a throwaway type if the sulphur removed from the flue gas is discarded. A process is regenerative if the sulphur is recoverable. Wet or dry refers to the phase in which the main reactions occur.

Table 8.7 Options for the control of SO_2 emissions (Cooper and Alley, 1986)

Option	*Suboption*	*Examples of processes*
Do not create SO_2	Switch to a lower sulphur fuel	
	Desulphurise fuel	Oil desulphurising Coal cleaning
SO_2 scrubbing		
Throwaway	Wet scrubbing	e.g. lime, limestone, dual alkali, forced oxidation (with gypsum disposal)
	Dry scrubbing	e.g. lime spray drying, lime injection
Regenerative	Wet processes	e.g. absorption with water, aqueous carbonate
	Dry process	e.g. activated carbon adsorption, copper oxide adsorption

Additional information is available from a wide range of sources on air pollution abatement techniques. General texts outlining the principals of air pollution control include Parker (1978), Cooper and Alley (1986), Stern (1976), Young and Cross (1980), Noll and Duncan (1973), Hesketh (1974) and Bretschneider and Kurfurst (1987).

8.4.4 Odour control

Odours are generally not localised and are a function of the whole process and its ancillaries (Valentin, 1980). Therefore all stages of the process should be considered, as important odour sources may be omitted from consideration. Odours may be emitted from various sources associated with a process and include (Fig. 8.4) (Valentin, 1980):

- The quantity and age of the raw material used in a process may be the cause of an odour;

Figure 8.4 Sources of odour generation requiring control in a process

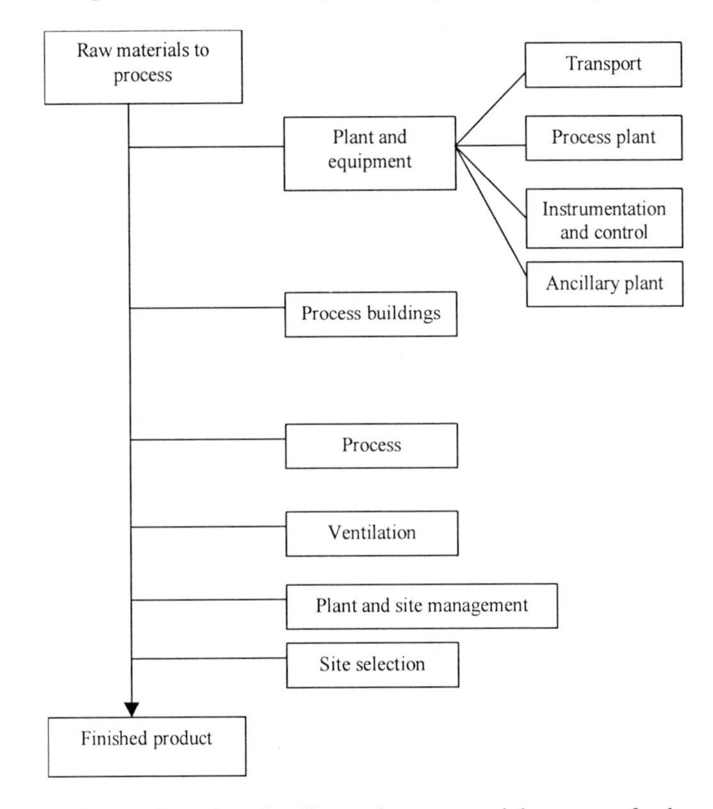

- The manufactured product itself may be a potential source of odour or by-products from the finished product;
- Plant and equipment:
    - Vehicular transport and/or material handling may be a source of odour emission. Open carrying systems have the greatest potential to cause odours;
    - Process plant is an important source of odour generation and leakage;
    - Odour generation is often provoked if temperatures, flows, component quantities, pressures, reaction times, etc. deviate from their desired or optimum values;
    - If ancillary plant is not adequately sized then odours may be emitted;
- The planning of buildings, particularly those dealing with biological materials, may require special care and expertise to prevent odour emissions. Areas of concern include storage and handling of materials, which can deteriorate rapidly on storage to become highly odorous. Effective building construction can prevent the egress of odours. Simple measures such as the use of separate areas, minimum entry points, negative pressure, etc. will also help minimise odours. An appropriate drainage and effluent system will similarly reduce odours;
- Careful design of the process may help to minimise odours;

- The principal aim of a ventilation system is to control odour egress such that it can be dealt with by dispersion or via a stack and to treat only the odorous air from specific collection points;
- Adherence to effect operating procedures will minimise odour emissions and generation. Odour generation is often due to poor housekeeping problems;
- The siting of particularly odorous processes away from ASR will minimise the likelihood of odour complaint.

Wright (1994) has identified four stages in defining odour control requirements:

- Process elements contributing to the malodorous problem should be identified such as raw materials, storage areas, process units and associated vents, waste material handling and treatment, leaks and fugitive emissions and controlled discharge points.
- The size and nature of the impact should be broadly estimated from an examination of the complaints received and their nature. In addition consideration should be given to the effectiveness and appropriateness of any existing *in situ* abatement equipment; the containment of emissions for abatement and the nature of any fugitive emission(s).
- Outline the scope and type of abatement approach to be adopted. Measurement work may be required to quantify the emissions and indicate appropriateness of abatement technique. Information may be required on odour strengths for each source, volume flow rate for each source, chemical analysis and pH, humidity and temperature of each source.
- Install and commission abatement equipment. An assessment of the efficiency of the system may be needed to ensure compliance of emissions standards.

The abatement of odorous emissions requires the removal of odorous compounds from a gas stream or their conversion into low or non-odorous compounds (Valentin, 1980). Table 8.8 details the criteria that may affect the choice of odorous emission abatement.

8.4.5 Control of dioxin emissions

Studies into the combustion processes that form PCDD/Fs have suggested a framework for their control (Hagenmaier *et al.*, 1987; Fangmark *et al.*, 1995; Addink *et al.*, 1998). Issues relating to the formation of dioxin emissions have recently been reviewed from the research literature (Environmental Resource Management Ltd, 2000a) and include:

- Incomplete combustion of organic wastes leads to the formation of organic fragments which serve as organic precursors to the dioxin/dibenzofuran molecule;
- Waste provides a source of chlorine and of metals. The latter are incorporated into fly ash, which carries over to the cooler (250 to 400°C) post combustion zone of the incineration system;

- Organic precursors adsorb onto the surface of the fly ash in the post-combustion zone and, following a complex sequence of reactions which are catalysed by metals (primarily copper) in the fly ash, lead to the formation of PCDD/Fs along with other chlorinated trace organics.

Table 8.8 Odour abatement criteria (after Wright, 1994)

Criteria	*Issues*
Containment	Ensure containment of odorous gas where possible (e.g. for the process, storage and transfer of materials, etc.). Limit fugitive emissions by ensuring integrity of building and negative pressure to prevent egress of odours
Particulates and aerosols	The odour concern may be principally or wholly particulate associated. Therefore removal of particles may improve the efficiency of the abatement equipment
Concentration and pollutant gas flow rate	Odorous compounds may be organic or inorganic and high odour strength is not necessarily associated with high chemical concentrations. Generally high odour concentration is associated with small air volumes and medium or low concentrations are liable to be associated with larger airflows. For high concentrations with low airflows incineration or adsorption may be an effective abatement measure. Multi-stage scrubbing is often more cost effective for higher volumes of pollutant air. Low odour concentrations are likely to require low costs for abatement such as bio-filtration or scrubbing
Temperature	The temperature of the emitted gas will effect the choice of abatement option. For example bio-filters and carbon absorbers are unlikely to be effective if the gas temperature is higher than 40°C. Greater exhaust gas temperature emissions may reduce the operations cost for an incineration option
Moisture content	The moisture content of an emitted gas will affect the suitability of the chosen abatement technique. Moisture content is likely to reduce the cost effectiveness of adsorbers owing to the preferential adsorption of water vapour. A higher moisture content in the effluent gas stream may be beneficial for bio-filtration as it reduces the amount of irrigation required to keep the bed sufficiently moist
Chemical composition	The chemical composition of the pollutants can affect the suitability of abatement options. Organic compounds may not be suitable for abatement by a water based scrubber but may supply a significant fuel fraction for an incinerator

The USEPA (1987b) introduced the concept of Good Combustion Practice (GCP), the term being defined as 'those combustion conditions which lead to low emissions of trace organic pollutants'. Following comprehensive studies of incinerators, the USEPA concluded that low organic emissions could be achieved by a combination of combustion control techniques and appropriate gas cleaning technology. The rationale for the application of GCP was that PCDD/Fs are the product of incomplete combustion. Hence, optimisation of combustion conditions to as close as possible to the theoretical ideal of complete combustion, coupled with appropriate end of pipe control strategies, should lead to reductions in emissions. The USEPA recommendations for GCP can be categorised:

- Minimisation of organic emissions to atmosphere through optimum design of the combustion process;
- Operation of the process within its design specifications, with control systems to prevent exceedences of the design criteria;
- Monitoring and verification of combustion performance, with continuous surveillance of key design and operating parameters.

From a regulatory perspective, the most appropriate means of implementing GCP is to outline a set of general rules that can be applied to all types of combustion process. Five classes of criteria can be identified:

- Design criteria such as a minimum gas phase residence time of 2 seconds in the combustion zone;
- Operational criteria such as a minimum furnace temperature of 850°C, or maintenance of a minimum excess oxygen level in the combustion gas;
- Measurement and control of surrogates such as CO and particulate emissions such as maintaining CO emission below 50 mg/m^3;
- Control regimes (restrictions of waste; failsafe, interactive control systems; automatic shutdown procedures, etc);
- Monitoring regimes (measurement and recording of combustion temperature; continuous monitoring of excess O_2, CO emissions and combustion efficiency, etc).

8.5 TECHNIQUES FOR THE CONTROL OF PARTICULATE EMISSIONS

The principal techniques available for the collection of particles include gravity settler, bag filters, cyclones, wet scrubber and electrostatic precipitators (Copper and Alley, 1986). Control devices are selected to achieve the required removal efficiency which is defined as (Tchobanoglous *et al.*, 1993):

$$E_e = ((W_{inlet} - W_{outlet}) / W_{inlet}) \times 100\% \quad (8.1)$$

8.5.1 Gravity settler

A gravity settler is a large vessel in which the effluent gas velocity is slowed, thus allowing particles to settle out by gravity.

8.5.2 Cyclone

A cyclone removes particles by causing the emission from the process to flow in a spiral pattern inside a chamber. Owing to centrifugal force, the large particles tend to move outwards and collide with the wall of the vessel. The particles slide down the sides of the cyclone into a collection vessel and the abated gas then flows out the top of the chamber. The principal reason why cyclones are widely used is because they are inexpensive, have no moving parts and can be constructed to withstand harsh operating conditions (Cooper and Alley, 1986).

Table 8.9 Approximate collection efficiencies (%) of particle abatement equipment for varying particle size (Muir, 1992)

Type of equipment	*Efficiency*			
	10 μm	*5 μm*	*2 μm*	*1 μm*
Inertial collector	30.0	16.0	7.0	3.0
Medium–efficiency cyclone	45.0	27.0	14.0	8.0
High–efficiency cyclone	87.0	73.0	46.0	27.0
Low resistance cellular cyclone	62.0	42.0	21.0	13.0
Tubular cyclone	98.0	89.0	77.0	40.0
Irrigated cyclone	97.0	87.0	60.0	42.0
Self-induced spray deduster	98.0	93.0	75.0	40.0
Spray tower	97.0	94.0	87.0	55.0
Wet impingement scrubber	>99.0	97.0	92.0	80.0
Disintegrator	99.0	98.0	95.0	91.0
Venturi scrubber–medium energy	>99.9	99.6	99.0	97.0
Venturi scrubber–high energy	>99.9	99.9	99.5	98.5
Electrostatic precipitator	>99.5	>99.5	>99.5	>99.5
Irrigated electrostatic precipitator	>99.5	>99.5	>99.5	>99.5
Shaker-type fabric filter	>99.9	>99.8	99.6	99.0
Pulse-jet fabric filter	>99.9	99.9	99.6	99.6

In a typical cyclone (Fig. 8.5) the particulate-ladden emission enters the top of the chamber and is forced downwards spirally due to the shape of the cyclone and the tangential entry of the emission. Centrifugal force and inertia cause the particles to move outward and collide with the wall of the chamber, and then drop to a collection point. Near the bottom of the cyclone chamber, the gas stream reverses its downward spiral and moves upward in a smaller inner spiral and exits the top of the chamber. Particle removal efficiency varies considerably with particle size and with cyclone design.

Table 8.9 shows that particle removal efficiency can be as much as 98% for large particles although the removal efficiency diminishes in the smaller micron range (<10 μm).

Figure 8.5 Cyclone units (courtesy of Cordah Limited)

8.5.3 Fabric or bag filters

The basic purpose of a fabric or bag filter (Fig. 8.6) is to remove solids from aerosols (Palmer, 1974). The filter operates on the same principle as a vacuum cleaner. Air carrying dust particles is forced through a fabric. As air passes through the material, the particles accumulate on the cloth resulting in a cleaner air stream. The particulates are periodically removed from the fabric by shaking or by reversing the airflow. The major design parameters for a fabric filter are the filter area, material and the method of cleaning.

8.5.4 Electrostatic precipitators

An electrostatic precipitator (Fig. 8.7) applies electrical force to separate particles from the emission. A high voltage drop is established between electrodes, and particles passing through the resulting electrical field acquire a charge. The charged particles are attracted to and collected on an oppositely charged plate, and the cleaned gas flows through the device. Periodical rapping to shake off the layer of dust that has accumulated cleans the plates. The efficiency of an electrostatic precipitator is a function of the flue gas characteristics, principally moisture and temperature, and the electrical resistivity of the particles. Typically the efficiency of electrostatic precipitators is greater than 99.9% for fine particles (<10 μm). Case

Figure 8.6 Fabric of bag filter system

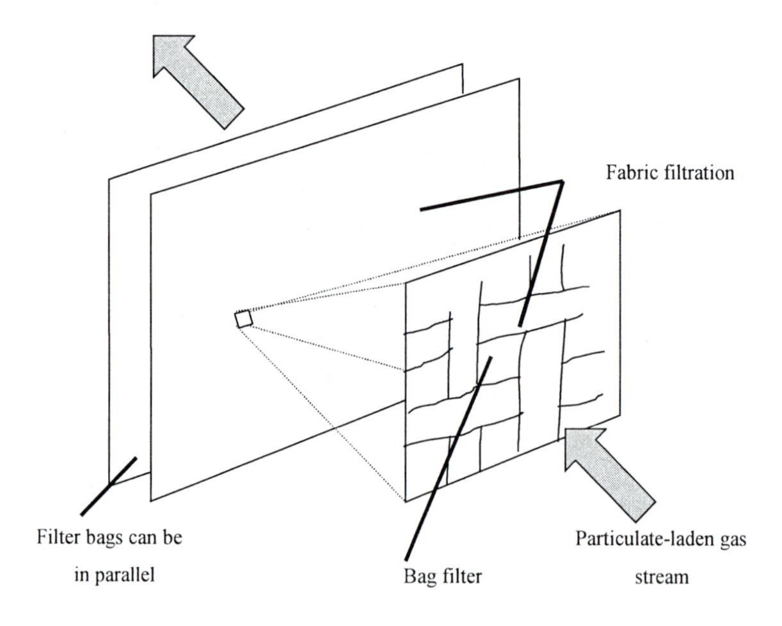

Figure 8.7 Electrostatic filter system

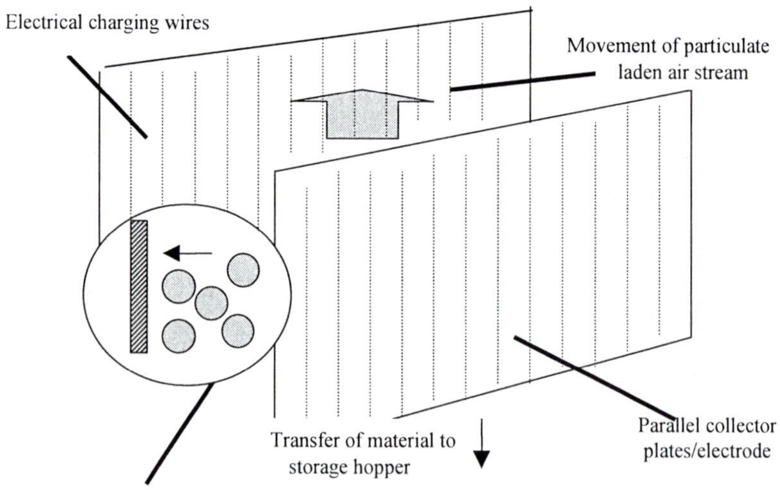

Electrically charged particles subjected to an electrical field are attracted towards the electrodes creating the field

Study 8 (Chapter 10, Section 10.9) provides an example of emission control for a woodchip manufacturing plant using an electrostatic precipitator.

8.5.5 Wet scrubber

A wet scrubber employs the principles of impaction and interception of particles by droplets of water (Section 8.4.1.2). Larger particles are easily removed from the gas stream by gravity. The solid particles can then be independently separated from the water, or the water can be otherwise treated before re-use or discharge.

The overall efficiency of collection of a system composed of two or more devices in series is neither simply the sum nor the product of the efficiencies of each device. Each device efficiency is based on the mass loading of particles entering that device, but the overall system efficiency is based on the total mass collection as a fraction of the total mass entering the first abatement device (Cooper and Alley, 1986). Table 8.10 details the advantages and disadvantages of particulate abatement equipment.

Table 8.10 General comments on the advantages and disadvantages of particle abatement techniques (Cooper and Alley, 1986)

Techniques	*Advantages and disadvantages*
Mechanical collectors (gravity settler and cyclone)	• Less expensive than other methods • Often only moderately efficient • Better for large particle collection • Often pre-cleaners for more efficient final control device
Fabric filters	• High efficiency • More costly than mechanical methods • Usually limited to dry, low temperature conditions • Handles many different types of dust
Electrostatic precipitators	• Handle large volume flow rates at low pressure drops • Very high efficiency • Costly and relatively inflexible to changes in process-operating conditions
Wet scrubbers	• May achieve high efficiencies • Some gaseous pollutants removed with particles • May be costly to operate owing to a high pressure drop • Produce a wet sludge requiring disposal

8.5.6 Equipment selection

Muir (1992) has provided selection criteria for particle abatement equipment. Table 8.11 provides a screening device for the selection of appropriate equipment

Table 8.11 Primary and secondary factors for particle control (Muir, 1992)

Equipment	*Particle size μm*			*Gas temperature °C*				*User preference*			
	> 10	*1 – 10*	*< 1*	*>400*	*250 – 400*	*>dewpoint - 250*	*Near dewpoint*	*Dry product*	*Low initial cost*	*Low operating cost*	*Minimum technical complexity*
Cyclones	✓	C	B	✓	✓	✓	C	✓	✓	✓	✓
Wet washers[a]	✓	C	B	C	✓	✓	✓	B	✓	✓	✓
Wet washers[b]	✓	✓	C	C	✓	✓	✓	B	C	U	C
Dry EP	✓	✓	✓	C	✓	✓	C	✓	U	✓	C
Wet EP	✓	✓	✓	B	C	C	✓	B	U	✓	C
Aggregate filters	✓	✓	C	✓	✓	✓	C	✓	C	C	C
Fabric filters	✓	✓	✓	B	C	✓	C	✓	C	C	C
Fibrous filters	✓	✓	C	B	C	✓	C	B	✓	✓	✓

	Dust properties							*Gas conditions*						
	High inlet burden	*Erosive*	*Sticky*	*Light fluffy*	*Difficult to wet*	*Pyrophoric*	*Resistivity problem*	*Constant pressure drop*	*Varying flow*	*Explosive combustible[c]*	*Corrosive*	*Suitable for high pressure*	*Minimum ancillary equip.*	*On-line regeneration*
Cyclones	✓	✓	B	B	✓	C	✓	✓	C	C	C	✓	✓	✓
Wet washers[a]	✓	✓	✓	✓	C	✓	✓	✓	C	✓	C	✓	C	✓
Wet washers[b]	✓	C	✓	✓	C	✓	✓	✓	C	✓	C	✓	C	✓
Dry EP	C	✓	C	C	✓	C	B	✓	C	B	C	C	C	✓
Wet EP	C	✓	✓	✓	C	✓	C	✓	C	B	C	C	C	C
Aggregate filters	C	✓	B	✓	✓	✓	✓	C	✓	C	C	✓	C	C
Fabric filters	✓	C	B	✓	✓	B	✓	C	✓	C	C	C	C	C
Fibrous filters	B	✓	C	✓	✓	C	✓	C	C	C	C	✓	✓	B

✓ = *Can generally cope with the process requirements if well designed*
C = Special attention required in plant design and operation to prevent problems
B = Could lead to severe operational difficulties; seek alternatives to avoid the problem
U = On purely economic grounds alternatives generally favoured if suitable
a) Low energy
b) High energy
c) Seek advice for all gas cleaning problems associated with explosive or combustible materials
EP = electrostatic precipitator

and details the necessary primary and secondary factors to be considered. These include:

- Collection efficiency versus particle size: The most common method for characterising the performance of a dust collector is by means of a grade efficiency curve which is a plot of the overall collection efficiency of the abatement process. Factors affecting the collection efficiency of abatement include particle density, shape and degree of agglomeration; operating pressure drop of the system and the physical condition of the emission stream (i.e. temperature, humidity, etc.). The grading efficiency of various arrestment equipment is given in Table 8.11.
- Gas temperature: The effluent gas temperature can affect the choice of arrestment plant. For example, fabric filters are sensitive to particle temperatures where elevated temperature may cause the bag filter to catch fire.

Table 8.12 Recommended air pollution measures for agricultural practices (MAFF, 1992)

Problem/activity	*Control measure*
Odours • Agriculture • Farm buildings • Slurry or manure stores • Animal feed products • Silage clamps	Housekeeping control measures: collect and transfer slurry to a suitable store; keep concrete areas around buildings clean; remove and dispose of dead animals, birds and foetal remains; maintain drains and repair broken or badly lain concrete to prevent liquids from ponding; ensure bedding materials are stored in dry conditions; manage drinking systems to avoid overflow and spillage and suitably site farm buildings
	Ventilation of buildings: ensure ventilation to control humidity and temperature and limit concentrations of poisonous gases
	Biological treatment: wet scrubbers, bio-filter system
Slurry spreading	Housekeeping control measures: take precautionary measures in adverse weather conditions; avoid high density of spraying (e.g. less than $50m^3$/ha or 50 t/ha); lightly cultivate land after spreading; avoid overfilling of spreader; avoid spreading near to sensitive receptors and use appropriate equipment
Treatment of livestock wastes	Treatment methods: mechanical separation (solid/liquid separation); biological treatment (aerobic and anaerobic); additives (oxidising agents, deodorants, etc.); electrolytic methods
Smoke pollution • Burning of waste • Burning/disposal of dead animals	Housekeeping control measures: waste minimisation; recycling of materials; use other methods for disposal of waste materials; incineration

8.6 CONTROL OF EMISSIONS FROM AGRICULTURAL PRACTICES

Agricultural practices which involve housed livestock, storing wastes or the spreading of wastes are those most likely to cause odour problems (principally from ammonia). Dark smoke or smoke nuisance from agriculture and horticulture can be caused by the burning of crop residues, packaging, plastics, tyres, waste oil or animal carcasses in the open or unsuitable equipment (MAFF, 1992). Agricultural activities can give off various gases, which may cause atmospheric problems. Table 8.12 details the recommended air pollution measures for the control of emissions from agricultural practices (MAFF, 1992).

8.7 FLARING

Occasionally large volumes of VOC releases may occur from industrial processes (e.g. refineries). For emergency and upset release conditions flaring systems have been used for safety venting purposes (Fig. 8.8). The flare system burns emissions at a safe height above the process area. Flares also minimise radiation effects at ground level. Incomplete combustion may give rise to smoke and unburnt hydrocarbon emissions. This can be overcome by injecting steam into the flare's combustion zone to provide turbulence and to entrain air. Concerns, however, have been raised about the emissions of unburnt hydrocarbons from flares and regulators

Figure 8.8 Flaring operations at an oil and gas terminal in Peru (courtesy of Cordah Limited)

are therefore seeking better control of mechanisms (e.g. thermal incinerators or vapour recovery systems) (Cooper and Alley, 1986).

8.8 CONTROL OF EMISSIONS FROM MOTOR VEHICLES

In countries which have removed Pb from petrol, concentrations of lead in air have been reduced to a level at which they are no longer a discernible problem (Fig. 8.9). The use of lead-free petrol has also increased the use of catalytic converters in motor vehicles, which has therefore helped to reduce the emissions of other pollutants. Catalysts substantially reduce emissions of hydrocarbons, NO_x and CO; they do however increase emissions of CO_2, and have no effect on emissions of particles. Since 1993 all new petrol engine cars in the EU have to be fitted with catalytic converters. Prior to the introduction of cars fitted with catalytic converters, diesel powered vehicles were considered cleaner than petrol powered cars. However, legislation for some countries requires that they meet the same limits for hydrocarbons, NO_x and CO as petrol-driven cars. Diesel fuel contains no lead but is an important source of particulate matter, PAH and SO_2. The introduction of lower sulphur diesel fuels has, however, reduced emissions.

Figure 8.9 Decline in Pb in air levels (μg/m^3) (City of Glasgow Council, 1999)
(Data from Environmental Health Department, Glasgow City Council)

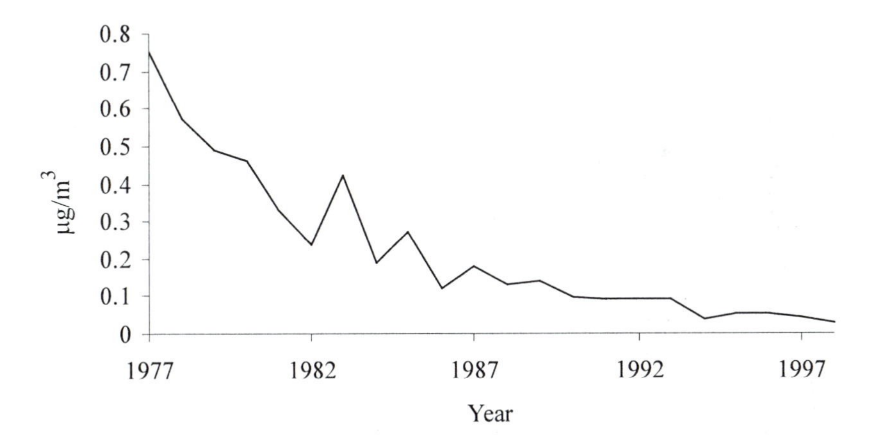

Countries, such as Singapore, control vehicle emissions by specifying the emission standards of vehicles entering the country. In Singapore vehicles imported for local use are required to meet specified emission standards (Singapore Government, 2000). In addition, in use vehicles are required to undergo mandatory periodic inspections. During the inspections, exhaust emissions are tested for compliance against specific standards. The quality of fuel used by vehicles is controlled. Unleaded petrol was introduced to Singapore in 1991 and leaded petrol was phased out in 1998. To reduce emissions from diesel vehicles, permissible sulphur content in diesel was reduced from 0.5% to 0.3% by weight in 1996 and this was further reduced to 0.05% in 1999. In the absence of their own emission standards countries often adopt other established standards. For example, Singapore applies either the EU, US or Japanese motor vehicle emission standards (Table 8.13).

Table 8.13 Summary of vehicle emission standards used in Singapore (Singapore Government, 2000)

Type of vehicle	*Emission standard*	*Implementation*
Petrol	All new petrol-driven vehicles are required to comply with 91/441/EEC Emission Directive or the Japanese emission standard (Article 31 of Safety Regulations for Road Safety).	1.7.1994
Diesel	All new diesel-driven passenger cars are required to comply with 93/59/EEC Emission Directive	1.7.1997
	All new light commercial vehicles <3.5 t are required to comply with the 93/59/EEC Emission Directive	1.7.1997
	All new heavy duty vehicles >3.5 t are required to comply with 91/542/EEC Stage 1 Emission Directive	1.7.1997
Motorcycle/scooter	All new motor cycles are required to comply with US Code of Federal Regulations (US 40 CFR 86.410.-80) emission standard	

In many countries there has been a progressive policy of tightening of emission standards for cars and lorries in line with EU directives (Chapter 7, Section 7.9.4.2) and UNECE standards. Much more will, however, need to be done to ensure that reductions in vehicle emissions are not offset by the rapid increase in vehicle ownership and use. This is an aspect in which action by local and regional government can make a significant impact on localised air quality. Improved public transport, 'park and ride' schemes, traffic restrictions, planning guidelines and encouragement to cycle and walk are some of the measures that can be taken (also see Table 8.15). Requiring vehicle owners to maintain their vehicles regularly will also ensure that fuel is burnt efficiently and economically, and will therefore be less polluting. Other measures that are available for motor vehicle emission control include:

- Provision of safe cycling and pedestrian routes;
- Regular emissions testing of private and public transport vehicles;
- Encouraging local business to use car pooling or car sharing schemes;
- Encouraging less use of motor cars during weather conditions likely to lead to an episode of air pollution;
- Encouraging better fuel quality and better planning of the built environment aimed at reduced mobility and improved access to shops, jobs.

An example of the management of motor vehicle emissions is seen in Greece (European Environment Agency, 2000a). In Athens, two types of air pollution are of principal concern: the high concentrations of particles and photochemical. Several measures have been taken in an effort to improve the existing air quality conditions in the city with regard to photochemical smog. A retirement plan for old vehicles, introduced in 1991 in an effort to renew the motor vehicles fleet with new cars equipped with catalytic converters, has resulted in 260,000 old cars being retired in Greece, of which 150,000 are in Athens. The national inspection programme for the control of emissions from motor vehicles was initiated in 1994, consisting of regular inspections of all private vehicles once per year and taxis and light trucks twice per year. In addition there has been an expansion of the subway system. Two new lines of a total length of 18 km have been constructed. The project was expected to have concluded by 2000. It is anticipated that daily automobile trips in the centre of the city will be reduced by 250,000.

A project for the full replacement of the old buses with new ones equipped with anti-pollution devices has also been initiated. In addition to the replacement of buses, several other measures have been taken in an effort to make public transport attractive to Athenians:

- Rescheduling of bus routes to accommodate passenger needs;
- Dedicated bus lanes to improve running conditions;
- The odd-even traffic regulation system to control cars entering the city;
- Banning of all types of motor vehicles (excluding public buses) from the commercial centre of Athens;
- In the event of adverse meteorological conditions full banning of traffic from the centre of Athens.

In a further effort to reduce traffic, the working schedule of public and private employees was modified during the summer months of 1994-1996. Employees were divided into four major categories with different start and completion times.

8.9 COST EFFECTIVENESS

The introduction of mitigation measures in pollution control should not only be effective from a pollution control perspective but also cost effective. To assess the cost effectiveness of an abatement measure, two elements are involved. The first assesses the likely reductions in emissions or air quality benefits of the control measure, along with the consequential non-air quality effects. The second assesses the costs of implementing the control measure, while also considering associated costs or possible revenue streams (NSCA, 2000b). The cost effectiveness of a control measure typically involves the dilemma of how much does it cost to reduce an emission of a prescribed substance to an acceptable/regulatory level?

Generally standard methods should be applied when evaluating cost issues (NSCA, 2000b). Cost can be calculated using the following expression:

$$PVC_0^k = \sum_{t=o}^{T^k} [NRC_t^k + ERC_t^k + NERC_t^k] [1 + r]^{-1} \quad (8.2)$$

Where PVC is the present value of the total cost stream for environmental protection measure k in year zero; NRC the non-recurring cost of environmental protection measure k in period t; ERC the energy recurring costs to operate environmental protection measure k in period t; NERC the non-energy recurring costs to operate environmental protection measure k in period t; T is the operating life of environmental protection measure k; and r is the appropriate discount rate.

Guidance from the European Environment Agency (2000b) is available on how cost data should be expressed and how it should be interpreted for stationary

Table 8.14 European Environment Agency summary guides for defining and documenting data on costs of possible environmental protection measures (European Environment Agency, 2000b)

Summary of guidelines
1. Pollutant definitions and assumptions regarding scope of pollutant categories should always be given wherever there is any possibility of ambiguity
2. Sufficient detail of the pollution source should be given to enable comparison with similar processes and to avoid ambiguity. It is recommended that published source sector classifications should be used wherever possible
3. Sufficient detail of the environmental protection measure should be given to avoid ambiguity to define its performance characteristics, and to clarify any special circumstances limiting applicability of the measure
4. It is essential that reported costs are defined: what is excluded, how they have been attributed or apportioned. It is recommended that costs are also explained in physical terms such as quantity of materials, and as unit prices
5. As a minimum, all data should have a background discussion of the key uncertainties
6. The year in which the following data apply should always be given: • cost data • currency exchange rates • data describing • control technologies (efficiency, applicability) and process technologies • emissions to the environment
7. The sources and origins of all data should be recorded as precisely as possible so that data may be traced at a later date if necessary
8. As a minimum, any discount/interest rates used should be recorded
9. If cost data are adjusted for inflation or changes in price through time, then the method used should be recorded and any index used should be recorded and referenced
10. If determining annual cost data, the approach that has been used to derive the annual costs should be recorded, along with all underlying assumptions

abatement technologies. The same principles apply technical measures for control of mobile sources. There are, however, some important differences for such measures (NSCA, 2000b). Table 8.14 details how cost data should be expressed and how it should be interpreted for stationary abatement technologies (European Environment Agency, 2000b).

8.10 AIR QUALITY MANAGEMENT

To ensure improved air quality requires the deployment of effective assessment and management practices. AQA is an environmental management tool, which enables impacts to be effectively evaluated and managed. AQA is an integral component of the environmental management process. It involves not only the identification, monitoring, prediction and evaluation of critical variables (e.g. source emissions and meteorological conditions), but also potential changes to air quality as a result of emissions to air from a source and the mitigation and control of these emissions, whether they are from controlled or fugitive sources (Harrop, 1999).

AQA can be used as a screening device for setting priorities for pollution control, or it can be used to evaluate alternative options at an early stage in the project/process cycle as well as to aid the identification of the most suitable site location to further mitigate potential impacts. AQA may also be applied at a strategic planning level as well as a localised project level. This can be seen when collectively assessing the impacts of emissions within the context of neighbouring land uses, regional planning policy and air shed management. AQM essentially involves designating the level of pollution deemed acceptable in terms of a set of ambient air quality criteria (Section 7.2), and then controlling the pollutant emissions to ensure that the criteria are not exceeded (Weber, 1982). AQM has been defined elsewhere as the regulation of the amount, location and time of pollutant emissions to achieve some clearly defined set of ambient air quality standards (AAQS) or goals (de Nevers *et al.*, 1977).

Elsom (1987) describes the component parts of an AQM system (Fig. 8.10). These components of the system are necessary before appropriate and effective pollution control can be introduced. A great concern of the AQM strategy is its use of AQS. Although AQS result in pollution concentrations being reduced to an acceptable level in an area of concern, air quality in an area of no relative concern, however, is allowed to increase up to the standard. The consequence is a degradation of air quality. Another criticism is that it may result in widely varying control requirements and pollution control expenditures for competing companies in regions of differing air quality. Inevitably, a strategy can vary in duration and differ from country to country depending on resources and the strategy's aims and objectives (Elsom, 1987). Nevertheless, Elsom (1987) has identified that there are common elements, which make it possible to distinguish three other types of strategy from that of AQM. These include:

- Emission standards;
- Economic issues;
- Cost benefit.

Figure 8.10 Stages involved in an AQM strategy (from Elsom, 1987) (courtesy of Blackwell Publishers Ltd)

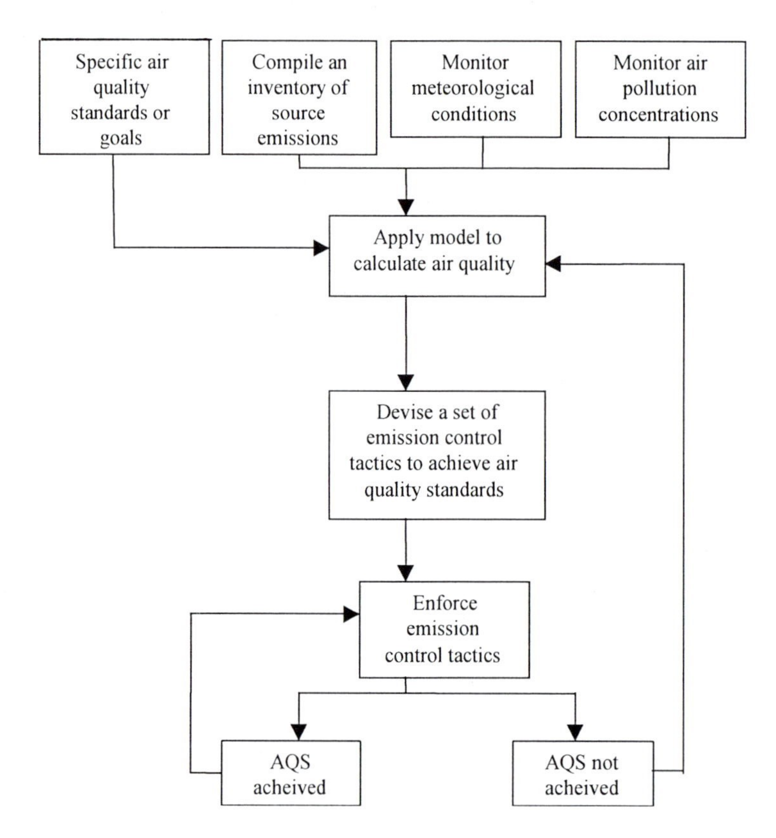

There are principally two types of emission standard. One relates to the concentration of the emission from a process (e.g. concentration of particulates in an effluent gas), and may be related to a particular control device (e.g. bag filter) that the emitting operator should install to meet the required emission standard. The other standard relates to the type and the impurities of the fuel used in the process, for example the sulphur content of the fuel or the lead in petrol content. Economic strategies (e.g. emission charges, pollution permits) provide a financial incentive for emission sources to entice the polluter to use the most cost effective means of pollution control. Essentially the premise of emission charges is to make polluters pay a levy related to the amount of pollution they emit (i.e. the Polluter Pays Principle). A cost-benefit strategy attempts to quantify the costs of environmental damage associated with air pollution and the costs of known ways of pollution control (Section 8.9). It then adopts the pollution control option(s) which minimise the sum of the pollution damage and the pollution control costs.

8.10.1 AQM process

In many countries the preliminary stages of the AQM process have been initiated to some degree, particularly for the criteria pollutants. However, it is further stages of the AQM that are the most demanding. Having set air quality objectives there is a requirement to undertake monitoring (Chapter 5) and modelling (Chapter 6) studies of specific pollutants to assess their individual contribution to the air quality concern. Should specific criteria be exceeded then there is a requirement for an Action Plan to reduce, mitigate and control air pollution to acceptable levels. Certain measures are required depending on measured air quality level (NSCA, 2000b):

- Identify zones where air pollution levels exceed or are likely to exceed the recommended AQO, plus the margin or tolerance (i.e. the percentage of the AQO by which the value may be exceeded), plans and programmes to achieve the AQO with a specified time frame must be drawn up and implemented and the public given details of them;
- In zones where the levels are lower than the recommended AQO these levels should be maintained and the best ambient air quality compatible with sustainable development maintained;
- Where alert thresholds for recommended AQOs are exceeded, steps must be taken to inform the public and administrative body of concern.

Individual countries have introduced their own NAQS. Fig. 8.11 details a generic AQM strategy, which is the basis for many plans.

Figure 8.11 Generic air quality management plan (courtesy of the School of Health and Life Sciences, King's College London)

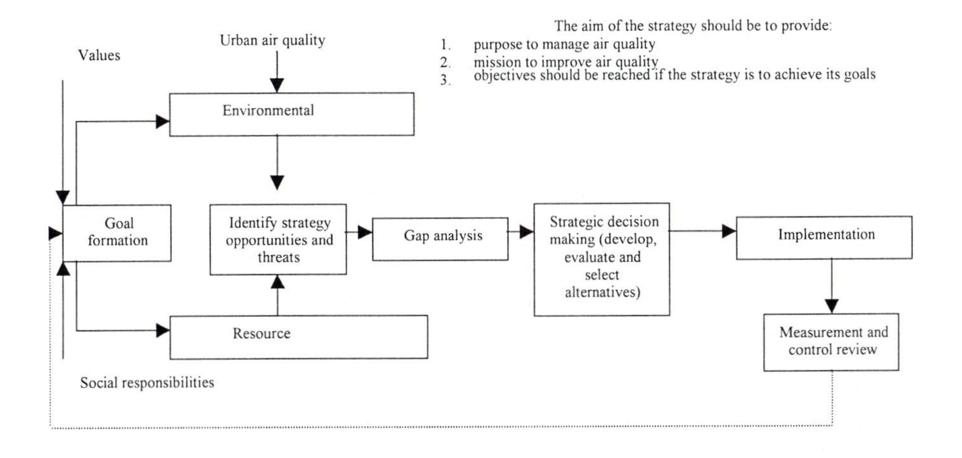

8.10.2 Air quality management in the UK

The UK Government began working towards its present strategy for air quality management in 1990 with the publication 'This Common Inheritance'. In 1994 they issued 'Improving Air Quality', a consultation paper on air quality standards and management which discussed measures which might be used to address these issues and secure a progressive long-term improvement in air quality. In 1995 the Government adopted the concept of LAQM by setting out their approach in 'Air Quality - Meeting the Challenge'. Finally in 1997 a NAQS for the UK set out a framework of standards and objectives for the air pollutants of most concern which will lead to reductions in the number and extent of episodes of poor air quality.

The NAQS focuses upon the concerns of eight pollutants and sets air quality criteria to be met by 2005. The Strategy is built upon two broad trends which have merged to create a platform for the formulation of a more strategic and integrated approach to air quality management. The first was the elaboration of the principles of sustainable development and the second was the progress at national, European and international levels in our understanding of air pollution and in the development of new legal instruments to tackle it. Under the 1995 Environment Act, Local Authorities are also required to review and assess air quality within their areas and to assess whether any of the AQOs specified in the NAQS are not likely to be met by 2005. In this event, a further more detailed assessment will be required for potential non-attainment areas. Any areas in which air quality will not or is unlikely to meet the AQO must be designated an AQMA. Within 12 months of designation the Local Authority should draw up an action plan for each designated area detailing measures and target dates in order to meet the NAQS objectives. The UK Environment Act, 1995, required the Secretary of State to produce a NAQS to provide a framework for air quality control through national strategies and policies for air quality and the establishment of AQS. The NAQS (DETR, 2000e) was first published in 1997 and was revised in January 2000. The NAQS sets objectives for eight main pollutants and includes benzene, 1,3-butadiene, CO, Pb, O_3, NO_2, PM_{10}, and SO_2 (Chapter 7, Section 7.2.3).

A minimum of two rounds of air quality review and assessment are required in order to assess compliance with the AQO. One at the outset of the NAQS and a second to be carried out towards the end of the time scale for meeting the objectives, either 2003, 2004 or 2005, depending on the pollutant. Each Review and Assessment may consist of Stage 1, Stage 2 and/or Stage 3 reports. Not every Local Authority would be required to proceed to each stage. The UK Government has provided guidance on a phased approach to the Review and Assessment process. The intention is that Local Authorities in the UK should only undertake as much work as necessary, commensurate with the extent of the air quality problems in their area. The three stages comprise:

- First Stage Review: Requires the collation of information on any existing or proposed significant source of pollutants of concern within the Local Authority's area or within the vicinity which may affect air quality within their area. The potential exposure of the population to these pollutants, within the averaging times established within the AQO, should be assessed. At the end of the First Stage Assessment, the Local Authority should be able to demonstrate that all potentially significant pollutant sources have been considered; that sufficient information has been collated on these sources; and that the need to proceed to a further stage within the assessment has been fully determined.

- Second Stage Review: The Second Stage of the review focuses on the areas where the highest likely occurrence of pollutant concentrations exists. From these areas, the Local Authority is required to identify general locations, which may not meet the AQO by the end of 2005. In addition, an estimation of ground level concentrations within these areas should be made at the points of greatest risk of exceedence (e.g. combinations of road side and/or industrial sources), using existing air quality monitoring data where available and carrying out basic modelling exercises. These estimations should allow the Local Authority to be able to predict air quality both now and in 2005.
- Third Stage Review: The Third Stage comprises an accurate and detailed Review and Assessment of the current and future air quality. This requires more advanced air dispersion modelling and monitoring techniques. At the conclusion of this stage, a Local Authority should be able to predict whether AQO would be met or not, and where exceedences may occur, taking into account the relevant exposure limits of the individual pollutants. Tools recommended for use within the Stage Three Assessment include detailed emissions inventories and validated dispersion modelling, backed up by a continuous automatic air quality monitoring exercise (Case Study 7, Section 10.8).

8.10.3 Designation of air quality management areas

Should AQO be exceeded then there is a requirement for an AQMA to be delineated. Owing to the transboundary nature of air pollution, the declaring of an AQMA may not be a straightforward task. Studies have shown that the interpretation of the extent of AQMA can be varied (Woodfield, 2000). Concerns raised in the designation process included:

- Uncertainty in the contour information – concerns on the accuracy of the modelled information to derive the AQMA;
- Exposure versus exceedence - concerns were raised about the need to consider exceedence of AQO together with the level of actual individual exposure to exceedences to derive the AQMA;
- Receptor definition – concerns were raised over what is a receptor or ASR (e.g. those receptors that required to be considered with regard to exposure to a pollutant);
- Demarcation of the AQMA boundary – concerns were raised over the exact positioning of the AQMA boundary;
- Designation of AQMA for more than one pollutant – concerns were raised over the issue of how to encompass two or more predicted pollutant exceedences.

The use of air quality modelling and monitoring techniques will help in the designation process of AQMA site boundaries. The principal concerns are the limitations of both techniques. Monitoring is spatially constrained (unless numerous sites are available) and modelling is predictive. To aid the practitioner the DETR (2000g) has provided guidance; techniques are provided to help identify areas of concern.

Table 8.15 Available actions to manage air quality (NSCA, 2000b)

Action	*Action*
Road traffic related action - zoning	• Walk to school plans
• 20 mph residential traffic zones	*Road traffic related action/traffic management*
• Traffic free residential areas	
• Low emission zones	• Speed regulation
Road traffic related action/public transport	• High occupancy vehicle lanes
	• UTC systems (e.g. SCOOT)
• Public transport priority schemes	• Parking controls
• Bus quality partnerships and contracts	*Air traffic related action*
	• Reduce aircraft emissions
• Subsidised public transport	• Travel plans (employees and air travellers)
• Park and ride	
• Light rail or tram systems	*General policy measures*
Road traffic related action/charging and enforcement	• Section 106 Agreements
	• Development plans (e.g. Structure Plan)
• Congestion charging or toll roads	
• Non-residential parking levy	• EMS/ISO 14000
• Roadside emission testing	*Industrial emissions*
Road traffic related action/infrastructure	• Changes to process technology
• Pedestrianisation	• Feedstock and/or fuel change
• Improved cycling and walking provision	• Emission abatement equipment
	• Changes in operating pattern
• Traffic calming	• Relocation of process
• Road system redesign	*Domestic emissions*
• Bypass and road building	• Smoke control zones
Road traffic related action – Schemes	• Home insulation schemes
• Car pool schemes	• Information on bonfires
• Travel plans	• Bonfire ban
• Car scrapping schemes	

UTC = urban traffic control.

SCOOT is a proprietary telematics package that controls traffic lights and optimises traffic flow.

8.10.4 Air quality action plans

Having identified AQMA there is the need to implement air quality action plans (AQAP) to assist in the delivery of proposed AQO. AQAPs ultimately provide the mechanism by which interested parties state their intentions for working towards

Figure 8.12 The Process for Developing AQAP (NSCA, 2000b) (courtesy of the National Society for Clean Air and Environmental Protection)

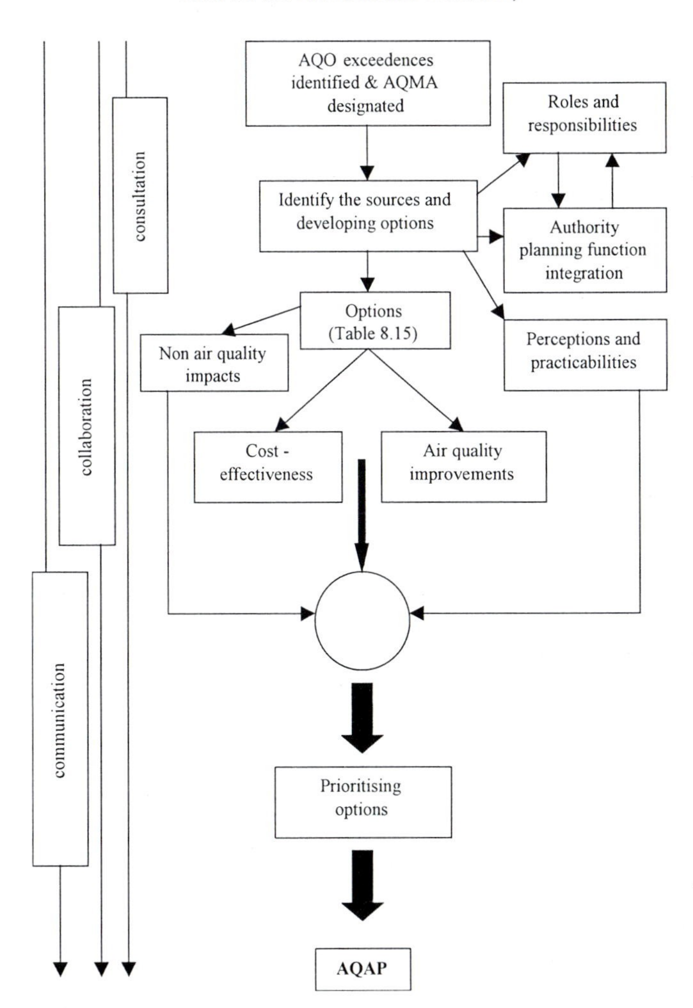

the AQO through the use of the powers they have available. Fig 8.12 shows the process of developing an AQAP. When developing a plan there is a need to include an assessment of the options available, an understanding of the wider non-air quality impacts of such options and the likely improvements offered by them (NSCA, 2000b). The length of an AQAP will vary depending upon the degree to which air quality improvements are required and the complexity of the options and actions considered necessary. Studies (NSCA, 2000b) have suggested that AQAPs should begin by assembling a list of tasks to consider, and a committed team who will undertake work on various aspects of the plan. Table 8.15 provides a range of options to manage, control and improve air quality to meet required AQOs. A full description of options is provided by NSCA (2000b).

CHAPTER NINE

Site Inspection and Project Management

9.1 INTRODUCTION

Several issues relevant, but often considered ancillary, to AQM may be encountered by practitioners that require particular diligence. One concerns the eliciting of useful information to help manage a problem more effectively; the other relates to managing the actual project that embraces the air quality problem.

A site inspection of an activity will provide useful information that can be utilised to improve its performance and management. Section 9.2 provides a generic methodology used to carry out an inspection (Cordah Limited, 2001). Section 9.3 details general project management skills that are pertinent to managing an AQA study.

9.2 PROCESS INSPECTION

An effective form of pollution mitigation is to anticipate and manage the problem. This can be done by systematic inspection of the process to identify problems and concerns. The essential components of an inspection are the:

- Opening meeting;
- Inspection techniques;
- Interview techniques;
- Document review;
- Verification techniques;
- Inspection findings and the close out meeting.

9.2.1 Presentation to management

For all site inspections there should be an opening meeting where the inspection team makes a presentation to management and other senior personnel responsible for the facility/site. The purpose of the meeting is to:

- Introduce members of the inspection team to the company's management;
- Review the scope, objectives and inspection plan and agree on an inspection timetable;
- Provide a short summary of the methods and procedures to be used to conduct the inspection;

- Establish official communication links between the inspection team and the company;
- Confirm that the resources and facilities needed by the inspection team are available;
- Confirm the time and date of the close out meeting;
- Promote the active participation of the company staff;
- Review relevant site safety and emergency procedures for the inspection team;
- Determine whether auditors will be guided around the site or left unattended.

9.2.2 Orientation tour

Following the opening meeting the inspectors should tour the facility. This will familiarise them with the activities on site and give them a general understanding of the processes. It will also be an opportunity for them to assess the site's relationship with its exterior environment.

9.2.3. Techniques

The main techniques include site inspections, interviews and documentation review. These techniques are used to verify that the environmental and process control systems in place are performing adequately. The control systems are those systems that are implemented by site management to control processes that impact on the environment. Control systems may be either management control (procedures and work instructions) or technical controls (abatement technology).

9.2.3.1 Site inspections

Inspections are important because they provide evidence that the management system and other assessment criteria are either working or not. It is also important for the inspectors to acquaint themselves with what is actually happening on the site (Fig. 9.1). Inspections can help to:

- Identify potential on-site problem areas and hazards;
- Identify the effects of change;
- Verify that remedial actions have been carried out adequately;
- Identify hardware deficiencies and incorrect personnel actions;
- Identify good practices and performance.

The formal stages of an inspection are essentially:

- Preparation;
- Inspection;
- Recommend immediate actions;
- Recommend follow-up actions;
- Prepare report.

Figure 9.1 Inspection team (courtesy of Cordah Limited)

As with anything, preparation is essential prior to carrying out an inspection. If inadequately prepared the inspector could miss vital information, waste valuable management time and be made to look a bit of a fool. If well prepared they can approach the inspection with a positive attitude and make it a motivating experience for themselves, but more importantly for those on the site. So, in preparation they should:

- Approach the inspection with a positive attitude;
- Establish and know what they are looking for (what are the critical areas or activities which they might want to concentrate on);
- Look at previous reports;
- Make checklists;
- Establish the tools and materials they may need.

When carrying out an inspection the following study approach may be adopted:

- Work to the checklist and use a floor/process plan;
- Keep a sharp eye out for non-obvious items and new hazards;
- Look for unnecessary things;
- Describe and locate items;
- Recommend or take any immediate or temporary actions as appropriate.

The overall recommendations will become clearer as the overall inspection progresses but the observations of the inspection should be developed into possible remedial actions. Inspectors should:

- Consider the implications of the observation in terms of potential significance of environmental effect;
- Consider possible management and hardware control possibilities;
- Consider a preferred option.

The inspector should also be looking out for examples of good housekeeping practice.

9.2.3.2 Interviews

Interviews should be pre-arranged to ensure that personnel could be away from their normal work site for the duration of the interview. The names of individuals should not generally appear in the inspection report. Interviews should include all levels of management, supervision, and operations personnel, including contractors on the site. Owing to the time scale of the inspection, sampling techniques should be used in order to cover the appropriate range of personnel. Inspection team members should take written notes of all relevant conversations so that these may be shared with other team members. Inspectors should possess personal attributes and skills conducive to carrying out interviews effectively, which include diplomacy, tact and ability to listen.

Clear objectives should be established, prior to interviewing, regarding the information required from the interviewee. The interviewer should secure facts by asking the important questions of what, which, who and why.

Open-ended questions should be encouraged in order to achieve the full participation of the interviewee. It should be made clear that the purpose of the interview is to establish facts and not, for example, to apportion blame for ineffective control.

9.2.3.3 Documentation review

Examination of documents is a key component of the inspection process. As with interviewing personnel, it is another means of collecting evidence to determine whether the activity is in compliance with the performance criteria against which they are being assessed. The inspection team would typically review:

- Records;
- Written procedures;
- Documents;
- Policies;
- Minutes of meetings.

The documents will be reviewed against the inspection criteria e.g.

- EMS standards (documentation requirements);

- Company standards (documentation requirements);
- Process operation criteria.

When reviewing documentation an inspector should be able to find their way around the process i.e. follow an inspection trail. If there were an incident on site, inspectors should be able to follow through the incident report, investigation and corrective action paperwork.

9.2.4 Verification techniques

As described above the techniques used when conducting an inspection include:

- Inspection;
- Interviews;
- Documentation review.

Appropriate, and sufficient, information must be collected, analysed, interpreted and documented as inspection evidence in an examination and evaluation process to determine whether inspection criteria are being met.

It is important that information obtained from these three techniques is independently verified e.g. information gathered during interviews should be verified by acquiring supporting information from independent sources, such as observations (inspections), records and results of existing measurements (documentation review). Is what is said is being done, actually being done in practice? Non-verifiable statements should be identified as such. Indications of non-conformity to the inspection criteria should also be recorded. In addition, inspectors should examine the basis of relevant sampling programs and procedures for ensuring effective quality control of sampling and measurement processes, that is verify reliability of data.

9.2.5 Inspection findings

The inspection team should review all of the inspection evidence to determine whether the inspection criteria have been conformed with. The inspection team should then ensure that those inspection findings that are of non-conformity are documented in a clear, concise manner and supported by inspection evidence. Inspection findings should be reviewed with the responsible company manager with a view to obtaining acknowledgement of the factual basis of all findings of non-conformity.

9.2.6 Close out meeting

After completion of the evidence collection phase and prior to preparing the inspection report, the inspectors should hold a meeting with the company's management team and those responsible for the functions inspected.

The main purpose of this meeting is to present inspection findings to the company in such a manner as to obtain their clear understanding and acknowledgement of the factual basis of the findings. Necessary improvements and means of achieving them should also be discussed with the site management team. Disagreements should be resolved, if possible before the inspector issues the report. Final decisions on the significance and description of the inspection findings ultimately rest with the inspector, although the company may still disagree with these findings.

9.3 PROJECT MANAGEMENT

AQA management involves information management and its communication. Unfortunately there are many opportunities for mis-communication (Harrop and Nixon, 1999). Three fundamental aspects of project management are context and procedure, technical management and fiscal control and budgets. To this can be added the importance of report writing.

In the absence of clear direction from the project proponent and/or client, an AQA project may not cover all the relevant issues while pursuing other less important matters. In such circumstances, requirements of the various interested parties may not be met. At the outset, therefore, terms of reference (TOR) should be prepared defining the key issues to be covered, the decisions to be taken and the options to be investigated so as to minimise omissions, and the possibility of the introduction of costly additional issues in the latter stages of the project. Project management issues are detailed elsewhere (Harrop and Nixon, 1999) and summarised below.

An AQA study may take a long time to produce, therefore the project manager will be responsible for the continuity of the project despite possible project team changes, changes in the tender brief, etc. The greater the duration of the project, the greater the potential for managerial problems. An initial starting phase of the project management process is the understanding of the manager's role in the context of the project and TOR. In some situations the context and procedure are well defined and are established in regulations, law or policy. In such circumstances the project manager's role is normally defined. What must not be forgotten is that the project management process provokes question asking, as well as problem solving. Both the questions and the answers can lead to a different focus for a project, perhaps even clarify its purpose and need. The project manager therefore serves an important role in asking questions and resolving them.

AQA studies are often conducted by interdisciplinary teams, that is a group of two or more persons trained in different fields of knowledge with different concepts, methods, and data and terms, organised to address a common problem with continuous inter-communication among participants from different disciplines (Dorney and Dorney, 1989). A study team for a specific study can be considered as a temporary entity, which has been assembled and possibly specifically appointed, for meeting the identified purpose of conducting an AQA for a proposed project. The team may be assembled with formal authority, responsibility, and accountability, but a more typical approach is the delineation of an informal authority within that team, with the team basically being subjected to the management of the team leader (Cleland and Kerzner, 1986).

The project manager is the individual who provides leadership for the team itself when directed toward accomplishing the end purpose, with the end purpose being the successful execution of the study (Cleland and Kerzner, 1986). The project manager should exhibit a number of specific qualities; required attributes should include (Cleland and Kerzner, 1986):

- Demonstrated knowledge and skill in a professional field;
- Positive attitude in support of the conduction of the environmental impact study;
- A rapport with individuals;
- An ability to communicate with both technical and non-technical persons;
- Pride in their technical speciality area;
- Self-confidence;
- Self-starter;
- A reputation as a person who gets things done;
- The ability to deal successfully with the challenge of doing quality work;
- The willingness to assume responsibility for the overall study and team leadership.

Several key characteristics should also be considered in the selection of the project manager. These characteristics include experience in serving as project manager, management/leadership skills and substantive area of expertise.

A number of considerations are related to the management of a study team and project. The project manager should consider several management techniques and develop approaches to utilise them for the successful operation of the specific team. For example, Cleland and Kerzner (1986) suggested the following important factors, which would be basic to the successful management of a study team:

- A clear concise statement of the mission or purpose of the team;
- A summary of the goals or milestones that the team is expected to accomplish in planning and conducting the study;
- A meaningful identification of the major tasks required to accomplish the team's purposes, with each task broken down by individual;
- A summary delineation of the strategy of the team relative to policies, programs, procedures, plans, budgets and other resource allocation methods required in the conduction of the environmental impact study;
- A statement of the team's organisational design, with information included on the roles and authority and responsibility of all members of the team, including the team leader;
- A clear delineation of the human and non-human resource support services available for usage by the study team.

A fundamental technique for team operation is the carrying out of periodic team meetings with planned agendas. It is a primary role of the project manager to develop schedules and to establish priorities with regard to manpower and other resources allocated to particular activities within the study. It should be recognised that modifications will probably be needed in scheduling and budgetary allocations as the study progresses. In addition to team meetings, the project manager must allow individual team members working in their own particular areas to carry out agreed assignments, and then subject the work products, or at least the ideas resulting from the work, to team review. The pattern of meeting, individual work, and a follow-up review meeting is a useful concept in the operation of a study team. While it is theoretically possible, it is unlikely that the study team will completely work together on every aspect of a study. The communication among specialists may not always be smooth and productive. This circumstance can be healthy, as it promotes debate and interaction, but it can also be negative. However, the project manager's job is to keep the debates in perspective.

9.3.1 Report writing

Canter (1996) identifies a number of basic principles to report writing:

- Always have in mind the audience of the report;
- Use simple and familiar language;
- Ensure that the presentation of the report is well-structured;
- Make the report visually attractive.

The report should follow a logical process and be prepared in a consistent manner, which can aid in communicating information to both technical and non-technical audiences. To further aid the communication of the study Canter (1996) stipulates a number of useful general writing suggestions to prevent mis-interpretation of the report findings and a failure in the communication process. The suggestions include:

- Do not use clichés and catchwords;
- Endeavour to make the report succinct and clear, with minimal use of written texts and liberal use of visual display methods;
- Avoid creating a credibility gap as a result of too many technical errors and mistakes in the document;
- Include both pro and con information with regard to a proposed action (e.g. always avoid bias);
- Efforts should be made to provide as complete a document as possible within the time frame and monetary constraints associated with a given writing effort;
- Care should be given to prevent plagiarism of existing documents and avoid improper referencing;
- Provide a document that has continuity from one section to another;
- Avoid inconsistencies in writing and presentational style due to project team member inputs;
- Avoid generalities in the presented information, make sure that all presented data is included for a reason and is properly displayed and interpreted.

The application of QA procedures will certainly improve the technical management and preparation of the final report. The procedures will help focus the project team's attention on the delivery of a product that meets the requirements of the study, the fiscal budget of the study and satisfies the client's needs.

9.3.2 Financial control

One of the uncertainties in the conduct of a study is related to the appropriation of costs. There are no systematically developed cost algorithms that could be used for estimation purposes. Unfortunately there is a never-ending call for the study to be done as cheaply as possible. Such an approach can prove costly in the long-term should the development be called before a planning inquiry and the costs of legal fees be encountered. It is undoubtedly the preferred option to produce a quality product, but the project manager is inevitably forced to focus on the realities of producing a good report within the budget provided. It is the responsibility of the project manager to monitor the progress of the study against the determined budgetary constraints. Should a variation occur to the TOR then it is their responsibility to identify the variance, determine additional costs and resolve financial issues prior to the commencement of additional work. Confirming variations to the original contract will avoid embarrassment at a later date when trying to resolve invoicing and payment issues with the client.

CHAPTER TEN

Case Studies

10.1 INTRODUCTION

The following case studies incorporate and demonstrate key elements of AQA and management that have been detailed in the proceeding chapters of this book. The following case studies are detailed:

- Stirling Council, Scotland - screening air quality monitoring study. The study demonstrates screening baseline air quality monitoring techniques for SO_2, NO_2 and PM_{10};
- Indoor air quality health assessment. The study outlines an air quality monitoring study to determine CO levels in an indoor environment, a health assessment and the remediation techniques to rectify the problem are given;
- Falkirk Council, Scotland - emission inventory. The study outlines the methodology used to undertake an inventory of atmospheric emissions from point, area and line sources within a Local Authority area;
- Belfast, Northern Ireland - baseline air quality monitoring study. The case study details a baseline air quality data assessment and rationale for a baseline monitoring study;
- Newton Stewart, Scotland - air dispersion modelling study. The case study details an air dispersion modelling study of emissions from a proposed incinerator and outlines a stack height optimising study;
- Hong Kong, People's Republic of China - quantitative health risk assessment. A QRA study is presented on emissions from a proposed energy from waste facility;
- Falkirk Council, Scotland - AQM. The case study details the air dispersion modelling studies and the procedures adopted to undertake an AQM study for a Local Authority;
- Ayrshire, Scotland - emission control. The study details the emission control procedures adopted for the control of VOCs from an industrial facility;
- Spireslack OCCS, Ayrshire, Scotland - dust assessment. The case study demonstrates the techniques available to assess potential dust impacts from a proposed OCCS.
- Peterhead Power Station, Aberdeenshire, Scotland – air dispersion modelling study. The case study details the approach taken to model emissions from a predominantly oil-fired power station.
- Widening of the Tolo /Fanling highway (Hong Kong), People's Republic of China – road traffic air quality impact assessment. The case study details a road traffic modelling study.
- Nigg Oil Terminal, Ross-shire, Scotland - emission inventory of VOCs. The case study details the approach taken to quantify VOC emissions from an oil terminal.

- Shenyang, People's Republic of China – assessment of air quality data. The case study details the presentation and analysis of air quality data.
- Stirling Council, Scotland – a comparison of model predictions and baseline monitoring data.

10.2 CASE STUDY 1 - SCREENING AIR QUALITY MONITORING STUDY

In 1999 Stirling Council, Scotland, undertook a 12 month air quality monitoring programme for NO_2, SO_2 and PM_{10} (Cordah Limited, 2000b). NO_2 and SO_2 were monitored using passive sampling techniques (diffusion tubes (Chapter 5)). The tubes were changed at 4-weekly intervals for a period of 12 months. In total there were 15 NO_2 sites and 6 SO_2 sites (Table 10.1, Fig. 10.1). PM_{10} was monitored using an active sampling technique (semi-automatic), where a known volume of air was pumped through a gravametric filter for a known period of time. PM_{10} was monitored for a two week period.

Figure. 10.1 Stirling Council air quality monitoring sites for NO_2, SO_2 and PM_{10} (courtesy of the Environmental Health Department, Stirling Council) (base map reproduced by kind permission of Ordnance Survey © Crown Copyright NC/01/24240)

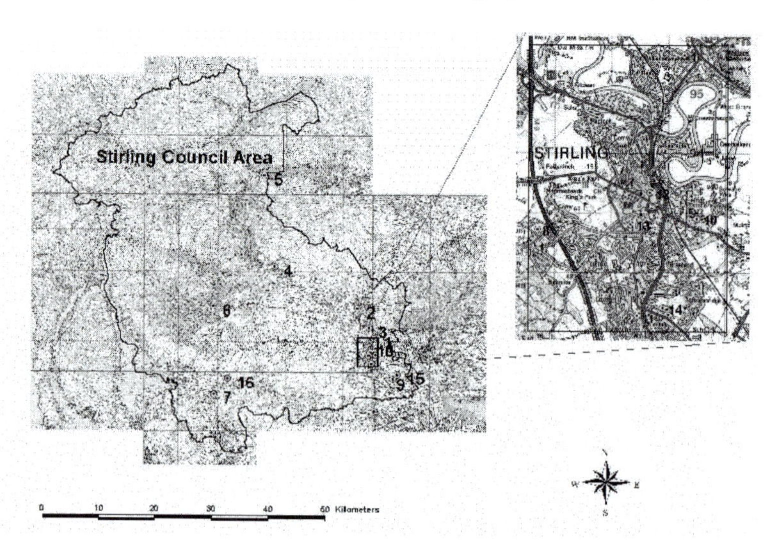

Tables 10.2 and 10.3 summarise the monitoring results for NO_2 and SO_2. The results show that the concentrations of NO_2 ranged from <1 to 54 $\mu g/m^3$. The highest values were for Callander (Site 4) due to the kerbside location of the site. The lowest values were at Lochearnhead (Site 5) which had low traffic volumes and no industrial processes. Annual means were considerably below the UK AQS of 40 $\mu g/m^3$ (Chapter 7). SO_2 concentrations ranged from <0.1 to 13.6 $\mu g/m^3$. The highest values were at Cambusbarron (13.6 $\mu g/m^3$) (Site 8). This may have been

Table 10.1 Site name and location of monitoring sites and pollutants monitored

Site name	NO_2	SO_2	PM_{10}
1. Causewayhead	✓		
2. Dunblane	✓		
3. Bridge of Allan	✓		
4. Callander	✓	✓	
5. Lochearnhead	✓	✓	
6. Aberfoyle	✓		
7. Killearn	✓		
8. Cambusbarron	✓	✓	
9. Main Street, Cowie/Redland Tiles	✓		
10. Springkerse	✓		
11. Laurencecroft		✓	
12. Port Street	✓	✓	
13. St Modans	✓		
14. Park Terrace	✓		
15. Lennox Avenue	✓		
16. Ochil View, Cowie	✓	✓	
17. Balfron	✓		
18. Corn Exchange Road			✓

Table 10.2 Mean NO_2 concentrations ($\mu g/m^3$)

Site	*Mean*	*Range*
1. Causewayhead	19.2	7-27
2. Dunblane	15.8	7-27
3. Bridge of Allan	25.9	15-40
4. Callander	23.2	16-54
5. Lochearnhead	2.3	<1-5
6. Aberfoyle	10.5	5-16
7. Killearn	10.5	5-17
8. Cambusbarron	17.4	9-27
9. Main Street, Cowie/Redland Tiles	17.1	6-25
10. Springkerse	19.9	8-32
12. Port Street	28.6	12-41
13. St Modans	15.6	3-35
14. Park Terrace	16.9	4-23
15. Lennox Avenue	14.0	4-27
16. Ochil View, Cowie	19.5	11-30
17. Balfron	10.0	N/A

due to domestic coal burning in the locality. The annual average ranged 0.9 to 4.7 $\mu g/m^3$. Although values could not be directly compared with UK AQS, all results were considerably below the EU guide value (Chapter 7). PM_{10} results showed that the values obtained from the monitoring study (4.4 μg/m^3) were considerably below the UK AQS of 40 μg/m^3.

Table 10.3 Mean SO_2 concentrations (μg/m^3)

Site	*Mean*	*Range*
4. Callander	1.56	<1-3.44
5. Lochearnhead	0.91	<1-1.96
8. Cambusbarron	4.71	<1-13.61
11. Laurencecroft	1.64	<1-2.8
12. Port Street	3.00	<1-5.42
16. Ochil View, Cowie	4.40	<1-2.66

10.3 CASE STUDY 2 - INDOOR AIR QUALITY HEALTH ASSESSMENT

A car wash facility was studied to assess CO levels and their effects on human health (Newton and Harrop, 1987). The car wash was situated in two adjacent railway arches, one arch acting as the motor vehicle entrance and the other as the exit (Fig. 10.2). On entering the premises the vehicles were valeted by employees prior to washing and polishing in the second arch. Throughout the entire cleaning cycle the motor engines were left idling. CO concentrations varied widely according to day and time, traffic flow through the premises and engine operating mode and frequently exceeded the UK Health & Safety Executive (HSE) OEL. The combination of queuing cars and poor ventilation generally provoked the worst conditions for atmospheric pollution concentrations. Previous studies (HSE, 1982) had found that under similar conditions in car parks, CO levels ranged from 50 to 700 ppm in and near cars.

Table 10.4 shows the results from the study. Prior to the installation of the ventilation system (Fig. 10.2) peak hourly levels frequently exceeded 100 ppm. The high 8 hour mean concentrations were calculated to be equivalent to levels of approximately 8.3 COHb% above normal background levels (0.1–1% in a non-smoker and 6–7% in a smoker). It was clear that CO levels in this poorly ventilated workplace (the only ventilation was through doors at one end of each arch) were potentially hazardous to human health as the HSE OEL was frequently exceeded on a number of days. Four options were available to resolve the problem:

- Substitution of process;
- Total enclosure of process;
- Local exhaust ventilation;
- Dilution ventilation.

The problem of elevated CO levels was eventually resolved by using dilution ventilation. The ventilation system that was finally installed was a somewhat hybrid solution following a number of modifications to earlier attempts. The final installation resulted in an almost fourfold reduction in the CO 8 hour concentration.

Table 10.4 Measured CO concentrations (ppm) in car wash facility (Newton and Harrop, 1987)

	Levels prior to air ventilation improvements[a]	*Levels after air ventilation improvements*[b]
CO_1	92	25
CO_2	47	13

CO_1 – average daily peak hourly concentration
CO_2 – average daily peak 8 hr concentration
a) based on 21 days continuous monitoring
b) based on 8 days continuous monitoring

Figure 10.2 Layout of study area and proposed ventilation system to rectify elevated CO levels (Newton and Harrop, 1987)

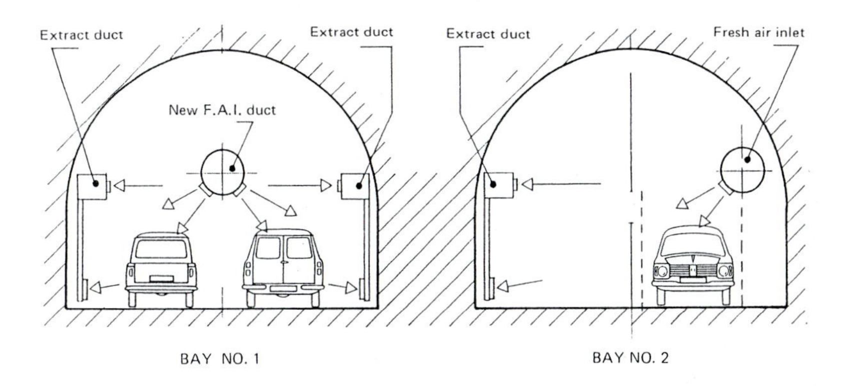

10.4 CASE STUDY 3 - EMISSION INVENTORY

In 1997, under Part IV of the Environment Act, the UK National Air Quality Strategy (NAQS) was published. The Strategy provided a national framework for the reduction of hazards to health from air pollution, outlining standards and objectives for the control and reduction of the main health threatening pollutants. Local Authorities were identified as one of the key contributors to meeting the Strategy's aims and were required to review and assess air quality in their area, taking account of the NAQS's objectives and, where necessary, drawing up an AQAP if proposed standards were being breached or at risk of being exceeded. Emission inventories are a fundamental aspect of the review and assessment process (Chapter 4).

This case study (Cordah Limited, 2000a) details an emission inventory for Falkirk Council, in Scotland, which was undertaken to aid the Council's development of a Local Air Quality Management System (LAQMS). The objectives of the study were:

- To identify the key emission sources of the primary pollutants within the Falkirk area (CO_x, NO_x, SO_x, VOCs and particulates);
- To provide a detailed emission inventory of these major pollutants by source area;
- To highlight data deficiencies where any such instances occurred.

Table 10.5 Sources of information for study area

Authority consulted	*Information required*
Falkirk Council	Number of houses per ward Number of houses utilising each fuel type per ward Number of landfill sites within the Falkirk area and the associated capacity data Commercial units and heating statistics within the Falkirk area
British Gas/Scottish Gas	Average amount of gas consumed per average household
The Coal Board	Average amount of coal consumed per average household
Oil Merchants	Average amount of oil consumed per average household
London Research Centre/UK Department of Trade and Industry	Emission factors for conversion of energy utilised to pollutants emitted for coal, gas and oil
Local Authority or SEPA	Source emission information
BP Chemicals	Ship movement statistics within Grangemouth docks

The emission inventories were based on the current situation, with the exception of road emissions which were projected to the year 2005. For all other categories of activity it was deemed reasonable to assume existing conditions in projecting to 2005 (DETR, 2000a). Area sources were categorised into those from domestic sources (including commercial units), landfill sources and in port shipping sources. Quantification of emissions from domestic sources necessitated the identification of the number of houses/commercial units utilising coal, oil and gas within a housing area (i.e. ward). Emission factors were used to convert the utilisation into actual emissions of pollutants. This process necessitated consultation with various bodies, which are detailed in Table 10.5. To estimate emissions from landfill sites, the number of sites within the district as well as site capacity data was required. Falkirk Council provided information on the number of houses in each ward. The total number of households within the Falkirk area amounted to over 60,000, of which approximately 65% were private and 35% were council houses. Falkirk Council was able to provide a percentage breakdown of the number of Local Authority houses and private houses with central heating and with gas central heating (Table 10.6).

Table 10.6 Percentage (%) breakdown of houses with heating and with gas central heating

Fuel type	*Housing sector*	
	Local authority	*Private*
Central heating	94.7	87.7
Gas central heating	70.1	84.8

Information on further fuel use was available for Local Authority houses only. A breakdown is given in Table 10.7. There was no oil use among these houses.

Table 10.7 Percentage (%) breakdown of fuel use by houses in the Local Authority sector

Fuel type used	*Percentage of total (%)*
Gas	69.0
Coal	4.0
Electricity	24.0
No central heating	1.4
Unknown	0.3

Gas producers were contacted to ascertain a figure on the average amount of gas utilised per household. Unfortunately this information was unobtainable. The UK DETR advisers, London Research Centre, were also contacted and were able to provide figures on gas consumed per grid square (km^2) (1996) across the Falkirk area. This information averaged at 20,574 kw/hr when totalled and this compared well with an estimated average annual utilisation of 19,050 kw/hr. Figures for those houses using coal were available for Local Authority houses (5%). There was no information available for those houses in the private sector, therefore this value was applied for houses in both sectors. Information on the number of houses utilising oil within the private sector was unobtainable. To estimate the amount of pollutants emitted to the atmosphere by the utilisation of each fuel type, the emission factors given in Table 10.8 were applied.

Table 10.8 Emission factors for domestic gas and coal utilisation

Fuel type	*Pollutant emission factor*			
	NO_x	SO_2	CO_2	*CO*
Gas[a]	0.000165473	0.0	0.200034118	0.0000294439
Coal[b]	0.00312	0.01600	2.83873	0.04500

a) The emission factors for the conversion of gas are in kg/kwh

b) Coal is assumed to be Solid Smokeless Fuel (SSF). The emission factors for the conversion of coal utilisation are in tonnes per tonne.

There were no emission factors available for landfill sites from the UK DETR or associated databases. This was primarily due to uncertainties arising over the quantities of waste already landfilled and the amounts and composition of waste disposed to landfill each year.

Table 10.9 shows the total mass emissions (gas and coal) from the domestic sector for each of the study pollutants. Information on fuel emissions from heating in commercial units within the Falkirk area was also unobtainable.

Table 10.9 Total mass emissions (t/km^2/a) from domestic sources

	Mass emissions				
	CO	*CO_2*	*NO_x*	*SO_x*	*Particulate*
Total emissions	432.3766	204915.1	160.3906	144.3552	1.36×10^{-5}

Activity data for regulated industrial processes in Scotland is held by SEPA, who were consulted throughout the project. While it was expected that some of the information from the Part A processes (the significant industrial process) would be available from Public Registers, a questionnaire was also constructed which aimed to obtain the information required for the emission inventory directly from industrial operators and to fulfil the requirements of the data to be inputted into an air dispersion model (e.g. ADMS-3 and ADMS-Urban (Chapter 6)). The information, and the units, requested in the questionnaire included:

- Mass emissions (t/a)
- Source emission height (m)
- Source emission diameter (m)
- Gas emission temperature (°C)
- Gas exit/efflux velocity (m/sec)
- Gas volume (m^3/sec)
- Ordnance Survey (OS) map grid references (m)

A complete list of Part A and B (minor) processes was obtained from SEPA for the Falkirk area. In total, there were 17 companies with Part A processes and 61 companies with Part B processes. All industrial processes were contacted by a standard covering letter and questionnaire. The Public Register was also consulted but the level of detail required for the model data was not met. Table 10.10 provides the emissions levels for the industrial processes.

Table 10.10 Inventory of total emissions from Part A and B processes

	Mass emissions (kt/a)				
	CO_2	*NO_x*	*SO_x*	*VOCs*	*Particulates*
Part A	9712.8	29.000	68.3	16.100	1.80
Part B	6.8	0.008	26.3	00.016	0.05
Total	9719.6	29.000	68.3	16.100	1.90

Line sources were categorised as those emissions from road and train sources. The information sought and the individual bodies consulted is detailed in Table 10.11.

Table 10.11 Source of information for line emissions

Authority consulted	*Information required*
Falkirk Council: Planning & Transportation Services	Traffic count statistics for motorways, A class roads and B class roads Traffic growth statistics Percentage breakdown of HGVs and LGVs by road type.
Scottish Executive	Percentage breakdown of HGVs by road type (motorways) Percentage breakdown of LGVs by road type (motorways)
London Research Centre	Emission factors for road traffic Emission factors for trains
Railtrack	Train count statistics Type statistics

Traffic count and traffic growth statistics were obtained from the Planning & Transportation Services of Falkirk Council. The Annual Average Daily Total (AADT) for 1997 was provided for links on all motorways and A class roads. Traffic count data for various links on B class roads was provided but the data for some links was from 1992 and no AADT data was available for a number of the B class roads. The average percentages of HGVs and buses on A and B class roads were provided by the Planning & Transportation Services but some of these figures dated back to the late 1980s. The percentage of HGVs and buses on the motorways for 1998 was provided by the Scottish Executive.

The methodology used was to divide all roads in Falkirk area into straight line segments and record the start and end point co-ordinates and the length of each segment. The AADT for each of the line segments was obtained from the corresponding link provided by Falkirk Council. If there were any data deficiencies for particular road segments then traffic counts from the nearest similar road section were used. The daily total of traffic for each line segment was divided into the number of HGVs plus buses and the number of LGVs. The emission factors were obtained from the London Research Centre (2000) with the HGV category divided into articulated and rigid vehicles. Therefore, since the number of each type of HGV in Falkirk was unknown the higher emission factors were used in order to obtain the worst case scenario. The composition of the LGVs in Falkirk was assumed to be totally petrol-engined. In order to obtain a prediction of total emissions in years 2000 and 2005 the AADT was multiplied by the national road traffic forecasts for total traffic supplied by Falkirk Council to obtain an estimate of the traffic figures for each year. Emission factors for motorways, urban and rural roads were used for each vehicle category and for years 1997, 2000 and 2005. Table 10.12 presents summary data highlighting the total emissions for LGV and HGV vehicles on A and B class roads.

Table 10.12 Total emissions (kts) for LGV and HGV class vehicles

Pollutant	*Total emissions by road type*				*Total emissions*	
	A Class		*B Class*			
	LGV	*HGV*	*LGV*	*HGV*	*A Class*	*B Class*
CO	18.100	0.250	1.200	0.0200	18.40	1.200
CO_2	245.800	109.100	30.100	8.0000	355.00	38.200
NO_x	1.700	11.200	0.140	0.1000	2.90	0.240
SO_2	0.040	0.035	0.005	0.0030	0.08	0.008
VOC	1.200	0.220	0.135	0.0200	1.40	0.200
PM_{10}	0.032	0.060	0.004	0.0050	0.09	0.008
Benzene	0.005	0.004	0.006	0.0004	0.06	0.007
1,3 Butadiene	0.020	0.002	0.002	0.0002	0.02	0.002

As may be seen from Table 10.13, point sources were the more significant source of emissions. Within the line source category, roads were the most significant source of emissions compared with train emissions. Table 10.14 provides a breakdown of the relative contribution from each source category towards overall emissions. Fig. 10.3 shows the geographical distribution of emissions in the Falkirk area.

Table 10.13 Comparison of mass emissions (t/a) from domestic, point and line sources

Source area	*Mass emissions over the Falkirk District*				
	CO_2	*CO*	NO_x	SO_2	PM_{10}
All source emission breakdown					
Domestic totals	204915.1	432.4	160.4	144.4	1.363×10^{-5}
Point totals	9719600.3		28956.4	68297.6	1867.5
Line totals	393116.6	19827.4	3173.2	131.4	97.5
Line					
Road totals	393116.6	19821.6	3122.8	82.8	96.7
Train totals		5.8	50.4	48.6	0.7
Total emissions	10317632.0	20259.7	32290.0	68573.4	1965.0

Table 10.14 Relative contribution from the various sources to total emissions

Source area	*Percentage contribution to total Falkirk emissions*				
	CO_2	*CO*	NO_x	SO_2	PM_{10}
Domestic totals	2	2	1	0.3	negligible
Point totals	94		89	99.5	95
Line totals	4	98	10	0.2	5

Figure 10.3 Geographical distribution of emissions in the Falkirk area (Cordah Limited, 2000c) (courtesy of the Environmental Health Department, Falkirk Council)

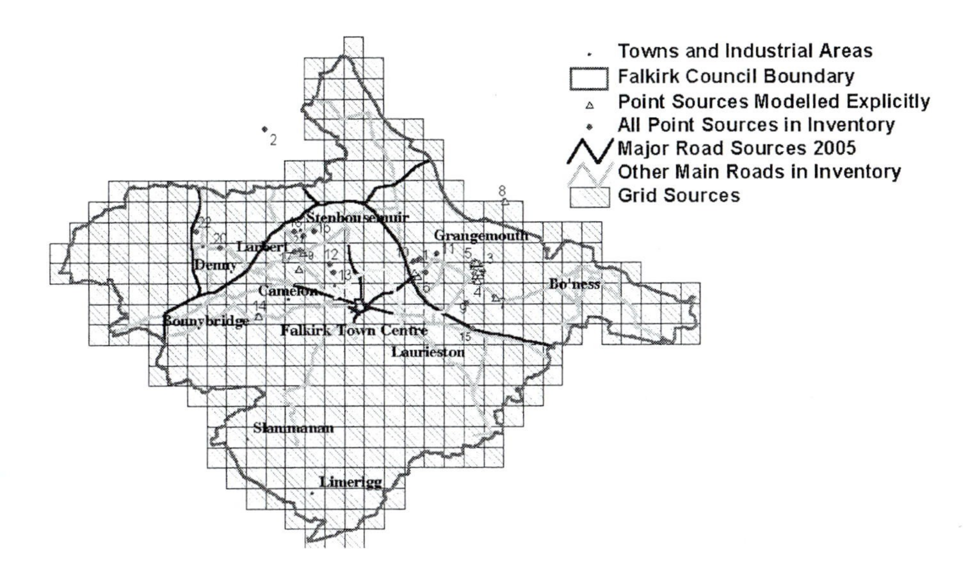

10.5 CASE STUDY 4 - BASELINE AIR QUALITY MONITORING STUDY

This case study details a baseline monitoring study undertaken for a proposed sewage sludge incinerator in Northern Ireland. Initially an evaluation of baseline air quality data for the study area in the harbour region of Belfast was undertaken. The purpose of the exercise was to evaluate existing air quality data and recommend additional monitoring requirements (Aspinwall & Company, 1994).

Baseline air quality data for Belfast at the time of the study in 1993 was limited primarily to SO_2 and smoke data measured as part of the UK Smoke and Sulphur Dioxide Monitoring Network where seven sites were operational. In January 1992, a new urban air quality monitoring network, funded by the then UK DoE became operational. Initially SO_2, CO, O_3, NO_2 and TSP were monitored. Lombard Street (Belfast 2) was the location of one of the continuous monitoring sites, another site (Belfast 1) measured SO_2 continuously. Two sites monitoring SO_2 and smoke (Belfast 33 and Belfast 41) and a NO_2 diffusion tube survey site (Dargan Road) were situated in close proximity to the development site in Duncrue Street. Fig. 10.4 shows the location of the air quality monitoring stations.

10.5.1 Sulphur dioxide and smoke

Table 10.15 shows the annual median SO_2 and smoke concentrations for the Belfast area. SO_2 levels measured at Site 42 (Shankill Road) exceeded the EC Directive on air quality limit and guide values for SO_2 and smoke (Chapter 7). The

Figure 10.4 Location of air quality monitoring sites around a sewage sludge incinerator development (not to scale)

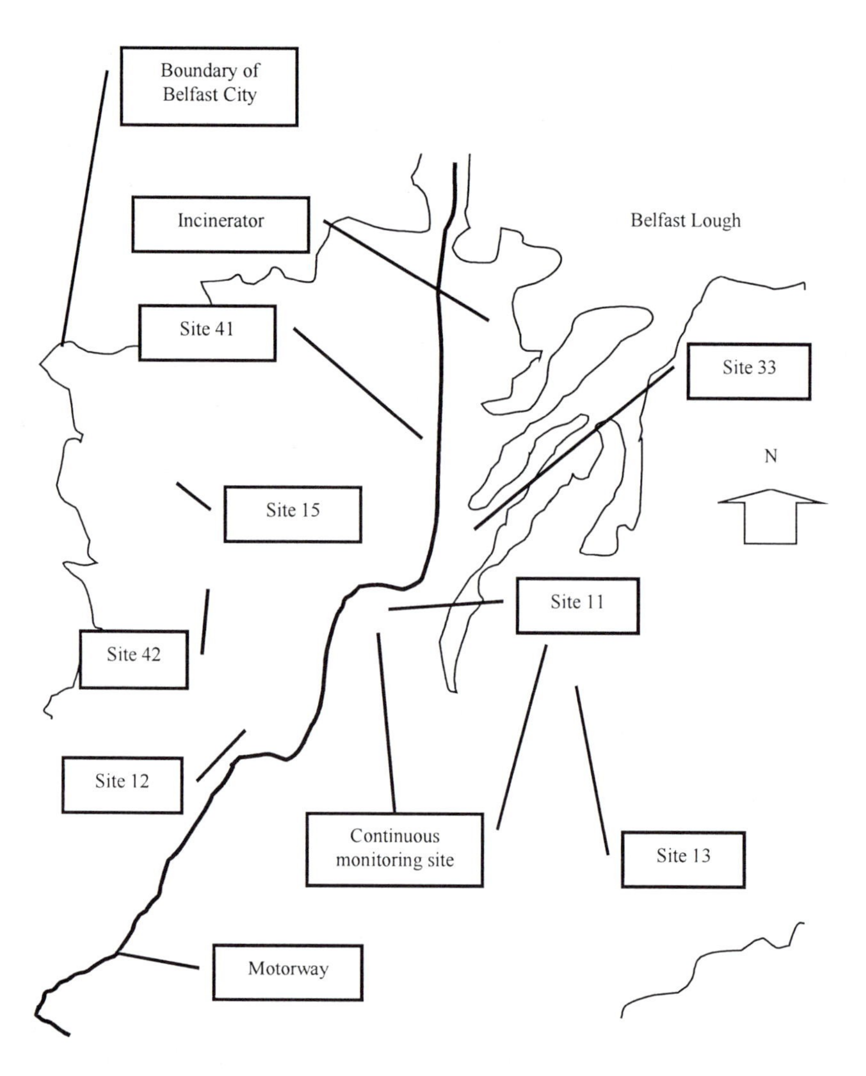

SO_2 98th percentile level of 340 μg/m^3 exceeded the limit value of 250 μg/m^3. SO_2 levels measured at Site 13 were marginally below the EC Directive 98th percentile limit value. SO_2 and smoke levels recorded at the two sites closest to Duncrue Street (Sites 33 and 41) were below the Directive's air quality limit value, although the 24 hour guideline was exceeded.

Table 10.16 shows that the general trend in ambient SO_2 and smoke levels was declining. The trend was more discernible for the period 1960-1992 than for the last 5 to 10 years, particularly for winter mean SO_2 and smoke concentrations. Results from the continuous monitoring station for SO_2 (Belfast 1), are summarised in Table 10.17 and showed that the then UK DoE air quality bands for good air quality occurred on 112 days and 343 days respectively for 1989/90 and 1990/91. Available data for the Belfast 2 monitoring station are shown in Table 10.18, however, the monitoring period was for less than a year and therefore comparison with the EU Directive AQOs was not possible. Since the EU Directive (80/799) for smoke and SO_2 came into operation in 1983 Belfast had exceeded the 98th percentile limit values for smoke and/or SO_2 every year until the study period year. Annual average smoke and SO_2 levels for North Down for 1991 (January to December) were 9.7 and 36.8 μg/m^3 respectively. Data for 1992 was incomplete. The annual average SO_2 levels for Carrickfergus for 1990-91 were 11 μg/m^3.

Table 10.15 Annual median (maximum daily) SO_2 and smoke levels (μg/m^3) in the Belfast locality 1991 - 92 for selected sites (Belfast City Council Environmental Health Department, 1993)

Site	*SO_2*			*Smoke*		
	Median	*Maximum daily*	*98th percentile*	*Median*	*Maximum daily*	*98th percentile*
Belfast 11	21	129	69	12	194	84
Belfast 12	39	251	178	12	264	156
Belfast 13	44	388	235	15	210	102
Belfast 15	25	169	105	15	283	137
Belfast 33	28	266	125	17	216	121
Belfast 41	28	311	198	10	162	104
Belfast 42	52	475	340	16	257	147

10.5.2 Nitrogen dioxide

A NO_2 diffusion tube survey in 1986 had recorded average levels in the Belfast area of 18.8-34.2 μg/m^3 at 9 sites. NO_2 levels ranged from 4.5-50.8 μg/m^3. Table 10.19 summarises the results from a NO_2 diffusion tube survey in Belfast in 1986 and 1991. Results showed that NO_2 concentrations had increased over the 5 year period by 11.7 to 101.7%. At the time of the study NO_2 levels were being monitored using NO_2 diffusion tubes at a number of locations in Belfast.

Results for 8 sites showed that levels ranged 22.0-73.5 μg/m^3 and averaged 50.8 μg/m^3 (Belfast City Council Environmental Health Department, 1993). A survey of NO_2 concentrations of 9 sites in 1988 in the Belfast Harbour Estate area of locations close to known emitters of NO_x monitored levels of 45.1-90.2 μg/m^3 (Industrial Science Division, Department of Economic Development, 1988). The highest concentrations were found north of Richardson Fertilizers and West Twin Power Station, both known sources of NO_x. A subsequent 4 week survey (Industrial Science Division, Department of Economic Development, 1990) in 1989 of 31 sites monitored an average level of 47 μg/m^3. Levels ranged from 1.9-120.3 μg/m^3. Average concentrations at each site ranged from 28.2-86.5 μg/m^3. A survey undertaken over a 3 month period in 1991 (Industrial Science Centre, 1991) of NO_2 levels around Kilroot Power Station and the Carrickfergus area observed a range of 1.9-26.3 μg/m^3 with a mean level of 7.5 μg/m^3.

Table 10.16 Ambient median SO_2 and smoke[a] levels (μg/m^3) for Belfast 1983 - 92 for selected sites (Belfast City Council Environmental Health Department, 1993)

Site	*Pollutant*	*Year(1983 – 1992)*								
		83 - 84	*84 - 85*	*85 - 86*	*86 - 87*	*87 - 88*	*88 - 89*	*89 - 90*	*90 - 91*	*91 - 92*
Belfast 11	SO_2 98th	116	211	133	213	190	*	187	249	69
	median	38	52	49	44	61	*	39	35	21
	smoke 98th	97	131	90	113	110	*	183	251	84
	median	17	19	20	20	23	*	16	13	12
Belfast 12	SO_2 98th	145	142	129	139	181	128	*	237	178
	median	49	50	50	44	53	44	*	59	39
	smoke 98th	163	243	265	162	165	156	*	251	156
	median	24	26	25	28	20	18	*	20	12
Belfast 13	SO_2 98th	127	226	201	345	201	152	342	246	235
	median	46	68	77	70	91	48	45	45	44
	smoke 98th	74	83	83	166	113	113	194	131	102
	median	21	21	23	32	29	22	22	19	15
Belfast 15	SO_2 98th	106	199	*	235	165	106	127	122	105
	median	41	57	*	48	70	37	31	31	25
	smoke 98th	47	72	*	149	195	124	199	165	137
	median	17	17	*	25	30	18	18	18	15
Belfast 33	SO_2 98th	125	258	145	193	201	143	108	158	125
	median	41	61	52	53	61	41	36	28	28
	smoke 98th	97	168	117	81	117	145	160	193	121
	median	26	26	17	21	20	16	16	20	17
Belfast 41	SO_2 98th					*	231	195	215	198
	median					*	42	28	34	28
	smoke 98th					*	175	141	168	104
	median					*	15	18	12	10
Belfast 42	SO_2 98th							*	374	340
	median							*	47	52
	smoke 98th							*	261	147
	median							*	19	16

** insufficient data*

a) The smoke concentrations are for the BS method as used in the UK. Concentrations given in the EC Directive relate to the OECD method; the relationship between concentrations determined using OECD and BS calibration is: OECD = BS divided by 0.85

98th - 98th percentile

The monitored air quality data for Belfast showed that the levels of air pollution breached the recommended EC directive for SO_2 and smoke. The EC 98th percentile air quality directive (80/779) for SO_2 of 250 μg/m^3 was exceeded in 1989/90, (509 μg/m^3) and 1990/91, (326 μg/m^3) at the continuous monitoring site. The EC Directive 98th percentile 'trigger' value for smoke (127 μg/m^3) was generally exceeded at 3 sites and thus imposed a more stringent 98th percentile

Table 10.17 SO_2 levels (μg/m^3) from the UK air quality monitoring network, Belfast (Broughton *et al.*, 1992)[a]

	1990/1991	*1989/1990*
Geometric mean	60.1	65.8
Arithmetic mean	94.4	103.0
Median	57.2	57.2
98th percentile	497.6	649.2
98th percentile[b]	509.0	326.0
Data capture (%)	89.0	55.0

a) Pollution year April 1990 - March 1991
b) 98th percentile of daily means

Table 10.18 Air quality levels (μg/m^3) (October to December 1992) for Belfast 2 continuous monitoring station (WSL, 1993)

Pollutant	*Average*	*98th percentile*	*Maximum 1 hr*	*Maximum 8 hr*	*Maximum 24 hr*
O_3	30.0	66.0	80.0	72.0	50.0
NO_2	47.0	116.6	201.2	131.6	97.8
CO[a]	1.2	6.3	20.7	11.2	6.3
SO_2	74.4	366.1	1101.1	554.8	328.9

(a) mg/m^3

Table 10.19 Average NO_2 levels (μg/m^3) for Belfast 1986 and 1991[a] for selected sites (Belfast City Council Environmental Health Department, 1993)

Site	*1986*	*1991*	*Change (%)*
Belfast 11	33.7	44.6	32.4
Belfast 12	33.8	37.8	11.7
Belfast 13	31.0	44.9	44.9
Belfast 14	25.7	30.8	19.7
Belfast 15	34.2	48.1	40.6
Belfast 18	19.7	32.5	64.8
Belfast 33	22.4	45.1	101.7
Belfast 39	30.8	41.7	35.4
Belfast 17	22.9	-	-
Belfast 40	18.8	28.0	49.0

a) Survey period July - December in 1986 and 1991
- No data

limit value for SO_2 of 250 μg/m^3. Of the seven sites monitoring SO_2 on a daily basis in Belfast (Table 10.15), Site 42 (Shankill Road) exceeded the 98th percentile limit value. The then WHO 1 hour guideline of 350 μg/m^3 was exceeded 274 times on 37 days in 1989-90 and 269 times on 52 days in 1990-91. Data for 1991-92 was unavailable. The then annual and daily WHO guideline for terrestrial effects of 30

$\mu g/m^3$ and 100 $\mu g/m^3$ and the UNECE annual and daily guidelines of 21.5 $\mu g/m^3$ and 75 $\mu g/m^3$ were exceeded in 1989-90 and 1990-91. The WHO CO 8 hour guideline of 10 mg/m^3 was also exceeded in 1992. The NO_2 data for Belfast was not suitable for comparison with the EC Directive (85/203).

10.5.3 Monitoring

Owing to the limited air quality data for the locality following the baseline data review a detailed air quality monitoring survey was conducted for the locality to supplement existing air quality data. Baseline monitoring was undertaken at the Duncrue Street site and also up wind and downwind of the site. The study involved the use of six accredited laboratories. Preliminary air quality monitoring was undertaken from March to June 1993. Following this initial study a longer period of monitoring for one year was instigated. The monitoring programme was agreed to by the then Northern Ireland Radiochemical and Alkali Inspectorate. The study comprised the following:

- SO_2 and NO_2;
- Dioxins and furans;
- HCl and HF;
- Heavy metals (Cd, Cr, As, Pb, Hg, Cu, Ni, Mn, Sn, Co, Tl, Sb, V, Zn);
- VOCs;
- TSP.

The additional data collected by the monitoring study allowed for a greater appreciation of air quality issues in the development area when used with existing baseline data. The results of the study showed that, where comparison could be made, levels were similar to concentrations monitored elsewhere in the Belfast area. However, elevated SO_2 levels were noted and were attributed to the sulphurous emissions from the lagoon close to the monitoring site. Monitoring results are shown in Table 10.20. The baseline air quality monitoring study was used to compliment air dispersion modelling studies, and both were used for an environmental impact assessment (Harrop and Nixon, 1999).

10.6 CASE STUDY 5 - AIR DISPERSION MODELLING STUDY

An air dispersion modelling study was conducted for gaseous and particulate emissions from a proposed energy from waste (EfW) plant. The air dispersion model used in the study was ADMS-3 (Chapter 6). The study assessed the effects of building downwash on plume dispersion and also the effects of surrounding topography (Energy From Waste Ltd, 1999).

The stack at the proposed plant had an exit height of 35 to 50m above ground level. A stack height optimisation study was undertaken to model four stack heights of 35, 40, 45 and 50m. Modelling parameters are given in Table 10.21. Pollutant emissions were modelled as a continuous plume with mass emissions in

Table 10.20 Baseline air quality monitoring results (μg/m³) (Aspinwall & Company, 1994)

	Mean	*Range*		*Mean*	*Range*
TSP	48.9	30-60	Mn	0.0135	0.0065-0.0232
SO_2	139.8	8.6-514.8	Ni	0.0153	<0.001-0.049
NO_2	51.5	32.0-71.4	V	0.039	0.012-0.074
HCl	3.49	1.7-6.3	Tl	<0.001	<0.001
HF	0.11	<0.05-0.31	Co	0.0011	<0.001-0.002
Cd	0.0019	<0.001-0.0043	Cu	0.015	<0.001-0.028
Hg	<0.001	<0.001	Zn	0.068	0.0041-0.142
As	0.0034	0.0011-0.0068	Sn	0.0016	<0.001-0.004
Pb	0.105	0.008-0.285	Sb	0.0094	0.0016-0.0348
Cr	0.01	<0.001-0.0186	VOC	144.0	40-280
Dioxins & furans	0.12				

grammes per second (g/sec). The modelling averaging time was set at 1 hour. The default specific heat capacity (1012 J/°C/kg) and molecular weight (28.96 g/mol) of air was used for the exit gas because dilution with air is almost instantaneous after emission of the buoyant plume pollutant gas mix from the stack. Ground level concentrations were predicted for a receptor grid with a 100m grid interval resolution. The modelling study predicted ground level concentrations within a 2 to 3 km grid square of the plant.

A ten-year statistical meteorological data set for the period 1988 to 1997 was input to the model. The meteorological data station is located at Eskdalemuir. The development site lay to the west of Eskdalemuir. Eskdalemuir was considered to be the most appropriate data set for the study area following discussions with the UK Meteorological Office. The alternative meteorological stations were considered to be unrepresentative of the locality. The wind direction is predominantly from the south and south west. A surface roughness value of 0.3m was considered representative of the area surrounding the installation which was principally agricultural in nature. The air dispersion modelling study assumed that all NO_x emissions were in the form of NO_2. The predicted values are given in Table 10.22 (for 1 g/sec) for a 50m high stack and Table 10.23 shows maximum predicted ground level concentrations for varying stack heights.

10.7 CASE STUDY 6 - QUANTITATIVE HEALTH RISK ASSESSMENT

The study (Aspinwall Enviros, 2000) involved identifying the principal air pollutants released from a proposed EfW facility in the People's Republic of China (Hong Kong) which, based on current knowledge, were of potential risk to human health via inhalation.

For the purposes of the assessment the substances of concern were classed under two categories, depending on whether they were hazardous by carcinogenic or other mechanisms. Some substances were included in both categories. Prior to undertaking the air quality health risk assessment an assessment was undertaken of

Table 10.21 Emissions data for the proposed EfW plant

Parameter	*EfW plant*	
	mg/Nm^3	*g/sec*
Particulates	10	0.3
VOCs	10	0.3
HCl	10	0.3
HF	1	0.03
CO	50	1.4
SO_2	50	1.4
NO_2	200	5.4
Cd and Tl (Total)	0.05	0.0014
Hg	0.05	0.0014
Sb, As, Pb, Cr, Co, Cu, Mn, Ni, V (Total)	0.5	0.014
Dioxins and furans (ng/Nm^3)	0.1	0.002 μg/sec
Stack exit height (m)	50	
Stack flue diameter (m)	1.90	
Stack gas exit temperature (°C)	145	
Gas velocity (m/sec)	15	

Table 10.22 Maximum predicted ground level concentrations (μg/m^3) in study area for a 50 m stack

	Predicted concentration[a]
100th percentile (maximum 1 hour)	19.7
99.9th percentile	15.5

(a) *Based on an emission rate of 1 g/sec, values should be multiplied by emissions given in Table 10.21 to derive specific pollutant concentrations*

Table 10.23 Maximum predicted ground level concentrations (μg/m^3) as a function of stack height

	Concentrations[a]			
	Stack 35m	*Stack 40m*	*Stack 45m*	*Stack 50m*
Long term mean (annual)	4.0	1.2	0.8	0.6
100th percentile (maximum 1 hr)	60.5	39.3	21.5	16.3
99.9th percentile	58.2	36.9	20.6	15.1
99th percentile	43.0	23.1	17.0	12.9
98th percentile	33.3	16.3	12.4	9.3

a) *Based on an emission rate of 1 g/sec, values should be multiplied by emissions given in Table 10.21 to derive specific pollutant concentrations*

existing baseline air quality levels in the locality of the proposed development. ASRs were selected in accordance with Section 2 of Annex 12 of the Hong Kong Technical Memorandum on EIA Process (HKEPD, 1997). In selecting ASRs, residential areas were identified, followed by hospitals, places of public worship, sport stadiums and other premises denoted in the Memorandum. Once these ASRs had been identified in the local vicinity of the proposed development, other potential ASRs were selected (i.e. golf course, parks). Finally, other discrete receptors expected to be heavily impacted by the development's emissions were also identified. Computer and physical air dispersion modelling studies were undertaken to predict ground level concentrations from the flue gas emissions of the development at the ASRs. Exposure intakes were calculated for individuals at the point of maximum ground level air concentrations and at identified worst affected discrete receptors. Default values for receptor characteristics are provided in Table 10.24 and reflected current usage for exposure assessments.

Table 10.24 Default values for receptor characteristics

Exposure parameter	*Unit*	*Scenario*
Body weight	kg	70
Inhalation rate	m^3/day	20
Life time	yr	70

Table 10.25 Maximum predicted long-term annual ground level concentrations (mg/m^3)

Pollutant	*Maximum long-term ground level concentration*
Dioxins	7.50×10^{-13}
Total heavy metals	3.70×10^{-6}
Cd	3.70×10^{-7}
Cr^{VI}	4.73×10^{-8}
Be	1.30×10^{-10}
Ni	7.80×10^{-8}
As	3.20×10^{-9}

Table 10.25 shows the predicted increases in maximum long-term (annual) ground level concentrations (mg/m^3) for the carcinogenic pollutants of interest. Specific data were unavailable for all of the heavy metal pollutants, they were therefore grouped together under the heading 'total heavy metals'. Table 10.25 also provides the predicted ground level concentrations for known heavy metals. Input data and assumptions for equations (7.1) to (7.3) are given in Tables 10.24 and 10.26. The data provided in Table 10.26 was sourced from the USEPA Integrated Risk Information System (IRIS) and from consultations with the USEPA IRIS Risk Information Hotline.

Table 10.27 provides the estimated lifetime risk due to exposure to emissions from the proposed development for selected operational scenarios. The total incremental risk from all assessed pollutants was 1.89×10^{-8}. This corresponded to a risk of 0.02 per million per year. Table 10.28 shows the calculated risks for proposed incinerator studies based on human inhalation of plant emissions (Harrop and Pollard, 1998). The risk from incineration plant was shown to be insignificant. Comparison with the HKEPD preliminary health risk guidelines (Table 7.16) showed that the health risks were well within recommended levels. The same observations were also seen when applying the WHO unit risk criteria. The calculated annual health risk was also therefore within the recommended levels set by both the USEPA and UK (Chapter 7, Section 7.4).

Table 10.26 Inhalation cancer potency factors

Pollutants	*Cancer potency factor (kg day/mg)*
Dioxins	1.5×10^{5}
Cd	6.3
As	15.1
Cr^{VI}	42.0
Be	8.4
Ni	0.84 - 1.68

Table 10.27 Increased lifetime risk of contracting cancer derived from USEPA IRIS cancer potency factors and WHO unit risk factors

Pollutant			*Lifetime risk*
Dioxins			3.20×10^{-8}
Cd			6.70×10^{-7}
As			1.36×10^{-8}
Cr^{VI}			5.68×10^{-7}
Be			3.10×10^{-10}
Ni			3.74×10^{-8}
Total risk per lifetime			1.32×10^{-6}
Total risk per year			1.89×10^{-8}
WHO criteria	*Unit risk ($\mu g/m^3)^{-1}$*	*Annual concentration ($\mu g/m^3$)*	*Total risk per lifetime*
As	1.5×10^{-3}	8.2×10^{-6}	1.2×10^{-8}
Cr^{VI}	4.0×10^{-2}	1.2×10^{-4}	4.8×10^{-6}
Ni	3.8×10^{-4}	2.0×10^{-4}	7.6×10^{-8}

Table 10.28 QRA predictions for proposed incineration plants in the UK based on inhalation of plant emissions (Harrop and Pollard, 1998)

Type of incinerator	*Level of annual risk (range where available)*
Waste-to-energy plant	1 in 33,000,000 to 55,000,000 (Bexley)
	1 in 26,000,000 (Belfast)
	1 in 26,900,000 (Thameside)
	1 in 15,900,000 (Knostrop)
Sewage sludge	1 in 2,400,000 to 5,500,000 (Belfast)
	1 in 12,500,000 to 58,000,000 (Calder Valley)
Clinical waste	1 in 9,000,000 to 38,900,000 (Knostrop)
Industrial waste	1 in 7,800,000 to 12,100,000 (Teeside)

10.8 CASE STUDY 7 - AIR QUALITY MANAGEMENT

In Scotland, Falkirk Council submitted a combined Stage 1 and 2 Review and Assessment report to the Scottish Office in 1998 to meet the requirements of the Environment Act 1995. The main conclusions of that report were:

- Monitored concentrations of benzene, CO and 1,3-butadiene were well within AQSs and that exceedences of the AQSs in the relevant future year were unlikely. However, it was decided to obtain more detailed monitoring data near potential sources for these pollutants in order to confirm the expected decrease in concentrations in the future. This monitoring data would be considered in the next round of Review and Assessment before 2003;
- There was a low risk of exceedence of the Pb objective in 2005 in an area where there was a potential for human exposure;
- There were exceedences of the annual mean objective for NO_2 and the 15 minute mean objective for SO_2 in areas where there was a risk of human exposure and therefore a risk of exceedences of these air quality objectives by the relevant future year;
- There were no exceedences of the then 24 hour rolling average objective for PM_{10}, but the AQSs had changed since presenting the initial reports and now referred to a 24 hour average and an annual mean concentration. PM_{10} was also only automatically monitored at one station and so it was necessary to obtain more data in order to assess the levels across the Falkirk Council area;
- It was decided to proceed to a Stage 3 Review and Assessment for NO_2, SO_2 and PM_{10}.

This case study outlines the current and predicted future air quality for the three pollutants and concludes by outlining one area within which it was possible that an AQMA may require to be designated for SO_2 (Cordah Limited, 2000c). A review of baseline air quality data for the study area and AQS revealed the following appraisal. Fig. 10.5 shows the location of the air quality monitoring sites in Falkirk.

Recent monitoring data from the diffusion tubes at 55 locations for the period April 1999 to March 2000 showed that there were two exceedences of the annual mean objective for NO_2 of 40 $\mu g/m^3$. These were at the A80 Northbound Carriageway at Banknock (55.4 $\mu g/m^3$) and in West Bridge Street in Falkirk (43.9 $\mu g/m^3$). The risk of human exposure was only applicable at the latter site in the Town Centre. Recent NO_2 monitoring data obtained from the automatic analysers for 1998 and 1999 at Falkirk Hope Street and Grangemouth Municipal Chambers showed that the annual average objective was not exceeded at either location for either year.

Recent SO_2 monitoring data obtained from the automatic analyser at Grangemouth Municipal Chambers for 1997-1999 showed that the number of exceedences of the 15 minute mean objective for SO_2 of 266 $\mu g/m^3$ dropped from 467 $\mu g/m^3$ in 1997 to 223 $\mu g/m^3$ in 1998 and increased slightly to 228 $\mu g/m^3$ in 1999. The dramatic reduction in exceedences between 1997 and 1998 corresponded with the installation of a second sulphur recovery plant at BP Power Station in the Grangemouth petrochemical complex. There were current

Figure 10.5 Location of monitoring stations in the Falkirk Council area (courtesy of the Environmental Health Department, Falkirk Council) (base map reproduced by kind permission of Ordnance Survey © Crown Copyright NC/01/24240)

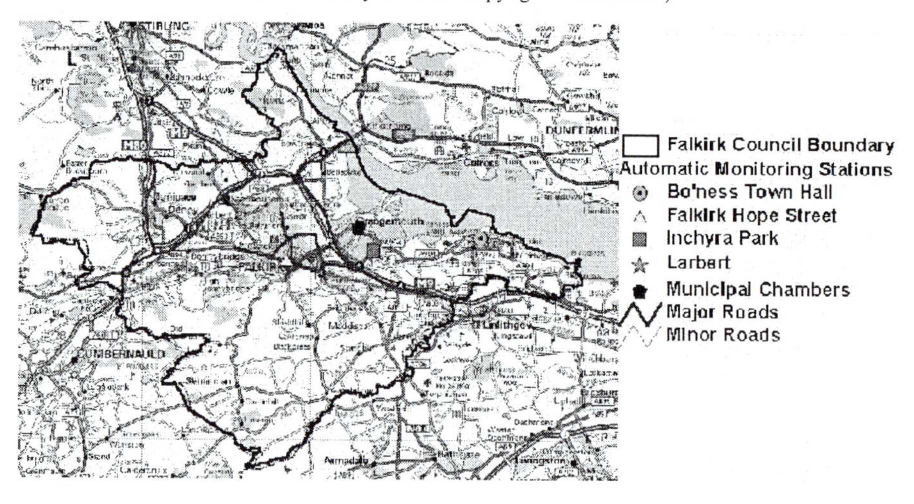

Mobile monitoring station in Falkirk

Continuous monitoring station in Falkirk

exceedences of the 15 minute mean objective of 266 μg/m^3 for SO_2. There were no current exceedences of the 24 hour mean objective of 125 μg/m^3 or the 1 hour mean objective of 350 μg/m^3 for SO_2.

To date (2000) there are no recorded exceedences of the annual mean PM_{10} objective of 40 μg/m^3 across the area. There were therefore no current exceedences of the 24 hour mean objective of 50 μg/m^3 or the annual mean objective of 40 μg/m^3 across the Falkirk Council area.

The description of the Falkirk area is detailed in Section 10.4. The emission inventory aspect of the study is also detailed. The emission inventory detailed the following key sources:

- Area sources - gas and coal domestic heating emissions;
- Point sources - EPA 1990 Part A and B processes;
- Line sources - LGV and HGV emissions from motorways, A and B class roads and passenger and freight train emissions.

The main data deficiency areas were from sub-sources within the area sources, notably landfill site emissions and heating emissions from commercial units and indeed from within the Part A and B processes themselves. The lack of data on those private houses using oil was also another gap. Gathering data from these areas is problematic and it was difficult to foresee how it could be overcome. Estimating emissions from heating systems within commercial and industrial units could be achieved through the questionnaire process. This, however, would be a time consuming process and it was questionable whether it would be a wise use of resources for what would undoubtedly be an insignificant source of emissions. Table 10.13 (Section 10.4) shows a comparison of the emissions from the various sources. As may be seen from Table 10.13, point sources were a more significant source of emissions. Within the line source category, roads were the most significant source of emissions compared with train emissions. The modelling study was carried out with four main objectives:

- To compare modelled predictions with monitored data obtained by Falkirk Council;
- To quantify the likely impact on concentrations within the Falkirk Council area of emissions from a major source outside the Falkirk Council area i.e. Longannet Power Station in the Fife Council area;
- To predict the likely air quality in the future to identify if exceedences of any of the AQSs outlined in the UK NAQS (DETR, 2000f) would be expected by the relevant future year for compliance;
- To assist in the decision making of the necessity for AQMAs within the Falkirk Council area.

The modelling component of the Stage 3 Review and Assessment was carried out using the dispersion model ADMS-Urban version 1.53 (Chapter 6). ADMS-Urban is categorised as an advanced model in the technical guidance note LAQM.TG3(00) (DETR, 2000f). ADMS-Urban is a new generation multiple source dispersion model. Specific features include the ability to treat both wet and dry deposition, building wake effects and complex terrain. ADMS-Urban allows for the use of point, line, area and volume sources as well as 1 km^2 grid sources and individual road sources and also includes options to model street canyons and the photochemical reactions between NO_x, NO_2 and O_3. The model can predict both long-term and short-term concentrations and is able to calculate pollutant concentrations for the averaging periods and percentiles set out in the NAQS. A number of validation studies have been carried out with ADMS-Urban (HMIP, 1995a; McHugh, Carruthers and Edmunds, 1997; Carruthers *et al.*, 1998; Cambridge Environmental Research Consultants Ltd, 2000). Data can be entered into the model directly, or the data can be stored in an emissions inventory database and linked to the ADMS-Urban model and a GIS. For the study, the model was linked to a geographical information system (GIS), ArcView, which was used to prepare the results on map data. Modelling scenarios were set up by a combination of the two methods. The modelling protocol adopted for the Stage 3 Review and Assessment is outlined below:

- Organise the emissions inventory into a Microsoft Access database and link it to the ADMS-Urban dispersion model;
- Carry out initial screening calculations of emissions from all sources within the emission inventory;
- Carry out one run using one full year of hourly sequential meteorology for all sources within the emissions inventory;
- Repeat the process with a different year of meteorological data to assess the annual variability in concentrations owing to changing meteorological conditions;
- Use the meteorological data giving the highest predicted concentrations in all subsequent runs to model worst case scenarios;
- Obtain information on new developments (e.g. roads, housing etc.) and compile an emission inventory for the calculations for the future emissions for different years;
- Use the results to identify areas of potential exceedences for relevant future years. Model those areas in greater detail where there was significant risk of human exposure;
- Carry out a validation exercise comparing predicted concentrations at specified receptors to the monitored data obtained from the same sites;
- Plot contours of concentrations in areas where AQS were predicted to be exceeded;
- Draw areas which correspond to likely AQMAs.

It was essential to validate predicted concentrations from the dispersion model with actual monitored data. Validation exercises were carried out in the study for the pollutant of concern. The method used in order to assess the uncertainty of the modelled results was taken from the guidance note LAQM.TG3 (00) (DETR *et al.*, 2000g). The monitoring sites were divided into categories representing their location classification e.g. roadside, urban background etc. The annual average modelled concentrations were compared to the annual average monitored results and the difference calculated. The square root of the mean sum of the squared differences between predicted and measured concentrations was calculated (the RMS value) to give the magnitude of the error of the predicted concentrations. The results for each category of site using the Derwent-Middleton correlation (Derwent and Middleton, 1996) are shown in Table 10.29. It can be seen that the modelled predictions of the annual mean NO_2 concentrations were well within 50% of the monitored predictions for all types of sites. It was therefore considered that the model was performing well in its predictions of NO_2 concentrations and modelled concentrations were therefore assumed to be reliable for the predictions of future air quality in the Falkirk Council area.

There were significantly less monitoring sites in the study area for SO_2 than there were for NO_2, therefore the method of validation used for NO_2 was not applicable. In an effort to calculate the accuracy of the modelled predictions of SO_2 concentrations, a number of comparisons were made. First, the calculated 99.9th percentile using 1998 meteorological data in the model at the location of the monitoring station at Grangemouth Municipal Chambers was compared to the peak 15 minute concentration measured there for 1998. The model predicted 99.9th percentile concentration of 15 minute averages at this location was 228 $\mu g/m^3$ and the maximum recorded 15 minute concentration was 1442.2 $\mu g/m^3$. These parameters were also compared for the Falkirk Hope Street site. The model predicted 99.9th percentile concentration of 15 minute averages was 166 $\mu g/m^3$ and the maximum recorded 15 minute concentration was 395 $\mu g/m^3$. The maximum predicted 99.9th percentile concentration was 550 $\mu g/m^3$, (which still indicated that the 15 minute objective concentration limit of 266 $\mu g/m^3$ was likely to be exceeded), but it occurred almost 900m to the east of the Grangemouth Municipal Chambers, very close to the main sources of SO_2 emissions.

On examination of the results it was seen that the meteorological condition predicting the highest concentrations in the model was a convective condition with a north-east wind which occurred very rarely. This was causing the plumes to spread rapidly in the vertical direction and have maximum ground level impact very close to the sources in the north-east direction. When observing the wind rose for Falkirk for 1999 (Fig. 10.6) it was considered that it would be more typical to have an east-north-east wind which, if modelled, would direct emissions closer to the monitoring station at Grangemouth Municipal Chambers. It was also known that wind speed and direction measured in the Firth of Forth or at coastline locations could differ quite significantly from those measured just a few hundred metres inland. It was therefore likely that there were coastline effects, which were not being considered in the model. Other differences may be due to the fact that when monitoring concentrations over short averaging periods, such as the 15 minute period required for SO_2, fluctuations in concentration were observed, particularly in convective meteorological conditions, due to the random nature of the turbulent motions in the atmosphere. In the current version of ADMS-Urban (1.53) it was not possible to model fluctuations, the 15 minute average calculation only takes account of one component of variability which is the meandering of the mean wind direction. If it were possible to model fluctuations in the model then the predicted maximum concentrations would have been higher than they were in the study.

In an effort to correlate the modelled predictions with known locations of exceedences of the 15 minute objective concentration of 266 $\mu g/m^3$, it was assumed that a recorded exceedence at the Municipal Chambers corresponded to a measured concentration of 266 $\mu g/m^3$ (although it was sometimes higher than this value).

Figure 10.6 Wind rose for Edinburgh Turnhouse (Data from Meteorological Office, 2000)

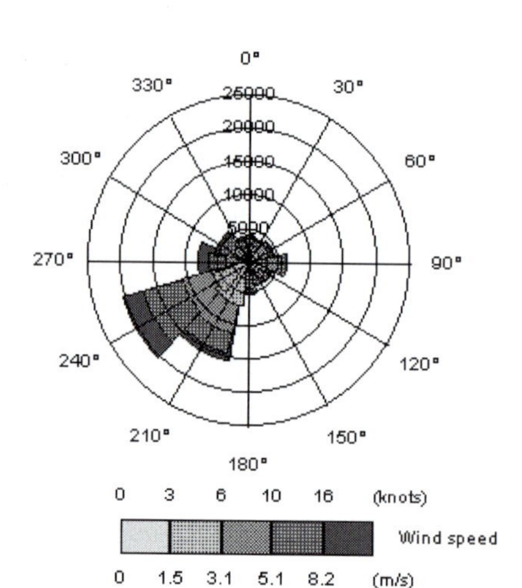

This was then compared to the modelled concentration at that location plus the background taken from the NETCEN background maps (AEA Technology, 2000b), of 238.6 μg/m^3, giving an error margin of 10.3%. Additional comparisons were made with the predicted and actual annual average concentrations measured at Grangemouth Municipal Chambers and Falkirk in order to obtain more information about model accuracy. At Grangemouth Municipal Chambers, the annual average concentration for 1998 was 16.7 μg/m^3 compared with a modelled prediction plus background of 20.1 μg/m^3. This was an over-prediction of 20%. At Falkirk Hope Street, the annual average concentration for 1998 was 17.2 μg/m^3 compared with a modelled prediction plus background of 16.1 μg/m^3. This was an under-prediction of 6%. The model predictions for the annual average were well within 50% of the monitored concentrations and so the model could be considered to be performing well in its calculations of the annual average at each location. Taking all of these comparisons into consideration and noting that the model can be under and over predicting then the largest margin of error calculated was 20%. This error margin was assumed as a worst case to be an under-prediction of the 99.9th percentile of 15 minute predicted concentrations of SO_2. It was applied when defining contours of predicted exceedences of the 15 minute objective in order to present a worst case geographic area within which exceedences were likely in future years.

There were significantly less monitoring sites for particulates than there were for NO_2, therefore the method of validation used for NO_2 was not applicable. In an effort to calculate the accuracy of the modelled predictions of PM_{10} concentrations, a number of comparisons were made. First, the calculated 99th percentile of 24 hour averages using 1998 meteorological data in the model at the location of the monitoring station at Grangemouth Municipal Chambers was compared to the maximum 24 hour concentration measured there for 1998. The model predicted

Table 10.29 Validation of NO_2 modelled concentrations

Site	*Derwent-Middleton correlation conversion NO_X to NO_2*				
	Modelled NO_X	*Modelled NO_2 using DM*	*Annual average monitored*	*Difference*	*Square of difference*
<u>*Roadside*</u>					
Falkirk Road/West Mains Ind. Est	82.25	34.10	43.13	9.03	81.56
Graeme High School Falkirk	85.77	34.51	42.02	7.51	56.46
Kerse Lane, Falkirk	93.31	35.31	58.26	22.95	526.64
Bellsdyke Road, Larbert	73.90	33.06	26.74	-6.32	39.94
A80 N. Bound C/way, Banknock	92.69	35.24	42.02	6.78	45.92
Kilsyth Road, Banknock	66.10	31.96	37.65	5.69	32.35
Laurieston	78.72	33.68	31.83	-1.84	3.40
Weir Lane, Falkirk	84.53	34.37	31.08	-3.29	10.79
Denny Cross	64.10	31.66	42.02	10.36	107.34
				Mean	100.49
				RMS	10.02
<u>*Urban centre*</u>					
Carnegie Drive Camelon	82.73	34.16	24.83	-9.33	87.03
Muirhall Road, Larbert	73.80	33.05	25.87	-7.17	51.48
				Mean	69.25
				RMS	8.32
<u>*Urban background*</u>					
Tinto School	83.32	34.23	26.32	-7.91	62.60
Cromwell Road, Falkirk	82.35	34.11	28.81	-5.31	28.15
Wellside Place	79.36	33.76	20.82	-12.94	167.37
Irving Parish Church, Camelon	86.94	34.64	24.66	-9.98	99.59
Glenview Avenue, Banknock	66.28	31.99	36.07	4.08	16.65
Hayfield Bankside	78.23	33.62	35.65	2.04	4.15
Thistle Avenue	96.84	35.65	33.43	-2.23	4.97
				Mean	54.78
				RMS	7.40
<u>*Automatic sites*</u>					
Urban Background					
Municipal Chambers	80.75	33.92	31.04	-2.89	8.34
				RMS	2.89
Kerbside					
Hope Street Falkirk	98.76	35.84	30.07	-5.77	33.25
				RMS	5.77

99th percentile concentration of 24 hour averages was 37.3 μg/m^3. Adding predicted background primary and secondary particulate concentrations gave 62.3 μg/m^3. The maximum recorded 24 hour concentration was 32.8 μg/m^3 TEOM, 42.7 μg/m^3 gravimetric equivalent. In this case the model was over predicting by 46%. Second, the model-predicted and actual recorded annual average concentrations were compared. The model-predicted annual average was 4 μg/m^3. Adding predicted background primary and secondary particulate concentrations gave 28.5 μg/m^3. The actual recorded annual average concentration was 14.7 μg/m^3 TEOM, 19.1 μg/m^3 gravimetric equivalent. In this case the model was over predicting by 49%. In both cases the model accuracy was within the tolerable 50% margin above or below the actual concentrations. It was therefore concluded that the model predictions were reliable for the purpose of estimating the likelihood of exceedence of the objectives in future years.

10.8.1 Screening study

The screening study was carried out for the seven meteorological conditions for south-west and east-north-east wind directions for NO_x, SO_2 and PM_{10}. The point sources were split into groups so that the likely contribution to air quality within the Falkirk Council area due to emissions from Longannet Power Station (Chapter 4) could be assessed in a range of meteorological conditions. This was only done for the east-north-east wind as emissions from Longannet Power Station would not impact on the Falkirk Council area in a south-west wind.

The predicted 1 hour concentrations of NO_x for convective, neutral and stable conditions (stability B, D & F) are summarised in Table 10.30 for both wind directions.

Table 10.30 Summary of screening results for NO_x modelling (μg/m^3)

Atmospheric stability class	*Wind direction*	*Maximum predicted concentration of NO_x*	*Maximum contribution from Longannet within the Falkirk Area*
B	SW	77	-
D	SW	73	-
F	SW	200	-
B	ENE	76	31.5
D	ENE	95	65.0
F	ENE	254	0

The NAQS objectives for NO_2 to be achieved by 2005, concern the annual mean and the maximum 1 hour concentration. The concentrations (Table 10.30) can be compared to the latter, for which the objective is 200 μg/m^3. The predicted concentrations are for 1 hour averages of total NO_x not just NO_2, but these results gave an early indication of areas where potential exceedences of the NO_2 objective may occur. In convective conditions the maximum ground level concentration occurred very near to the point sources at Grangemouth, but the road source contributions were more significant in stable conditions. The contribution from Longannet was greatest in neutral conditions when the wind speed was higher and is responsible for 68% of the maximum concentration predicted in the area. There were no contributions from Longannet in stable conditions because the stack and plume were penetrating the inversion of the boundary layer. Based on these screening results it was possible that the NAQS objectives for NO_2 might be exceeded but this was examined in more detail in the long-term calculations.

The predicted 1 hour concentrations of SO_2 for convective, neutral and stable conditions (stability B, D & F) are summarised in Table 10.31 for both wind directions. There were three UK NAQS objectives for SO_2 based on the maximum 1 hour mean and the 24 hour mean to be achieved by 2004, and the 15 minute mean to be achieved by 2005. The concentrations (Table 10.31) can be compared to the objective for the 1 hour mean which is 350 $\mu g/m^3$ not to be exceeded more than 24 times a year. This concentration was exceeded in convective conditions for the south-west wind direction with maximum impact very close to the sources. The more stringent objective to be achieved by 2005 is based on the 15 minute mean of 266 $\mu g/m^3$ not to be exceeded more than 35 times in a year. It follows that the 15 minute concentrations will be higher than the predicted 1 hour concentrations. Based on these screening results and the current point source emission rates, it was probable that the NAQS objectives for SO_2 would be exceeded in the future but this was examined in more detail in the long-term calculations.

Table 10.31 Summary of screening results for SO_2 modelling ($\mu g/m^3$)

Atmospheric stability class	*Wind direction*	*Maximum predicted concentration of SO_2*	*Maximum contribution from Longannet within the Falkirk Area*
B	SW	350.0	-
D	SW	228.0	-
F	SW	28.0	-
B	ENE	186.0	68.4
D	ENE	141.0	141.0
F	ENE	17.6	0

The predicted 1 hour concentrations of PM_{10} for convective, neutral and stable conditions (stability B, D & F) are summarised in Table 10.32 for both wind directions. There are two UK NAQS objectives for PM_{10} based on the annual mean and the 24 hour mean to be achieved by 2004. The concentrations listed in Table 10.32 are 1 hour concentrations. It can be assumed that the 24 hour mean and the annual mean will be less than the maximum 1 hour concentration, which for these results was only 13.8 $\mu g/m^3$. The annual mean limit is 40 $\mu g/m^3$ and the 24 hour mean limit is 50 $\mu g/m^3$ which must not be exceeded more than 35 times.

On examination of the results it could be seen that the main contributions to PM_{10} concentrations were from road sources with the contributions from the industrial sources at Grangemouth and Longannet Power Station clearly identifiable. In convective conditions the maximum ground level concentration

Figure 10.7 Predicted annual mean concentration of NO_2 over the whole of the Falkirk Council area (courtesy of the Environmental Health Department, Falkirk Council)

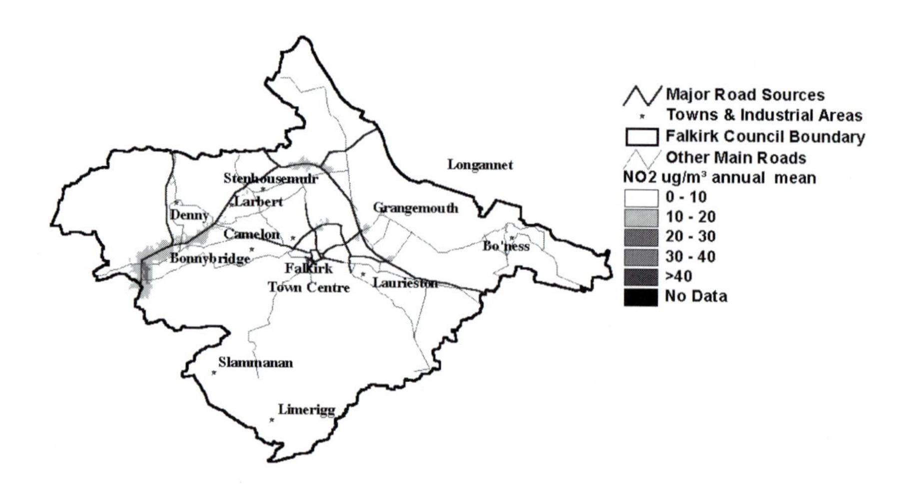

Figure 10.8 Predicted 99.8th percentile of concentration of NO_2 across the Falkirk area (courtesy of the Environmental Health Department, Falkirk Council)

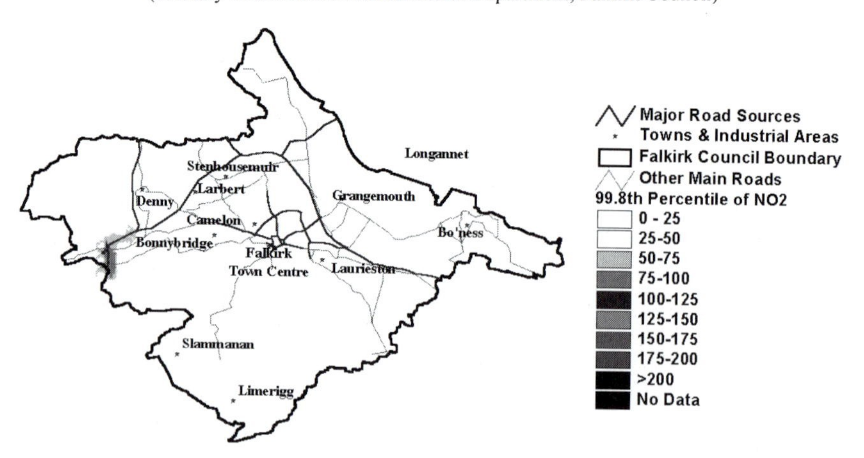

occurred very near to the point sources but in stable conditions the contribution from road sources was higher. There were significant contributions from Longannet in convective and neutral conditions with the east-north-east wind contributing 32% of the maximum concentration calculated in the area in convective conditions and 55% in neutral conditions. Based on the screening results and the current point source emissions, it was unlikely that the NAQS objectives for PM_{10} would be exceeded in 2004 but this was examined in more detail in the long-term calculations.

Table 10.32 Summary of screening results for PM_{10} modelling ($\mu g/m^3$)

Atmospheric stability class	*Wind direction*	*Maximum predicted concentration of PM_{10}*	*Maximum contribution from Longannet within the Falkirk Area*
B	SW	7.75	-
D	SW	8.26	-
F	SW	11.32	-
B	ENE	5.96	1.90
D	ENE	7.12	3.94
F	ENE	29.11	0

10.8.2 Long-term calculations

During the validation study it was concluded that the monitoring results were higher with 1998 meteorological data than with 1996 meteorological data from Edinburgh Turnhouse. Therefore, in order to model worst case scenarios, this 1998 meteorological data was used for all of the future predictions discussed in this section.

The Derwent-Middleton correlation (Derwent and Middleton, 1996) was used in the calculations of NO_2 concentrations. This was included in the modelling calculations by activating the chemistry scheme in the ADMS-Urban model (Cambridge Environmental Research Consultants Ltd, 2000). Fig. 10.7 shows the contour map of the predicted annual mean concentration of NO_2 over the whole of the Falkirk Council area. The predicted traffic flow data obtained from Falkirk Council was used in conjunction with the DMRB (DETR, 1999a) emissions factors for 2005 in the ADMS-Urban model for these calculations. Fig. 10.8 shows the predicted 99.8th percentile of concentration of NO_2 across the whole of the Falkirk Council area. The maximum predicted concentrations of NO_2 occurred along the A80 and M80 with a maximum predicted annual average concentration of 24 $\mu g/m^3$ and a 99.9th percentile of 128 $\mu g/m^3$. There were no predicted exceedences of either the annual mean objective or the 1 hour mean objective for NO_2 for 2005.

The calculations for predicting the future concentrations for SO_2 were based on the current emissions from the point sources. Whilst it was known that there were planned installations at BP and Longannet Power Station to reduce emissions of SO_2, it was not known what the expected reductions would be. Therefore, the predictions calculated by the model were based on the maximum expected concentrations and reflect the worst case scenario. The predicted modelled results are shown in Fig. 10.9 for the predicted 99.9th percentile of the 24 hour means and the 99.7th percentile of the 1 hour means respectively:

- The maximum recorded 15 minute mean concentration was 550 $\mu g/m^3$
- The maximum recorded 1 hour mean concentration was 170 $\mu g/m^3$
- The maximum recorded 24 hour mean concentration was 469 $\mu g/m^3$

Figure 10.9 Predicted 99.9th percentile of 24-hour means and 99.7th percentile of 1-hour means respectively (courtesy of the Environmental Health Department, Falkirk Council)

(a) Predicted 99.9th percentile 24-hour mean SO_2 levels

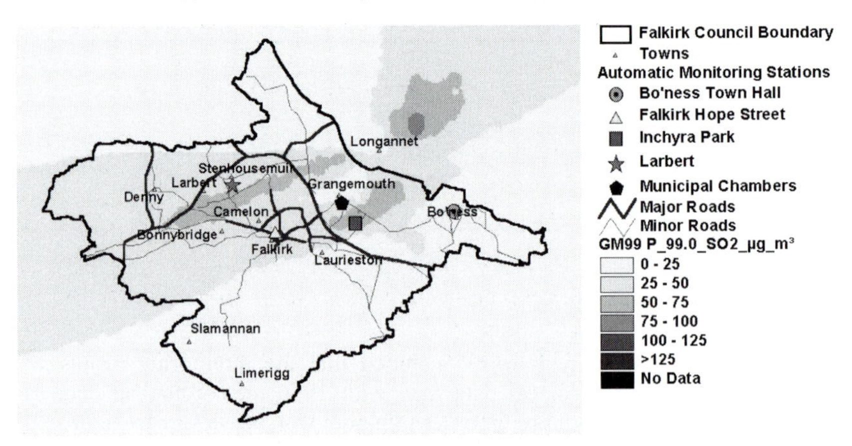

(b) Predicted 99.7th percentile 1-hour mean SO_2 levels

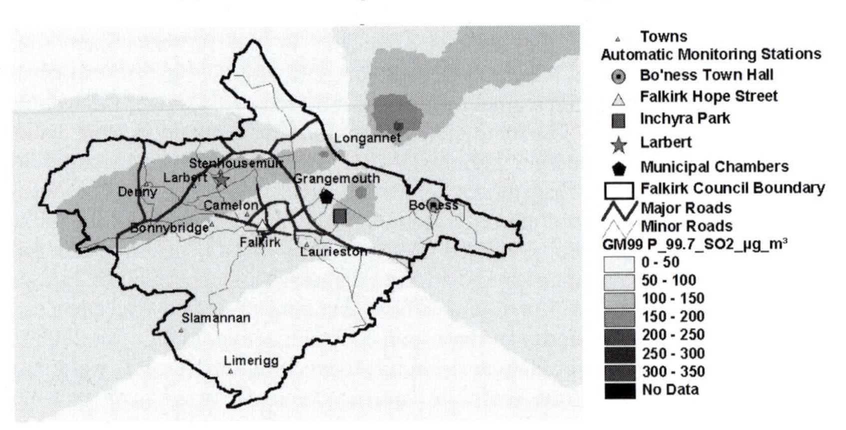

It was seen that the model predicted exceedences of the 15 minute objective of 266 µg/m^3 in the Grangemouth area. The model also predicted exceedences of 350 µg/m^3, the concentration associated with the 1 hour objective for SO_2 in a very localised area near the industrial sources at Grangemouth but there were less than the 24 exceedences permitted in the objective.

The predicted traffic flow data obtained from Falkirk Council was used in conjunction with the DMRB emission factors for 2004 in the ADMS-Urban model for these calculations in order to compare PM_{10} predictions with the objectives for 2004. The maximum predicted annual mean concentration was less than 5 µg/m^3. With the addition of forecast background concentrations of primary and secondary particulates there were still no predicted exceedences of the annual mean objective of 40 $\mu g/m^3$. The maximum predicted 24 hour mean concentration was less than 5 $\mu g/m^3$. With the addition of forecast background concentrations of primary and secondary particulates there were still no predicted exceedences of the 24 hour objective of 50 $\mu g/m^3$.

10.8.3 Areas of expected exceedences of air quality objectives

The initial study proposed an absolute worst case AQMA which was based on the SO_2 emissions data that were available at that time. It did not account for planned sulphur reduction technologies at BP Grangemouth or Longannet Power Station in Fife, data for which were available for the preparation of the Phase 3 study. Therefore, there was a need for more accurate data and further investigations were required before a decision could be made on whether or not it was necessary to declare an AQMA. A supplementary report was undertaken to address those issues and some of the responses to the report that were received by Falkirk Council from SEPA and the Scottish Executive.

The results of detailed modelling undertaken showed that the modelled predictions for SO_2 were below each of the NAQS objective concentrations during normal operating conditions at BP and Longannet. When BP was modelled on maximum load, the predicted 99.9th percentile of 15 minute concentrations and the predicted 99th percentile of 24 hour concentrations were shown to exceed the objective levels. However, it should be noted that the model calculation assumed that BP Power Station operated on maximum load for a whole year whereas in reality, the periods of maximum load would be short, intermittent and occur for a total period of up to one month in a year and it would be impossible to predict under which meteorological conditions they would occur. If the wind direction was at the prevailing south-westerly direction at the time of occurrence, there would be no exceedences within the Falkirk Council area.

The results for the 15 minute objective in normal operating conditions showed that the predicted concentrations were near to the objective level of 266$\mu g/m^3$ but did not exceed it. Model predictions compared well with monitoring data and the model results were accepted as reliable. Based on the results of the extensive work undertaken and data collected during the complete Review and Assessment process, Falkirk Council took the decision not to declare an AQMA within the Council area. However, Falkirk Council intend to maintain an extensive monitoring programme of air quality within the area and have identified areas for future work in order to maintain and improve on the air quality information available for the Falkirk Council area.

10.9 CASE STUDY 8 - EMISSION CONTROL

Eltimate Ltd (2000) (a subsidiary of Egger (UK) Holdings Ltd) operate a chipboard manufacturing plant in Ayrshire, Scotland. The plant utilises the latest continuous press technology, where the design incorporates the appropriate pollution control

Figure 10.10 VOC emission levels from a chipboard manufacturing plant (courtesy Egger (UK) Holdings Ltd)

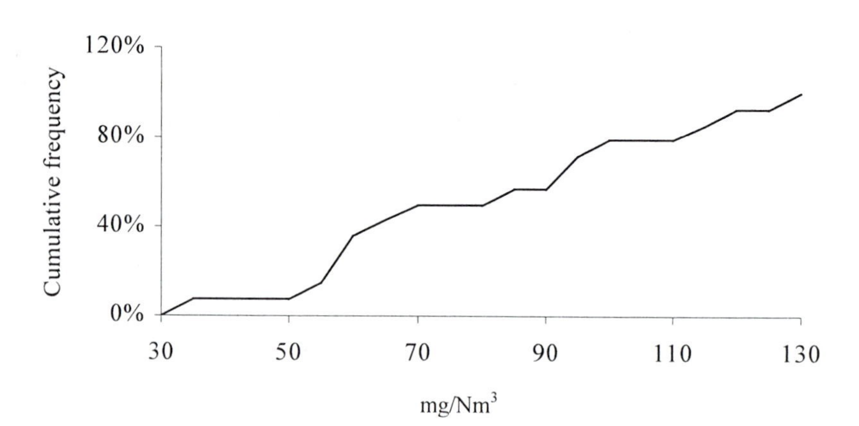

equipment to limit atmospheric emission levels to within the authorisation conditions set out in the UK Secretary of States Guidance Note PG 6/4 (94).

Chipboard is manufactured from thin dried flakes of wood blended with synthetic formaldehyde resins and compressed under heat to form flat panels. The six defined stages within the manufacturing process are wood particle preparation and storage; particle drying; particle separation and grading; resin application; hot pressing; and calibration and finishing.

The emissions from the plant are principally VOCs, particulates, formaldehyde and total aldehydes. Following preparation the moisture content of the material is reduced by a drying process from 50–120% to 2–5% at the drier exit. During the drying process, naturally occurring VOCs and aldehydes are released from the wood. These compounds, together with particulate matter and formaldehyde are then discharged from the driers, together with steam, prior to atmospheric discharge through a single stack. Emissions pass through a wet electrostatic precipitator system designed to reduce the levels of the discharge to well within the plant's operating consent conditions. Fig. 10.10 shows the measured VOC levels (mg/Nm^3) as a series of percentiles. The PG Note guidance level is 130 mg/Nm^3 (100th percentile). Data (Fig 10.10) shows that the level is not exceeded and the majority of levels are significantly below this level.

10.10 CASE STUDY 9 - SPIRESLACK OPEN CAST COAL SITE

This case study details an assessment of the impact of dust emissions resulting from excavation and restoration activities of a proposed OCCS in Ayrshire, Scotland (Scottish Coal, 1996a). The site is located approximately 4 km east of Muirkirk in Ayrshire. The sensitive receptors, identified in consultation with the developer, of particular concern were Glenbuck Home Farm and West Glenbuck adjacent to the site.

10.10.1 Method of assessment

Following consultations with interested parties it was agreed that a qualitative assessment approach would be adopted for the study owing to the isolated location of the site and the non-occurrence of complaints from an adjacent operational OCCS. It was therefore agreed that the following study approach would be applied. The study approach was broadly in accordance with that recommended by the UK DoE guidance (DoE, 1995a).

- Review of baseline air quality data for the area;
- Review of air quality levels at and around other similar OCCS;
- Assessment of those operational activities that would have the potential to emit dust;
- Comparison of air quality levels with recommended AQS;
- Identification of the location of sensitive receptors;
- Specification of suitable dust mitigation measures.

The principal air pollutant of concern was dust emitted from excavation, transportation and restoration activities. The air quality criteria of concern were those relating to human health and nuisance (Chapter 7). At the time of the study the EU Directive (80/779) (Council for European Communities, 1980) had set a limit for suspended particulates (smoke) at 80 $\mu g/m^3$ measured over one year as a median of the daily values. There was also a higher limit for the winter period only of 130 $\mu g/m^3$. The Directive gave a 98th percentile limit value for smoke of 250 $\mu g/m^3$ based on daily mean values taken throughout the year. The USEPA proposed an annual average standard of 50 $\mu g/m^3$ PM_{10} (previously 75 $\mu g/m^3$ as TSP, this concentration was roughly equivalent to 50 $mg/m^2/day$). The 24 hour mean, which should not be exceeded on more than 1 day in a year, was 150 $\mu g/m^3$ PM_{10} (previously 260 $\mu g/m^3$ TSP). The UK DoE Expert Panel on Air Quality Standards (EPAQS) had recommended a standard for PM_{10} of 50 $\mu g/m^3$ measured as a 24 hour running average (DoE, 1995b). However, the WHO had stated that no guideline should be set for particulate matter because there was no evident threshold for effects on morbidity and mortality (WHO, 1995b; 2000b) (Table 7.3). Table 7.18 summarises the criteria cited by the UK DoE for dust nuisance (DoE, 1995a).

10.10.2 Baseline air quality and meteorological conditions

Smoke levels were monitored at several locations in the vicinity of the site (Table 10.33). Average smoke levels were well within the EU Directive (80/779) limit for suspended particulates (smoke) of 80 $\mu g/m^3$.

TSP and PM_{10} levels monitored in the near locality (Enterprise Ayrshire, 1996) averaged 21.0 to 38.5 and 15.8 to 23.3 $\mu g/m^3$ respectively. PM_{10} levels were well within the recommended UK DoE EPAQS. Table 10.34 shows the dust deposition levels from other OCCS in the vicinity of the development location (Scottish Coal, 1996b). Monitoring was undertaken using directional dust deposit gauges (Chapter 5). Airdsgreen abutted the western boundary of the site and

Gasswater was 2 miles east of Cumnock. Dust at Airdsgreen was monitored at two sites. Average levels ranged from 116.1 to 130.0 mg/m^2/day. The average level for both sites was 123.1 mg/m^2/day. Dust was measured at 3 sites at Gasswater and average levels ranged from 38.7 to 55.6 mg/m^2/day. The average level for all sites was 47.5 mg/m^2/day. Average deposition levels were monitored using the 'Frisbee' technique at Gasswater and ranged from 21.8 to 25.2 mg/m^2/day.

Table 10.33 Average smoke levels (µg/m^3) for the Cumnock area (Cumnock & Doon Valley District Council, 1996)

Location	*1993*	*1994*	*1995*
Cumnock	23.0 (40.0)	19.8 (30.7)	16.5 (22.2)
New Cumnock	18.0 (26.1)	13.8 (18.7)	12.8 (13.6)
Dalmellington	15.3 (21.1)	16.2 (21.7)	13.2 (14.8)

The values given in the brackets were the highest monthly average value monitored for the respective year.

Table 10.34 Dust deposition levels (mg/m^2/day) for Airdsgreen and Gasswater OCCSs (1996)

Site	*Site 1*	*Site 2*	*Site 3*
Airdsgreen	116.1 (57.4-187.2)	130.0 (95.9-222.2)	-
Gasswater	48.3 (22.8-104.8)	55.6 (9.8-201.5)	38.7 (5.8-97.8)
Gasswater[a]	22.6	21.8	25.2

Values in brackets are the range of deposition values.
a) Measurements monitored using the 'Frisbee' technique.

Typical rural levels for dust deposition were 10-50 mg/m^2/day, although other sources gave 65 mg/m^2/day, more when harrowing fields, of 20-100 mg/m^2/day in rural locations (DoE, 1992). The monitored average levels at Airdsgreen and Gasswater were well within recommended dust nuisance criteria (Table 7.18) and were similar to typical levels found in rural areas. Available data showed that air quality levels for the Spireslack and surrounding locality were within recommended guidelines and standards.

The meteorological parameters of concern to the study were wind speed and direction, and rainfall. Wind speed and direction cause the entrainment and determine the transportation and dispersion characteristics of particulates. Rainfall acts as a natural dust suppressant. Rainfall (or more precisely precipitation) is one of the most variable of the meteorological elements, both in space and time. The River Ayr valley receives an annual rainfall level of approximately 1128 millimetres (mm) (Meteorological Office, 1996a). The potential evapotranspiration and rainfall levels for Cumnock (Meteorological Office, 1996b) showed a net moisture deficit in the months May to July (e.g. evaporation is greater than rainfall). Such weather conditions contribute to periods of the year that would have the greatest potential to cause dust emissions.

The speed and direction of the wind were measured at a limited number of reliable places in the vicinity of the site. For the purposes of the study, data from Prestwick Airport (Meteorological Office, 1995) and Cumnock (Cumnock Academy, 1990) were assessed. Fig. 10.11 shows the annual variation in wind

Figure 10.11 Wind rose for Prestwick Airport (Data from Meteorological Office, 2000)

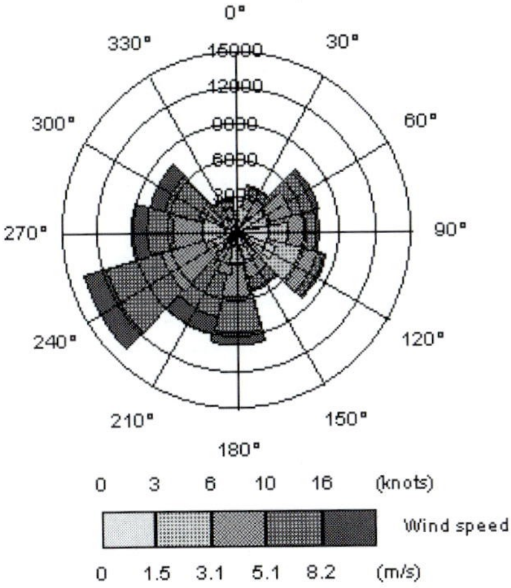

direction for the Airport. The principal wind direction was west-south-westerly, although the wind rose also showed a pronounced west and east orientation. Table 10.35 shows the directional variation of wind for the months where moisture deficit conditions occurred. The predominant wind direction was south-west to north-west. The predominant wind direction during dry conditions was away from the ASR close to the site. The development site was situated to the north of these premises. Northerly winds occurred on average 1.5% of the time and 1.1 to 2.2% of the time during moisture deficit conditions. North directional winds (NNW to NNE) occurred on average 6.2% of the time. The average wind speed for Prestwick was 9.4 knots (kt) approximately 4.6 m/sec. An average wind speed of 13 kt (6.7 m/sec) was considered a moderate breeze with the potential to raise 'dust and loose paper' (DoE, 1995a).

Meteorological data was also available from another site at Cumnock Academy. A comparison of the wind roses showed that both sites exhibited pronounced easterly and westerly winds. The wind rose for the Academy showed more pronounced easterly and westerly winds, this was partially due to the smaller number of wind sectors monitored. A summary of data for the Academy for 1973-1989 showed that on average there were 216 wet days per annum (e.g. 59.2% of the days per year had effective rainfall). The relative high percentage occurrence of 'wet days' in the locality acted as a natural dust suppressant. Prevailing wind directions, particularly during moisture deficit conditions, dispersed dust emissions away from sensitive receptors.

10.10.3 The proposed development

The proposal was to extract, by open cast mining methods, about 3 to 3.5mt of coal over a ten-year operational period. Approximately 40m^3 of overburden would be handled to facilitate the extraction of the coal. There would be approximately 80-90 coal lorry movements associated with the development per working day.

Briefly, operations for the site would consist of soils being stripped from the site and stored in mounds, and the initial box-cut is then created. Overburden would then be back-filled into the coaled out areas, thus beginning progressive restoration. The site would then be worked in a series of cuts. Overburden from the final void would then be back-filled into previously worked areas and soils progressively restored in the back-filled areas.

Table 10.35 Wind direction variation for months with moisture deficit conditions (percentage occurrence (%))

Direction	*Annual*	*May*	*June*	*July*
Calm	1.9	1.7	1.8	1.7
N	1.5	2.1	1.1	1.1
NNE	2.7	3.5	2.5	2.4
NE	5.0	7.9	7.1	4.7
ENE	6.2	9.0	8.7	4.9
E	6.0	6.1	6.2	4.5
ESE	7.3	6.8	6.6	5.1
SE	5.3	4.4	3.7	3.9
SSE	5.2	3.9	3.7	4.1
S	8.1	5.9	4.8	5.0
SSW	7.7	4.9	3.5	4.7
SW	9.6	6.7	6.6	9.8
WSW	11.7	10.9	13.6	17.1
W	8.3	8.9	11.8	12.7
WNW	6.8	9.0	10.7	10.2
NW	4.7	5.9	6.0	6.6
NNW	2.0	2.5	1.6	1.4

10.10.4 Potential emissions

The major air quality issue during the preparation, excavation and restoration of the site were dust emissions. Therefore, dust control and mitigation measures were adopted and rigidly enforced through the use of statutory powers and contractual requirements. All dust control measures were agreed with the regulatory body. All of the proposed development operations and activities had the potential to generate dust. The quantity of dust emissions from on-site works depended on the size of the area being worked and the intensity of activity, specific operations, prevailing meteorology and the intrinsic characteristics of the particulates. Previous research had shown that large dust particles (>30 μm), that make up the greatest proportion of dust emitted from mineral workings, would largely deposit within 100m of the source (DoE, 1995a).

At surface mineral sites, the most significant dust emission sources are generally mechanical handling operations and haulage of material on unsurfaced site roads. However, wind blow across disturbed site surfaces can also generate significant levels of dust. The principal OCCS activities of concern included:

Pre-operational civils: The construction of the site access was considered to be minor compared to the proposed main site activities. The duration of these works would only be for a limited period (e.g. a few weeks). The site access would be through the existing Airdsgreen access point and would therefore avoid residential premises. Dust emissions from this phase of the development were therefore not expected to be significant. The existing site infrastructure immediately adjacent to the site access would also be used. The nearest site haul road in the proposed OCCS would be approximately 375m from a sensitive receptor. It was proposed to use the existing road as a secondary access for site workers and light service vehicles. The road would be a made road (e.g. metallic) and would be swept regularly to minimise dust emissions. Other internal haul roads used by site vehicles would be concealed out of general view, as they would be north of the elevated strip of land that ran alongside the nearby A70 road.

Should dust emissions prove excessive during dry weather conditions (e.g. moisture deficit periods), watering of surfaces would be implemented to control emissions. An effective watering programme (twice daily) was previously estimated to reduce dust emissions by up to 50% (USEPA, 1995a; 1998).

Soils handling: The handling of soils was to be carried out under dry weather conditions in accordance with recommended guidance. However, such material generally had a relatively high moisture content and normally its removal and subsequent replacement was a short-term operation taking only a matter of a few weeks to complete at the commencement and restoration phases of site operations. Consequently impacts were therefore likely to be of short duration and minimal. The remote location of the site also contributed to a reduction of potential impacts.

Two residential premises, however, were located at the site boundary, Glenbuck Home Farm and West Glenbuck, therefore specific attention was required to minimise the impacts of potential dust emissions from these activities at this site. The soil storage areas would be predominantly to the north of the sensitive receptors to minimise the potential for dust nuisance.

A small storage area would be approximately 175m from West Glenbuck although prevailing winds would disperse dust emissions away from the premises. The same proposed soil tip would be 300m from Glenbuck Home Farm. The outer layers of the soil mounds would quickly form a crust, which would subsequently act as a seal to the underlying material. Long-term soil storage mounds were usually seeded to grass and a significant sward would develop within a few weeks. Both events would improve particle stability and prevent dust emissions. Unnecessary trafficking of plant over the soils would be avoided through careful design of haul roads within the site.

Overburden excavation: Overburden was to be excavated using hydraulic excavators and dump trucks. The actual excavation and loading of the trucks was likely to be relatively free from dust generation owing to the inherent high moisture content of the material. However, it was possible that the haul roads would be a potential source of dust emissions, therefore dust control measures including haul road grading, speed control and watering of dry surfaces would be undertaken to minimise emissions. The initially formed feature would be allowed to crust to minimise the potential for dust emissions. The spraying of exposed surfaces of mounds would help maintain surface moisture. Minimising handling would also reduce emissions. An existing overburden mound was to be situated to the northern sector of the site away from sensitive receptors.

An overburden area was to lie on the eastern boundary of the Airdsgreen site (western boundary of the Spireslack site). Dust measurements for Airdsgreen (closed to the overburden site) (Table 10.34) showed that average levels of 130 mg/m^2/day were within recommended air quality criteria.

Coal extraction: Coal excavation was to be undertaken using small hydraulic excavators loading into small articulated dump trucks. This operation would be relatively dust free owing to the inherent moisture content of the material. The average moisture content of the coal was about 8-11%, which reinforced the potentially dust free nature of the material. The main source of potential dust emissions from these activities would be the movement of trucks across the site. Dust control measures would include those measures identified above and a wheel wash facility prior to site exit. Any coal stock-piles would be kept damp by the application of water with sprays as and when necessary. Dust would be suppressed from crushers and screens by application of water through sprays immediately prior to feeding, if required.

Where appropriate, the distance between sensitive areas and dust-generating activities was maximised, and incorporated within the site design. In the absence of a quantitative dust impact assessment, guidance (DoE, 1995a) recommended a stand-off distance of 100 to 200m between sensitive areas and the medium to high dust-generating activities of a continuous nature. The nearest excavation point to the sensitive receptors was approximately 175m. The potential for severe impacts was greatest within 100 to 200m.

Blasting: Blasting might occur up to twice per day on average. Dust emissions are difficult to control during such activities. However, removing dusty material from the blast area prior to detonation would help to minimise dust emissions. Many drilling rigs used air to flush the boreholes free from debris to minimise dust emissions. No nuisance had been previously reported relating to blasting activities about the Airdsgreen site workings.

On and off-site haulage: Dust emissions from vehicle movements on unpaved surfaces are a function of vehicle speed, vehicle weight, number of wheels per vehicle, surface texture and moisture. Particles are lifted and dropped from rolling wheels, and the road surface is exposed to strong air currents in the turbulent wake behind vehicles. The haulage road would be sprayed with water to minimise dust emissions. A tractor brush would be available to sweep the access/internal roads. In addition, vehicle exhaust emissions would be routed upwards away from road surfaces to prevent resuspension of dust. Modern vehicles would be fitted with radiator fan deflector plates to minimise dust resuspension.

Motor vehicle and plant emissions on-site were not considered to be significant. Plant equipment would be regularly serviced to ensure that engines were properly operating.

Restoration: Restoration of the site might have caused dust emissions through material handling and displacement activities. Principal emissions would be from the haulage of the material and its tipping/discharging. Dust mitigation measures identified above to minimise emissions due to truck movements were imposed. Minimising the drop height of tipping activities would also reduce emissions.

At the time of the study precise excavation, filling and restoration operational methods were unavailable and therefore dust emissions from specific material handling activities were difficult to evaluate and quantify. Emissions were, however, likely to be an insignificant contribution to those that have been assessed.

10.10.5 Mitigation measures

Suitable control measures for dust at the OCCS were recommended and these are detailed in Section 8.3.1. Operations likely to cause the generation of dust within the site boundary were to be monitored continuously by site management and all findings, including prevailing meteorological conditions, recorded. The developer deployed effective methods of suppressing undue amounts of dust arising from consented infill operations. In addition, the developer provided on site, and maintained at all times, an adequate supply of water for spraying equipment, to ensure a sufficient supply of rapid filling of mobile spraying units, to have available a sufficient number of units, and to ensure that the rate of application would be sufficient for the purposes of effectively wetting the ground.

In addition to the above mitigation measures, all long-term storage mounds closest to the sensitive receptors were seeded to grass at the earliest opportunity and adequately maintained. The provision and maintenance of a windsock also benefited the management of the site to help in assessing operational activities in conjunction with wind speed and direction, and sensitive receptor locations.

Dust control and mitigation measures were rigidly applied to minimise dust impacts during all excavation and restoration activities. A complaints response system was also established with a contact name and telephone number. This enabled the public to call if a dust incident occurred and for appropriate action to be taken on-site to reduce dust emissions. In the unlikely event of a particular activity on the site creating a specific dust nuisance in adverse weather conditions, the site management would stop operation and only allow it to be resumed when satisfactory control was achieved. The site management had absolute authority to undertake this action if necessary. Operational staff would regularly inspect the site to ensure that dust mitigation measures were effectively adhered to.

10.10.6 Environmental consequences

Dust emissions from on-site activities were unlikely to have a significant effect on sensitive receptors provided dust mitigation measures were rigorously applied and closely monitored. The remoteness of the site coupled with the proposed operational practices (e.g. location of storage and excavation areas, haul roads, etc. relative to sensitive receptors) together with prevailing meteorological conditions also contributed to the reduction of potential dust impacts. Nevertheless, owing to the close proximity of an ASR regular monitoring was undertaken to ensure that dust mitigation measures were being effectively implemented.

10.11 CASE STUDY 10 - POWER STATION AIR DISPERSION MODELLING STUDY

An AQA (Scottish Hydro-Electric, 1997) was conducted for gaseous emissions from Peterhead power station, located at Boddam near Peterhead, Scotland (Fig. 10.12). The power station was equipped with multiple fuel burners, capable of burning heavy fuel oil, North Sea (Miller Field) gas, North Sea gas (methane), or natural gas liquid (NGL). The study modelled the NO_x and SO_2 emissions resulting from three different fuel combustion scenarios. The scope of the study was:

- Review of AQS;
- Review of local conditions, including meteorological parameters for the locality to establish predominant and prevailing meteorological conditions that may influence the dispersion of emissions from the power station;
- Air dispersion modelling studies to predict:
  - annual ground level concentrations;
  - 100th percentile levels (maximum 1 hour concentrations);
  - 98th percentile levels of NO_x and SO_2;
  - annual deposition levels for SO_2 and NO_2 for comparison with critical load levels for acidity, for soil, vegetation and freshwater, recommended by the UK CLAG.

Predictions were undertaken using the ADMS model. Concentrations were predicted for a receptor grid for the local area. The study used sequential meteorological data from the nearest available meteorological station to the site.

10.11.1 Meteorological data

A ten-year meteorological data set for the period 1986 to 1995 was input to the model. The meteorological station was located at Aberdeen (Dyce). The surface roughness length at the station was assumed to be 0.15 m. Roughness affects air turbulence, but is usually secondary to surface temperature variations in magnitude of effects on air turbulence. It was assumed that conditions at the meteorological station were representative of the power station. This was reasonable, since topography, surface roughness and climate were similar. The meteorological data included surface sensible heat flux and boundary depth values calculated using the same algorithms as those used by the model.

Figure 10.12 Location map of Peterhead power station (not to scale)

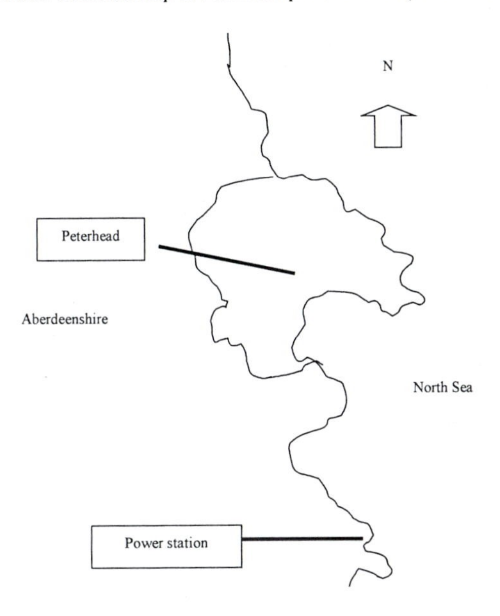

Figure 10.13 Wind rose for Dyce, Aberdeenshire (Data from Meteorological Office, 2000)

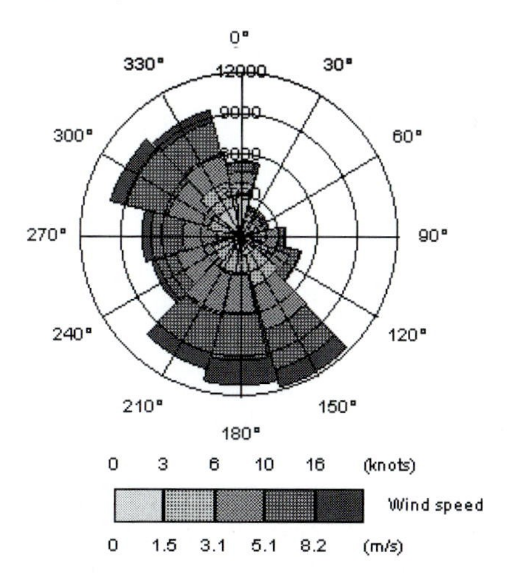

The total number of hours used in the meteorological data statistical analysis was 85,597, which was 97.7% of the total number of hours available in the ten-year measurement period. A wind rose was constructed from the meteorological data (Fig. 10.13). Wind direction was bimodal, being predominantly from the south, with a strong secondary direction from the north-west.

10.11.2 Modelling parameters

ADMS is a PC-based model of dispersion in the atmosphere of passive, buoyant, or slightly dense, continuous or finite duration releases from single or multiple sources. The model used an up-to-date parameterisation of the boundary layer structure based on the Monin-Obukhov length and the boundary layer height. Since the boundary layer is defined in terms of measurable physical parameters (obtained from meteorological readings), a better representation of changing dispersion characteristics with height is obtained, compared to the generalised Pasquill-Gifford stability classes (Chapter 6).

The stack at Peterhead power station had one flue with an internal diameter of 9.14m and an exit height of 170m. Three scenarios were modelled:

- Case 1 - emissions from exclusive burning of heavy fuel oil;
- Case 2 - exclusive burning of gas;
- Case 3 - exclusive burning of NGL.

NO_x and SO_2 emissions were modelled as gaseous reactive pollutants emitted in a continuous plume with mass emissions in grammes per second. The averaging time was set at 1 hour. Complex effects of buildings, hills or coastline were not included, since the stack height was considerably more than the height of surrounding buildings or local topography.

The model default specific heat capacity (1012 J/°C/kg) and molecular weight (28.96 g/mol) of air was used for the exit gas because dilution with air is almost instantaneous after emission of the buoyant plume pollutant gas mix from the stack. For wet deposition the washout coefficient, Λ, varied with precipitation rate, P, and default values of $A = 10^{-4}$ and $B = 0.64$ were used to calculate the washout coefficient from the relationship.

$$\Lambda = AP^{B} \qquad (10.1)$$

10.11.3 Modelling results

The ADMS model output was in the form of long-term average mean concentrations of pollutants at ground level, long-term average mean deposition fluxes and long-term average percentile concentrations. Table 10.36 lists the maximum values predicted by the model within the study area. Long-term mean of concentrations contour plots, 98th percentile contour plots, and deposition contour plots all reflected the wind rose pattern of pollutant distribution in two predominant directions: SSE and NNW. The 100th percentile contour plots showed a concentric circle pattern.

In Case 1, the maximum long-term mean of concentration indicated for NO_x was 1.7×10^{-6} g/m^3, equivalent to 1.7 μg/m^3. The maximum long-term mean SO_2 concentration was 5.4 μg/m^3. The geographical distribution of NO_x concentrations in Case 3 (which had the highest NO_x emission) was the same.

Table 10.36 Modelling results ($\mu g/m^3$)

	Case 1		*Case 2*		*Case 3*	
	NO_x	SO_2	NO_x	SO_2	NO_x	SO_2
Annual mean	1.70	5.4	1.00	0.030	1.90	-
98th percentile	16.00	50.0	11.00	0.350	22.00	-
100th percentile	700.00	2200.0	350.00	11.000	600.00	-
Wet deposition[a]	0.64	2.0	0.34	0.010	0.64	-
Dry deposition[a]	0.03	0.1	0.02	0.001	0.04	-
Total deposition[a]	0.64	2.0	0.34	0.010	0.64	-

a) $\mu g/s/m^2$

10.11.4 Assessment of predicted concentrations

NO_2 diffusion tube data was provided for sites near Peterhead. The Arbuthnot House site in the centre of Peterhead recorded the highest readings, with an average over 6 months of 26.8 $\mu g/m^3$. Concentrations were recorded over the period April to October 1996. The results were probably influenced by vehicle exhaust emissions. The diffusion tube data can be approximated to the 98th percentile EU limit and guide values by the relationship (Association of London Chief Environmental Health Officers and Greater London Council Scientific Services Branch, 1985/86; Campbell *et al.*, 1992) given in equation 5.1 (Chapter 5).

Using this relationship, and assuming the annual mean approximated the mean observed over six months, the 98th percentile measured NO_2 concentration was 64.3 $\mu g/m^3$. The concentration included a component for the Peterhead power station. The 64.3 $\mu g/m^3$ value was compared to the EU Directive limit value of 200 $\mu g/m^3$ and the guide value of 135 $\mu g/m^3$. The diffusion tube NO_2 level was below the EU Directive criteria. The Aberdeen kerbside mean annual NO_2 concentration for 1994 was 43.2 $\mu g/m^3$, with a recorded maximum 56.4 $\mu g/m^3$ (Stevenson and Bush, 1996). Kerbside concentrations were likely to be elevated above typical ambient concentrations because of the proximity of motor vehicle exhausts. The kerbside readings were diffusion tube data. Using the relationship quoted above for relating annual mean values to 98th percentile values, the 98th percentile-measured NO_2 concentration for Aberdeen was 104 $\mu g/m^3$. The contribution from Peterhead power station was likely to have been minimal, given the 42 km distance from Aberdeen. Comparing 104 $\mu g/m^3$ to the EU Directive limit and guide values, the measured background levels were below the EU Directive criteria.

SO_2 was monitored at several locations in the locality to the power station. Annual average SO_2 levels for 1995-1996 for Peterhead, Longside, Fraserburgh, Balmedie, Hatton and Fyvie were 9.7-14.0, 12.9, 14.0, 9.7, 8.6 and 10.9 $\mu g/m^3$ respectively (DoE, 1997b). Aberdeen was the nearest other location to the power station with background air quality data. In Aberdeen, an annual mean SO_2 concentration of 18 $\mu g/m^3$ was recorded during the period April 1994 to March 1995 (AEA Technology, 1996). Again, the contribution from Peterhead power station was likely to have been minimal. The annual mean SO_2 concentration of 18 $\mu g/m^3$ was considerably below the EU guide value of 40 to 60 $\mu g/m^3$. Aberdeen recorded a 98th percentile SO_2 concentration of 31 μg/m^3 during the same period. Data for locality was considerably below the EU limit value of 120 μg/m^3.

10.11.5 Comparison of predicted deposition levels with critical load levels

The mean annual wet deposition of nitrate between 1989 and 1991 in the Peterhead area was reportedly in the range 0.24 to 0.36 gN/m^2 (DoE, 1994a). Mean annual dry deposition for the period July to December 1991 was reportedly in the range 0.12 to 0.24 gN/m^2 (DoE, 1994a). These values were expressed in nitrogen equivalents. Case 1 maximum NO_2 predicted wet and dry deposition values were 0.64 and 0.03 μg/s/m^2, which equated to 5.8 gN/m^2/yr and 0.3 gN/m^2/yr, respectively; or 57.6 and 3.0 kgN/ha/yr. However, the high deposition values were predicted immediately above the stack, and were unrepresentative of deposition in the surrounding area. Contour plots indicated that the values reduced rapidly away from the stack, so that 1 km from the power station, total deposition values are less than 0.2 μg/s/m^2 or 19.2 kgN/ha/yr, and there was limited deposition 3 km away from the stack in any direction.

10.11.6 Environmental consequences

Worst case NO_x emissions were predicted for Case 3, when NGL was used exclusively as fuel. There were no SO_2 emissions from the combustion of NGL. The worst case SO_2 emissions were predicted for Case 1, when heavy fuel oil was burned exclusively. Case 1 emissions of NO_x were almost identical with Case 3 emissions. The lowest modelled emissions of both NO_x and SO_2 were predicted for Case 2, where natural gas was burned exclusively. Despite the use of worst case data combinations and modelling techniques, predicted air quality remained very good after the addition of worst case emissions from the power station. Predicted air quality levels did not exceed the strictest EU air quality criteria. Compared to CLAG critical load levels for NO_2 values, the impact of acid deposition from NO_2 was likely to be within acceptable limits.

10.12 CASE STUDY 11 - WIDENING OF THE TOLO/FANLING HIGHWAY BETWEEN THE ISLAND HOUSE INTERCHANGE AND FANLING (HONG KONG), PEOPLE'S REPUBLIC OF CHINA

The study (Enviros Ltd, 1999) evaluated the likely air quality impacts associated with the operational phase of the widened Tolo Highway between Island House Interchange and Lam Kam Road Interchange, and the Fanling Highway between Lam Kam Road Interchange and Wo Hop Shek Interchange, Hong Kong (Fig. 10.14). The study focused on the future road traffic emission impacts (NO_2 and RSP) and considered the existing Air Pollution Control Ordinance (APCO), the Technical Memorandum on Environmental Impact Assessment Process (TM-EIA), representative ASRs and potential sources of air pollutants. The APCO established a number of AQOs which detail the Hong Kong statutory limits for a range of

Figure 10.14 Map of study area for road development study (not to scale) (courtesy BMT Asia Pacific)

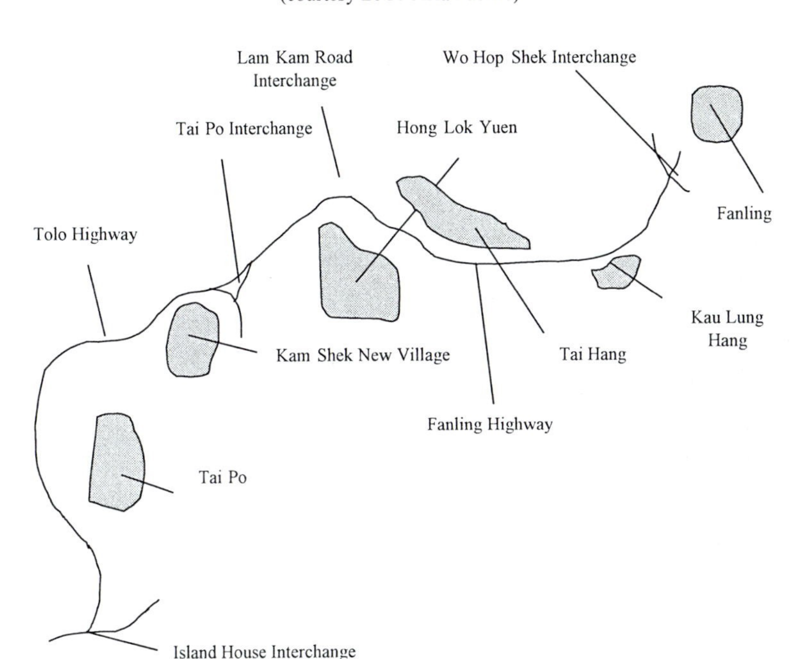

pollutants, including NO_2 and RSP (Table 7.10). The study was undertaken to evaluate potential residual impacts and determine their acceptability.

The potential for the generation of construction dust was addressed qualitatively and recommendations on the appropriate remedial actions to minimise any potential impacts were provided. This was done to ensure compliance with Air Pollution Control (Construction Dust) Regulations (Section 43, Cap.311 of APCO) and to ensure effective control of any potential dust impacts.

10.12.1 Description of study area

At the time in line with accepted vehicle-related air modelling procedures in Hong Kong, no baseline air quality measurements were taken as part of the study. To identify suitable background concentrations for the modelling work, reference was made to the results recorded under a long-term monitoring programme carried out by the HKEPD. The Tai Po Air Quality Monitoring Station was the closest station to the study area and the most recent available monitoring data from the station was utilised as being representative of background air quality conditions. Annual averages of TSP, RSP and NO_2 for 1998 are summarised in Table 10.37. RSP and NO_2 levels were used in the air dispersion modelling study. Measured concentrations met the applicable Hong Kong AQOs (Table 7.10).

Table 10.37 TSP, RSP and NO_2 levels for 1998 from the Tai Po air quality monitoring station (μg/m^3)

Pollutant	*Highest 1 hr*	*Highest 24 hr*	*Highest monthly*	*Annual average*
TSP	-	156	98 (Jan)	68
RSP	354	129	72 (Jan)	50
NO_2	245	124	67 (Nov)	51

Note: Annual Average AQO limits for TSP, RSP and NO_2 are 80, 55 and 80 μg/m^3, respectively (Chapter 7, Table 7.10).

HKEPD monitoring results for 1998 indicated an improvement over 1997 for the annual average concentrations of both RSP and TSP, while that for NO_2 was essentially unchanged. With the completion of the road widening works, the traffic volume along the roadways would increase and this might lead to an increase in the ambient levels of NO_2 and RSP. However, with the impending conversion of both taxis and mini buses to LPG, improvements in fuels (i.e., ultra-low sulphur diesel), the adoption of the Euro 3 and in future, Euro 4 standards for vehicle engines, as well as the retrofitting of catalytic oxidisers and particulate traps on goods vehicles which do not meet the applicable Euro standard, it was anticipated that the overall future impacts would actually be lower than current impacts. However, this would ultimately depend upon the vehicle numbers, mix of vehicle types, along with the general state of repair of the engines in question.

The spatial impact zone for the assessment of air quality impacts was defined as 500m from the boundary of the work site. Air quality impacts were assessed for the operational phase of the project only and related to potential vehicle emission impacts from road traffic.

Table 10.38 List of outline zoning plans and outline development plans referenced

Outline Zoning Plan and Outline Development Plan Number	*Date*
Kau Lung Hang – Outline Zoning Plan S/NE-KLH/1	24.6.1994
Draft Kau Lung Hang – Outline Zoning Plan S/NE-KLH/2	6.8.1999
Proposed Amendment on the Draft Kau Lung Hang – Outline Zoning Plan S/NE-KLH/1	12.9.1997
Fanling / Sheung Shui – Outline Zoning Plan S/FSS/7	27.2.1998
Fanling / Sheung Shui – Outline Zoning Plan S/FSS/8	2.7.1999
Tai Po – Outline Zoning Plan S/TP/9	27.1.1998
Proposed Amendment on the Tai Po – Outline Zoning Plan S/TP/7	21.11.1997
Proposed Amendment on the Tai Po – Outline Zoning Plan S/TP/9	13.11.1998
Tai Po – Outline Zoning Plan S/TP/10	23.3.1999
Draft Tai Po – Outline Zoning Plan S/TP/11	20.8.1999
Tai Po New Town Outline Development Plan D/TP/A (Provisional)	20.3.1990
Fanling / Sheung Shui Outline Development Plan D/FSS/A	20.3.1990

In accordance with Annex 12 of the TM-EIA, domestic premises, hotels, hostels, hospitals, clinics, nurseries, temporary housing accommodations, schools, educational institutions, offices, factories, shops, shopping centres, places of public

Table 10.39 Selected ASRs between Pak Wo Road and Hong Lok Yuen Road

ASR	*Description*	*Land use*[a]	*Dist. from kerb (m)*
SR1	Avon Park	R	90
SR2	Fanling Government Secondary School	Ed	48
SR3	Dawning Views	R	65
SR7	Southwest Tong Hang	R	56
SR9	Wo Hop Shek 2	R	28
SR11	Kiu Tau	R	10
SR17B	Tai Hang 3	R	14
SR20	Hong Lok Yuen 2	R	34
SR22	Wai Tau Tsuen 1	R	22
SR75	Wong Kong Shan	R	78
SR77	Yuen Leng 2	R	56

a) Residential uses (R); Educational uses (Ed)

Table 10.40 Selected ASRs between Hong Lok Yuen Road and Tai Po Tai Wo Road

ASR	*Description*	*Land use*[a]	*Dist. from kerb (m)*
SR23	Wai Tau Tsuen 2	R	32
SR25	Kau Liu Ha 2	R	22
SR28	Northwest Shek Kwu Lung	R	17
SR29	Parc Versailles	R	96

a) Residential uses (R)

Table 10.41 Selected ASRs between Tai Po Tai Wo Road and Tat Wan Road

ASR	*Description*	*Land use*[a]	*Dist. from kerb (m)*
SR31	Shek Kwu Lung 2	R	47
SR33	Shek Kwu Lung 3	R	26
SR34	Pun Chun Yuen	R	88
SR35	Buddhist Tai Kwong Middle School	Ed	54
SR36	Ma Wo 1	R	17
SR55	Dynasty View 2	R	53
SR56	Monastery at Ma Wo	T	16
SR78	Dynasty View 3	R	75

a) Residential uses (R); Educational uses (Ed); Temple/Place of Worship (T)

Table 10.42 Selected ASRs between Tat Wan Road and Island House Interchange

ASR	*Description*	*Land use*[a]	*Dist. from kerb (m)*
SR39	The Paragon	R	52
SR43	Wan Tau Tong Estate – Wan Lam House 2	R	29
SR45	HK Teacher's Association Secondary School	Ed	66
SR47	Wang Fuk Court – Wang Cheong House 1	R	42
SR54	Riverrain Bayside	R	65
SR57	King Nga Court – King Yuet House 1	R	95
SR61	Tak Nga Court 2	R	114
SR62	Ha Wun Yiu	R	41
SR64	Shan Tong New Village 1	R	27
SR66	P.L.K. Tin Ka Ping Primary School	Ed	119
SR68	Island House Park – Bicycle Track	Rec	40
SR69	Island House Park – Garden	Rec	85
SR70	Kwong Fuk Estate – Kwong Lai House	R	42
SR71	Tai Po Waterfront Park	Rec	63
SR72	Tai Po Waterfront Park	Rec	68
SR73	Island House Conservation Studies Centre	Ed	125
SR74	Yuen Chau Tsai – Tennis Court	Rec	58
SR100	KCRC Staff Quarter at Tai Po Kau	R	54

a) Residential uses (R); Educational uses (Ed); Recreational uses (Rec)

Table 10.43 Meteorological parameters

Meteorological parameter	*Daytime*	*Night-time*
Wind direction[a]	45°	45°
Wind speed	1 m/s	1 m/s
Atmospheric stability class	D	F
Mixing height	500 m	500 m
Wind direction standard deviation	18°	12°
Ambient temperature	25.5°C	25.5°C
Aerodynamic roughness	100 cm	100 cm

a) Wind direction input prompt of CALINE4 model requires the general wind direction in Hong Kong, i.e. north-easterly.

worship, libraries, courts of law, sports stadia or performing arts centres are considered as ASRs. A thorough review of all the Outline Zoning Plans, Outline

Development Plans, and the Register of all the recent re-zoning applications in the Technical Services section of the Planning Department was carried out to identify both existing and committed land uses in order to identify potential sensitive receivers. The Outline Zoning Plans and Outline Development Plans referred during the study are listed in Table 10.38. More than forty ASRs were identified and are detailed in Tables 10.39 to 10.42. Table 10.43 details the meteorological parameters used in the CALINE4 (Chapter 6).

10.12.2 Construction phase air quality impacts

During the construction phase of the project, there was the potential for dust generation arising during demolition, earthworks, stockpiling of materials and vehicle movements. If unmitigated, these activities may cause the TSP and RSP air quality criteria to be exceeded. The potential dust generation sources during the works were as follows (also see Section 10.5 and Section 8.3.1):

- Demolition work including breaking concrete;
- Earthworks, including excavation, soil stripping, re-grading;
- Site clearance, including removal of vegetation and topsoil;
- Unloading and handling of excavated materials;
- Truck movements on unpaved haul roads;
- Wind blown dust from stockpiled materials;
- Deposition of dust from haulage trucks onto local roads.

In addition, there were also gaseous emissions from construction vehicles and Powered Mechanical Equipment (PME) such as air compressors and generators. However, emissions from construction vehicles were unlikely to result in any adverse impacts owing to their limited number. Moreover, with good on-site housekeeping practices and regular maintenance of the PME, gaseous emissions were unlikely to result in any adverse impacts. Therefore, construction dust was considered to be the most important air quality issue during the construction phase of the work. The TSP impacts caused by construction dust generation at the identified ASRs may have exceeded the hourly limit of 500 $\mu g/m^3$ if unmitigated. With the implementation of dust mitigation measures however, dust emissions would be effectively controlled and thus comply with the Hong Kong AQOs. Existing highways would be fully utilised to serve as access roads and haul roads during the construction phase of the project. Therefore, it was considered likely that there would only be a limited number of unpaved access/haul roads used. As such, the dust generation from construction vehicle movements was considered to be minor. Construction dust impacts were required to be within the 1 hour TSP criterion of 500 $\mu g/m^3$ and the 24 hour AQO 260 $\mu g/m^3$. Therefore, effective control measures were needed to meet the requirements of the Air Pollution Control (Construction Dust) Regulation. Typical dust control measures during the construction phases would include:

- Restricting heights from which materials are dropped as far as practicable, to minimise the fugitive dust arising from unloading/loading;
- All stockpiles of excavated materials or spoil of more than 50 m^3 should be enclosed, covered or dampened during dry or windy conditions;
- Effective water sprays should be used to control potential dust emission sources such as unpaved haul roads and active construction areas;
- All spraying of materials and surfaces should avoid excessive water usage;
- Vehicles that have the potential to create dust while transporting materials should be covered, with the cover properly secured and extended over the edges of the side and tail boards;
- Materials should be dampened, if necessary, before transportation;
- Travelling speeds should be controlled to reduce traffic induced dust dispersion and re-suspension within the site from the operating haul trucks;
- Vehicle washing facilities would be provided to minimise the quantity of material deposited on public roads.

To ensure construction dust impacts were controlled within the relevant standards as stipulated in Annex 4 of the TM-EIA, an environmental monitoring and audit programme was established and implemented by the contractor throughout the construction period.

10.12.3 Operational phase air quality impacts

It is standard procedure in Hong Kong, when assessing vehicle traffic-related air quality impacts, for NO_2 (1 hour) and RSP (24 hour) to be evaluated. The actual modelling was based on the worst case hourly traffic flows (year 2020) as prepared by the study team's traffic consultants. Modelling results were compared to the applicable AQO (1 hour NO_2 AQO of 300 $\mu g/m^3$ and 24 hour RSP of 180 $\mu g/m^3$) to establish compliance.

The evaluation of traffic-related air quality impacts resulting from the project was performed using the Gaussian dispersion CALINE4 model. In accordance with the requirements of the TOR, all roads, including the widened Tolo and Fanling Highways, their associated slip roads and the existing roads within a 500m radius of the project area, i.e., the proposed road-widening works limit, were modelled.

In order to verify the maximum potential air quality impacts of future traffic in the area, vehicle emissions from the worst case predicted traffic flow within 15 years of the commissioning of the widened highways were used. The worst case hourly traffic flows were predicted to be during the morning peak hour in 2020. The most updated vehicle emission factors available from the HKEPD were for 2011 and these were adopted for the study. It was understood that the vehicle emission factors for Hong Kong traffic would decline with time owing to improvements in vehicle engines, the increased use of catalytic converters and particulate traps, as well as improved fuel (i.e. LPG and low-sulphur diesel). However, as the future emission factors beyond 2011 were not available, the year 2011 fleet average emission factors of NO_x and particulate matter were used.

Background levels used in the study were an annual average of 51μg NO_2/m^3 and an annual average of 50 μg RSP/m^3.

Two types of direct noise mitigation, in the form of vertical and canopy type barriers, were proposed to abate adverse traffic noise impacts arising from the widened highways. In order to account for the presence of these barriers, the HKEPD required the study approach to raise the vehicle-related air pollutant release height to the top of the barriers. This represents a worst case scenario and was considered to be very conservative. In addition, the worst case wind direction option was selected for each run of the CALINE4 model.

For the calculation of NO_2 concentrations an approach taking into consideration the O_3 limiting factor on NO_2 emissions was adopted. Therefore the NO_2 option is used in the CALINE4 model with the NO_x emission factor being utilised. The following input parameters, as agreed with HKEPD, were used:

- Ambient O_3 concentration of 62 μg/m^3;
- NO_2 photolysis rate constant (1/second) of 0 (for the most conservative analysis).

Since only a limited number of road segments can be input into each CALINE4 data file, each run was divided into six separate input files. Because the total O_3 background concentration (62 μg/m^3) must be utilised in each CALINE4 modelling file, the actual NO_2 formation was overestimated in each run. As a result, the overall predicted NO_2 concentrations were overestimated.

Owing to a limitation of the CALINE4 model, only 1 hour pollutant concentrations could be predicted. In order to provide a RSP concentration that is comparable to the 24 hour RSP AQO, the RSP concentration was modelled separately for daytime and night-time periods and the daily concentration was calculated based on the daytime and night-time results. The meteorological conditions for the daytime modelling were considered to be applicable during 08:00-21:00 hours, while the night-time parameters were utilised for 21:00-08:00 hours.

The averaged hourly traffic flow figures during the two time periods were calculated based on the 24 hour traffic flow breakdown figures. These were used in the CALINE4 model for the daytime and night-time model. The 24 hour concentrations were then calculated by adding the predicted concentrations proportionally according to the relative durations, i.e., 13/24 of the daytime predicted concentrations plus 11/24 of the night-time predicted concentrations.

Owing to limitations within the CALINE4 model (source elevation input parameter range: −10 to +10 m; receptor elevation input parameter: must be positive), the pollutant concentrations at some ASR heights could not be modelled directly. When the vertical separation between the road segment and the ASR was greater than 10 metres and the road segment was at a higher level than the ASR, the ASR was modelled at 10 metres below that particular segment(s). By doing so, a conservative prediction was obtained and this is considered to be representative of the worst case scenario. Air pollutant concentrations were evaluated at the worst case elevations for the corresponding ASRs. The worst case elevations were estimated with reference to the shortest slant separations between the nearest roads and the ASRs.

It should be noted that the limitations of the CALINE4 model, as well as the lack of vehicle emission factors for 2020, led to the generation of extremely conservative results. As such, it is believed that the reported results were representative of the worst case scenario and, in certain instances, may overestimate the level of impact. The NO_2 and RSP modelling results for 2020 with barrier scenario, including background concentrations for the 42 ASRs are presented in Table 10.44. The modelling results showed no exceedences of either the NO_2 hourly AQO limit (300 $\mu g/m^3$) or the RSP 24 hour AQO limit (180 $\mu g/m^3$) for any of the representative ASRs at the worst case elevations and wind directions.

The predicted concentration levels of NO_2 and RSP at the identified representative ASRs comply with the applicable AQOs. Therefore no mitigation measures were considered necessary. No adverse residual air quality impacts were anticipated from the operation phase of the roadways.

10.12.4 Environmental consequences

A qualitative assessment on construction dust impacts identified that fugitive dust was the primary potential air pollutant during the road-widening works. Established dust suppression techniques such as regular watering of haul roads, covering/dampening any stockpiles and dampening dusty materials before transportation were proposed. Through the proper implementation of the recommended mitigation measures, dust generation would be controlled and would not exceed the acceptable criteria. This, however, was to be further verified through an environmental monitoring and audit programme which would be undertaken as part of the construction works.

The CALINE4 modelling results indicated that neither the hourly NO_2 or 24 hour RSP AQO would be exceeded during the operation phase of the project. Because the results were based upon both the worst case traffic flows (2020) and wind directions, it was concluded that the traffic-related air quality impacts would be insignificant.

10.13 CASE STUDY 12 - EMISSION INVENTORY OF VOCs

The scope of the study was to estimate VOC emissions from various operating aspects of an oil terminal at Nigg, Ross-shire, Scotland (Cordah Limited, 1999). According to the American Petroleum Institute (1987) hydrocarbon losses during tanker loading may be estimated using the following loading emission factor:

$$EL = 1\,lb/1000\ US\ gal\ loaded \tag{10.2}$$

Assuming an approximate density of 0.88 t/m^3 for the oil terminal crude and converting to SI units this becomes:

$$EL = 0.14\ kg/t\ crude\ loaded = 0.1361835 \tag{10.3}$$

Table 10.44 Predicted 2020 levels (μg/m^3) at the worst case elevations with noise barriers in place

ASR	*Description*	*NO_2[a]*	*RSP[b]*
SR1	Avon Park	150.7	84.5
SR2	Fanling Government Secondary School	177.3	82.8
SR3	Dawning Views	179.2	94.8
SR4	Fanling Centre 1	132.8	78.7
SR7	Southwest Tong Hang	258.8	123.0
SR9	Wo Hop Shek 2	201.5	102.5
SR11	Kiu Tau	288.8	149.5
SR17b	Tai Hang 4	242.9	112.6
SR20	Hong Lok Yuen 2	151.4	92.8
SR22	Wai Tau Tsuen 1	259.1	110.8
SR23	Wai Tau Tsuen 2	261.0	147.8
SR25	Kau Liu Ha 2	207.1	103.1
SR28	Northwest Shek Kwu Lung	240.4	142.1
SR29	Parc Versailles	167.9	86.2
SR31	Shek Kwu Lung 2	241.9	96.2
SR33	Shek Kwu Lung 4	237.4	122.3
SR34	Pun Chun Yuen	153.2	89.1
SR35	Buddhist Tai Kwong Middle School	227.4	109.3
SR36	Ma Wo 1	276.5	146.8
SR39	The Paragon	186.4	99.0
SR43	Wan Tau Tong Estate - Wan Lam House 2	158.7	83.6
SR45	HK Teacher's Association Secondary School	219.0	99.2
SR47	Wang Fuk Court – Wang Cheong House 1	237.6	108.3
SR54	Riverrain Bayside	182.0	86.2
SR55	Dynasty View 1	190.3	98.9
SR56	Monastery at Ma Wo	181.3	99.7
SR57	King Nga Court – King Yuet House 1	158.3	82.0
SR61	Tak Nga Court 2	146.5	76.5
SR62	Ha Wun Yiu	240.2	126.5
SR64	Shan Tong New Village 1	227.9	120.1
SR66	P.L.K. Tin Ka Ping Primary School	165.2	81.2
SR68	Island House Park – Bicycle Track	198.3	93.9
SR69	Island House Park – Garden	169.0	86.9
SR70	Kwong Fuk Estate - Kwong Lai House	188.5	90.7
SR71	Tai Po Waterfront Park 1	148.5	78.9
SR72	Tai Po Waterfront Park 2	119.9	73.0
SR73	Island House Conservation Studies Centre	225.5	105.9
SR74	Yuen Chau Tsai - Tennis Court	281.6	118.5
SR75	Wong Kong Shan	195.4	94.5
SR77	Yuen Leng 2	216.3	111.1
SR78	Dynasty View 3	157.9	88.8
SR100	KCRC Staff Quarter at Tai Po Kau	174.8	92.8

The NO_2 AQO hourly limit is 300 μg/m^3 and the RSP AQO 24 hour limit is 180 μg/m^3

a) *1 hr + Background*

b) *24 hr + Background*

The total crude loaded in 1998 at the terminal was 2,247,924t. Therefore, the estimated tanker hydrocarbon losses were 306.1 t/a. The American Petroleum Institute has developed emission factors at three levels of increased detail and accuracy for estimating emissions during crude oil loading and ballasting operations. The above calculations were classed as Level One for terminals, when little or no information is available on cargoes or conditions of vessels calling at the terminal. Details on data required to produce such calculations are presented in American Petroleum Institute (1987). A detailed valve survey of the terminal identified 248 total crude and gas valves on site (i.e. $\geq$3 inch). The ratio of components were:

- Valves 1.0
- Connections 3.5
- Other 0.3

Table 10.45 Emission factors for values (kg/hr)

Source	*Emission factors*
Valves	0.001326
Connections	0.000164
Other	0.007526

From American Petroleum Institute (1995) the emission factors for valves, connections and others components within light crude facilities are given in Table 10.45. Component fugitive emissions were calculated using the above factors and the valve count, the findings are presented in Table 10.46. Estimated component fugitive emissions were 9.031 t/a.

Table 10.46 VOC emissions

Component type	*Estimated emissions (kg/yr)*	*Percent by source*
Valves	2879	32
Connectors	1247	14
Other	4904	54
Total	9031	100

Fig. 10.15 shows crude oil storage tanks for a similar facility in Oritupano-Leona, Venezuela and the VOC vents.

10.14 CASE STUDY 13 - ASSESSMENT OF AIR QUALITY DATA

Chapter 5 details the air quality monitoring strategy for Shenyang. A review of 1998/99 air quality data for the air quality monitoring stations in the city revealed seasonal trends and identification of potential emission sources. SO_2 concentrations exceeded both the daily and annual air quality standards of 250 $\mu g/m^3$ and 100

Figure 10.15 Venting system for crude oil storage tanks, Oritupano-Leona, Venezuela (courtesy of Cordah Limited and Perez Companc SA)

μg/m^3 (Class 3 standard) respectively (Chapter 7). Fig. 10.16 shows that SO_2 concentrations in Shenyang exhibit large seasonal fluctuations. For all sites the maximum daily average concentrations exceeded 1000 μg/m^3. At several urban sites in winter (e.g. Tianyuan Street and Woolen Mill), SO_2 concentrations exceeded 2000 μg/m^3. The principal source for these elevated concentrations were low-level district and domestic heating emissions, however, Shenyang has a large copper smelting plant (although this is due for closure). The plant may give rise to some summer peaks of SO_2. Temperature inversions in the winter further contributed to the accumulation of pollution in the early morning and evening. The air pollution load in the winter is 2.6 times greater than during the summer (AEA Technology, 2000a). TSP concentrations also show a marked annual variation in concentration in Shenyang; however, this is less discernible than for SO_2. TSP concentrations exceeded both the daily and annual air quality standards of 500 μg/m^3 and 300 μg/m^3 (Class 3 standard) respectively. Other cities also showed a less discernible seasonal variation. Scatter plots of SO_2 and TSP for Shenyang show that concentrations were not strongly correlated, this would indicate that combustion sources (principally coal) were not the sole dominant source of TSP emissions. Other sources (e.g. wind blown, construction, transport, etc.) were likely to contribute to total TSP emissions.

Figure 10.16 SO_2 concentrations for Shenyang, People's Republic of China (mg/m^3)(AEA Technology, 2000a; Shenyang Urban Planning Project, 2000)

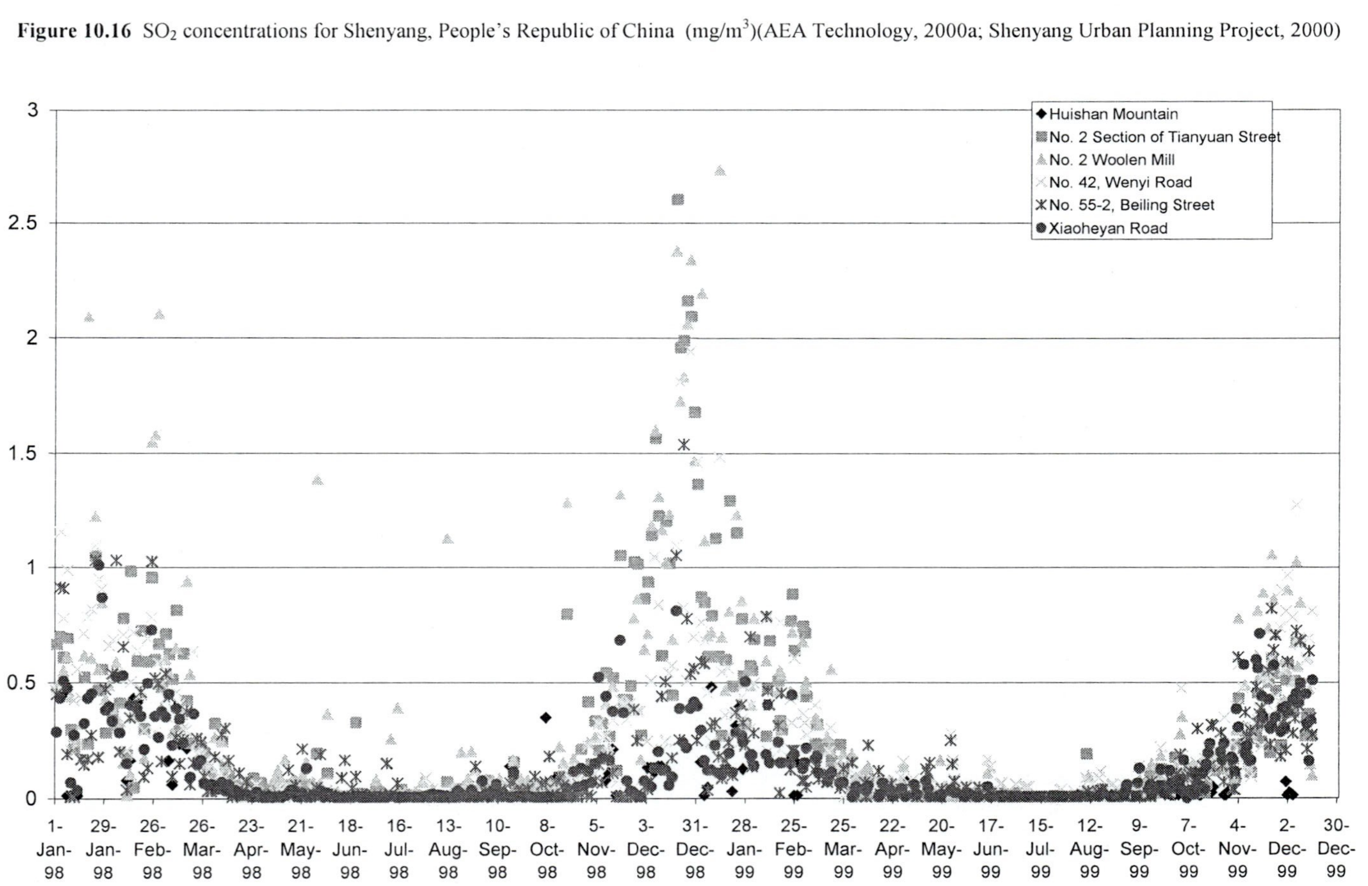

10.15 CASE STUDY 14 – A COMPARISON OF MODEL PREDICTIONS AND BASELINE MONITORING DATA

A Second Stage Air Quality Review and Assessment, to meet the requirements of the UK Environment Act, was prepared for Stirling Council, Scotland (Cordah Limited, 2000e). Concerns related to air quality issues near to a roundabout and therefore a second stage review was required to determine more accurately the risk of the objectives being exceeded. The road section was approximately 640m in length. The road section had a few residential buildings adjacent to it. The buildings along the road were all low rise. Running parallel to the road section was an industrial estate. The road section acted as a primary route into the industrial estate and as a result experienced a relatively high percentage (19%) of HGVs. The road was not a busy pedestrian thoroughfare. Results obtained from an air quality survey showed that NO_2 concentrations ranged from 8-32 $\mu g/m^3$ with an annual mean of 22 $\mu g/m^3$. Concentrations were below the annual NO_2 AQS of 40 $\mu g/m^3$.

Although the site was compliant with the AQS, the First Stage Review and Assessment (Cordah Limited, 2000d) determined, through the assessment procedure given in the DMRB, that the road running parallel to the monitoring site may exceed the AQS in 2005 due to projected high traffic flow. For the purpose of the Second Stage Review an additional four NO_2 diffusion tubes were set up along the road section of concern. The monitoring programme ran for three months. In addition, more detailed modelling was carried out using the DMRB model in order to compare calculated concentrations with monitoring data for the current situation and in order to predict future NO_2 concentrations.

Table 10.47 summarises the results from the monitoring study at each of the monitoring sites and corrects the data to estimated background levels of NO_2 in 2005 based on the three months of data. The guidelines (Stanger, 2000) recommend a method to predict concentrations for future years based on current data.

Table 10.47 Monthly mean NO_2 concentrations ($\mu g/m^3$)

Site	*Month 1*	*Month 2*	*Month 3*	*Mean*	*Range*	*Projections for 2005*
Site 1	30	42	22	31	22-42	24.5
Site 2	-	24	25	25	24-25	19.8
Site 3	28	25	22	25	22-28	19.8
Site 4	20	25	27	24	20-27	19.0

- Tube missing therefore no data available

Results showed that the concentrations of NO_2 along the road section range from 20 to 42 $\mu g/m^3$. All the values except one (42 $\mu g/m^3$) were well within the annual AQS of 40 $\mu g/m^3$. The mean values for the monitoring period range from 24-31 $\mu g/m^3$. Results were used to project the concentration of NO_2 in 2005. The results showed that, owing to a decrease of pollutants from car emissions, NO_2 concentrations would decrease, ranging from 24.5 to 19.0 $\mu g/m^3$. All the values were considerably lower than the annual AQS.

It is recommended that annual mean background concentrations should be corrected for 2005 using a methodology outlined in the guidelines (DETR, 2000g). The correction factors were derived from the estimated reduction in road traffic emissions in future years. A proportion of the background contribution (25%) was assumed to be 'non-traffic' related and was assumed to remain unchanged by 2005. Table 10.48 outlines the corrected background concentrations estimated for 2005 for NO_2 and NO_x.

Table 10.48 Corrected background concentrations estimated for 2005 for NO_2 and NO_x ($\mu g/m^3$)

Pollutant	*1996*	*2005*
NO_2	<9.55-19.1	<7.07-14.14
NO_x	< 7.64	<4.81

To predict the annual mean NO_2 concentrations the DMRB model was used. The model required input data on annual average traffic flow, annual average speeds, the fraction of HGVs and the distance from the centre of the road to the receptor. The model was run for two years of assessment. Firstly, the model was run for 2000, which allowed a comparison between monitored data from the 3 month monitoring period along the road section and the DMRB model. This allowed any difference between measured and predicted to be ascertained and considered when predicting NO_2 concentrations for 2005. Secondly, the model was used to predict NO_2 concentrations for 2005. Table 10.49 shows the results of the DMRB model for 2000 and 2005, which was run at a range of vehicle speeds from 5 to 30 mph. The road was sub-divided into 4 sections.

Table 10.49 DMRB results for concentrations of NO_2 at study road section for 2000 and 2005 ($\mu g/m^3$)

Road section	*Speed (mph)*					
	5	*10*	*15*	*20*	*25*	*30*
Section 1						
2005	73.5	55.4	46.9	42.0	36.3	36.3
2000	97.4	74.1	63.1	56.7	49.5	49.5
Section 2						
2005	87.4	68.5	59.4	53.8	50.0	47.4
2000	111.1	87.7	76.4	69.5	64.9	61.8
Section 3						
2005	87.1	68.3	59.2	53.6	49.8	47.3
2000	110.8	87.4	76.2	69.3	64.7	61.6
Section 4						
2005	84.4	65.3	56.2	50.6	46.9	44.3
2000	108.8	84.8	73.4	66.4	61.8	58.7

On the basis of historical monitoring within Stirling Council and new monitoring commissioned for the Second Stage Review and Assessment it was concluded that there were no current exceedences of the annual AQS for NO_2. A comparative study illustrated that the DMRB model over predicted the concentrations of NO_2 by between 50-303% compared to actual data obtained through the monitoring programme. Using the methods recommended to predict data for the future from the monitoring programme the values obtained were less than half the annual mean objective and unlikely to exceed the AQS. Therefore, in consideration of all work undertaken in the study, it was concluded that it was not necessary to proceed to a Stage 3 Review and Assessment for NO_2.

Epilogue

AQA is a useful and invaluable technique to the environmental management process. It is a tool which aids the efficient use and management of air resources involving not only identification, prediction and evaluation of critical variables, but also assesses potential changes of air quality as a result of emissions and their mitigation, control and management to ensure compliance with recommended air quality criteria.

Breathing is obligatory. Without air there is no life. It is today's generation of people who have realised that the air they breathe is not an unlimited natural resource. It is a real and present concern requiring effective management to ensure a sustainable future. The management of such a precious resource requires methods and techniques based on sound science and their careful application towards finding practical solutions to ever complicated problems. Future development in AQM must therefore incorporate training and guidance of the practitioner and a better awareness by the general public of the scale of the issue and the means to resolve it.

All developing economies have brought with them air pollution problems. In turn a vicious cycle has been perpetuated from developed to developing nation. One learns from the other. Or do they? Inevitably the learning process brings air pollution concerns.

There is a need to develop a new philosophy and attitude to AQM. There is also a need for a proactive stance as opposed to the usual reactive response that has previously resulted. It is encouraging that in the last 25 years air quality and its management is now firmly placed on national and international political agendas. It is important that it is kept there.

As individuals we must consider the consequences of our actions and work to improve air quality for future generations. As a society of careless individuals we have recklessly degraded the quality of our air (Bach, 1972). The time has arrived for everyone to take responsibility and participate in the management of such an essential ingredient of life. It is important to remember a few thoughts of our forefathers and reminisce a little!

'You breathe in the wildness of the sky; you reach out to the free horizon. It makes a man big, it builds a man within.'

Stephen Graham (1927), The Gentle Art of Tramping.

The objective is to avoid the reoccurrence of the air pollution problems of yesterday and anticipate and manage tomorrow's problems.

Industrial air pollution in Coatbridge, Scotland c 1950s
(courtesy of The City of Glasgow Council and the Scottish Environment Protection Agency)

'this most excellent canopy, the air, look you, this brave o'erhanging firmament, this majestical roof fretted with golden fire, why, it appears no other thing to me but a foul and pestilent congregation of vapours'

Hamlet, Act 4 Scene 2

Appendix

A.1 TERMINOLOGY

AADT	Annual Average Daily Total
AAQS	Ambient Air Quality Standards
ADMS	Air Dispersion Modelling System
AIRS	Aerometric Information System
AMIS	Air Management Information System
AOT40	Accumulated Exposure Over A Threshold of 40ppb
APCO	Air Pollution Control Ordnance
APHEA	Air Pollution and Health: A European Approach
API	Air Pollution Index
AQA	Air Quality Assessment
AQAP	Air Quality Action Plans
AQI	Air Quality Index
AQM	Air Quality Management
AQMA	Air Quality Management Areas
AQMD	Air Quality Management District
AQO	Air Quality Objective
AQS	Air Quality Standards
ASMA	Alam Sekitar Malaysia Sdn Bhd
ASR	Air Sensitive Receptors
BAT	Best Available Techniques/Technology
BATNEEC	Best Available Techniques Not Exceeding Excessive Costs
BCR	Community Bureau of Reference
BOP	Best Operating Practice
BPEO	Best Practicable Environmental Option
BPM	Best Practicable Means
BRI	Building Related Illness
BS	British Standard
CAA	Clean Air Act
CEC	Council for European Communities
CEN	Comite Europeen de Normalisation
CHIEF	Clearing House for Inventories and Emission Factors
CIWEM	Chartered Institute of Water and Environmental Management
CL	Confidence Limit
CLAG	Critical Loads Advisory Group
CLRTAP	Convention on Long-range Transboundary Air Pollution
CNS	Central Nervous System
COMAH	Control of Major Accident Hazard Regulations
COMEAP	Committee on the Medical Effects of Air Pollutants
COP	Conference of Parties
CORINAIR	CO-oRdination d'INformation Environnementale Air

DALR	Dried Adiabatic Lapse Rate
DETR	Department of the Environment Transport and the Regions
DMRB	Design Manual for Roads and Bridges
DOAS	Differential Optical Absorption Spectroscopy
DoE	Department of the Environment
EA	Environment Agency
EAL	Environmental Assessment Limit
EC	European Commission
EEC	European Economic Community
EfW	Energy from Waste
ELR	Environmental Lapse Rate
EMAS	Eco-Management and Audit Scheme
EMEP	European Monitoring and Evaluation Programme
EMS	Environmental Management Systems
EPA 1990	Environmental Protection Act 1990
EPAQS	Expert Panel on Air Quality Standards
ETC	European Topic Centres
ETC/AE	European Topic Centres Air Emissions
ETS	Environmental Tobacco Smoke
EU	European Union
FEV	Forced Expiratory Volume
FGD	Flue Gas Desulphurisation
GC	Gas chromatography
GCP	Good Combustion Practice
GEMS	Global Environmental Monitoring System
GIS	Geographical Information Systems
GWP	Global Warming Potential
HEW	Health Education and Welfare
HGV	Heavy Goods Vehicle
HI	Hazard Index
HKEPD	Hong Kong Environmental Protection Department
HMIP	Her Majesty's Inspectorate of Pollution
HPLC	High Performance Liquid Chromatography
HQ	Hazard Quotient
HSE	Health & Safety Executive
ISC	Industrial Source Complex
ISO	International Standards Organisation
IUFRO	International Union of Forest Research Organisations
IPC	Integrated Pollution Control
IPCC	Intergovernmental Panel on Climate Change
IPPC	Integrated Pollution Prevention and Control
IRIS	Integrated Risk Information System
LAPC	Local Air Pollution Control
LAQM	Local Air Quality Management
LAQMS	Local Air Quality Management System
LCP	Large Combustion Plant as defined under the Large Combustion Plant Directive (88/609/EEC)
LGC	Local Government Chemist

LGV	Light Goods Vehicle
LPEPB	Liaoning Provincial Environmental Protection Bureau
LPG	Liquid Petroleum Gas
LRS	Lower Respiratory Symptom
LTEL	Long-term Exposure Limits
MAFF	Ministry of Agriculture, Food and Fisheries
MAP	Major Air Pollutants
MEET	Methodologies to Estimate Emissions from Transport
MS	Mass Spectrometry
MEL	Maximum Exposure Limit
NAEI	National Atmospheric Emission Inventory
NAICC	National Air Issues Co-ordinating Committee
NAMS	National Air Monitoring Stations
NAMAS	National Measurement Accreditation Service
NAP	National Air Pollution
NAPSEA	Nomenclature for Air Pollution Socio-Economic Activity
NAQS	National Air Quality Strategy
NEIPTG	National Emissions Inventory Projections Task Group
NERI	National Environmental Research Institute
NETCEN	National Environmental Technology Centre
NET	National Emission Trends
NGL	Natural Gas Liquids
NIST	National Institute of Standards and Technology
NMEP	National Material Exposure Programme
NPRI	National Pollutant Release Inventory
NSCA	National Society for Clean Air and Environmental Protection
NSPS	New Sources Performance Standards
OAQPS	Office of Air Quality Planning and Standards
OCCS	Open Cast Coal Sites
OECD	Organisation for Economic Co-operation and Development
OES	Occupational exposure standard
OEL	Occupational Exposure Limit
OJ	Official journal
ONT	Odour Nuisance Threshold
OPA	Operator Performance Assessment
OS	Ordnance Survey
OSHA	Occupational Safety & Health Administration
OT	Odour Threshold
PAMS	Photochemical Assessment Monitoring Stations
PC	Personal Computer
PEF	Peak Expiratory Flow Rate
PER	Polluting Emissions Register
PFCs	Perfluorocarbons
PME	Powered Mechanical Equipment
POPs	Persistent Organic Pollutants
PORG	Photochemical Oxidant Research Group
PRTR	Pollutant Release and Transfer Register
PSI	Pollution Standards Index

QA	Quality Assurance
QC	Quality Control
QRA	Quantitative Risk Assessment
RCEP	Royal Commission on Environmental Protection
RH	Relative Humidity
SBS	Sick Building Syndrome
SCR	Selective Catalytic Reduction
SEPA	Scottish Environment Protection Agency
SEPB	Shenyang Environmental Protection Bureau
SIC	Standard Industrial Classification
SIP	State Implementation Plans
SLAMS	State and Local Air Monitoring Stations
SNAP	Selected Nomenclature for Sources of Air Pollution
SNCR	Selective Non-Catalytic Reduction
SORG	Stratospheric Ozone Review Group
SPMS	Special Purpose Monitoring Stations
SSF	Solid Smokeless Fuel
STEL	Short-term Exposure Limits
TAP	Toxic Air Pollutants
TEF	TCDD Equivalent Factor
TEQ	TCDD Equivalents
TEOM	Tapered Elemental Oscillating Microbalance
TM-EIA	Technical Memorandum on Environmental Impact Assessment
TOR	Terms of Reference
VI	Visibility Index
WHO	World Health Organisation
WMO	World Meteorological Organisation
WSL	Warren Spring Laboratory
UK	United Kingdom
UKAMN	UK Air Monitoring Network
UN	United Nations
UNCED	United Nations Conference on Environment and Development
UNECE	United Nations Economic Commission for Europe
UNEP	United Nations Environment Programme
UNFCCC	United Nations Framework Convention on Climate Change
US	United States
USEPA	United States Environmental Protection Agency
USSR	Union of Socialist Soviet Republic
UV	Ultra violet
5EAP	5th Environmental Action Programme

A.2 POLLUTANTS

Al	Aluminium
Ar	Argon
As	Arsenic
BaP	Benzo-*a*-pyrene

Be	Beryllium
BS	Black smoke
BTX	Benzene, Toluene and Xylene
C_6H_6	Benzene
C	Elemental carbon
Ca	Calcium
$CaCO_3$	Calcium carbonate
Cd	Cadmium
Cl^-	Chloride
Cr	Chromium
Cr^{VI}	Hexavaliant chromium
CFCs	Chlorofluorocarbons
CH_4	Methane
Co	Cobalt
CO	Carbon monoxide
CO_2	Carbon dioxide
CO_x	Oxides of carbon
Cu	Copper
DMS	Dimethylsulphide
Fe	Iron
F	Fluoride
HC	Hydrocarbons
HCFCs	Hydrochlorofluorocarbons
HCHO	Formaldehyde
HCl	Hydrogen chloride
H^+	Hydrogen ion
H_2	Hydrogen
He	Helium
HF	Hydrogen fluoride
HFCs	Hydrofluorocarbons
Hg	Mercury
HNO_3	Nitric acid
H_2O_2	Hydrogen peroxide
H_2S	Hydrogen sulphide
H_2SO_4	Sulphuric acid
K^+	Potassium
Kr	Krypton
Mg^{2+}	Magnesium
Mn	Manganese
MMVF	Man-made vitreous fibres
N_2	Nitrogen
Na^+	Sodium
NaCl	Sodium chloride
Ne	Neon
NH_3	Ammonia
NH_4^+	Ammonium
NH_4HSO_4	Ammonium bisulphate
NH_4NO_3	Ammonium nitrate

$(NH_4)_2SO_4$	Ammonium sulphate
Ni	Nickel
NMVOC	Non methane volatile organic compounds
NO	Nitric oxide
NO_2	Nitrogen dioxide
NO_3^-	Nitrate
NO_x	Oxides of nitrogen
N_2O	Nitrous oxide
N_2O_3	Nitrogen trioxide
N_2O_4	Dinitrogen tetroxide
N_2O_5	Nitrogen pentoxide
O_2	Oxygen
O_3	Ozone
PAH	Polyaromatic hydrocarbons
PAN	Peroxyacetylnitrate
Pb	Lead
PCB	Polychlorinated biphenyls
PCDDs	Polychlorinated dibenzo-para-dioxins
PCDFs	Polychlorinated dibenzofurans
PFCs	Perfluorocarbons
PM	Particulate matter
PM_{10}	Particulate matter 10 microns in diameter
$PM_{2.5}$	Particulate matter 2.5 microns in diameter
$PM_{1.1}$	Particulate matter 1.1 microns in diameter
RCF	Refactory ceramic fibres
RSP	Respirable suspended particulates
S	Sulphur
Sb	Antimony
Se	Selenium
SF6	Sulphur hexafluoride
Si	Silicon
Sn	Tin
SO_2	Sulphur dioxide
SO_4^{2-}	Sulphate
SO_x	Sulphur oxides
SPM	Suspended particulate matter
Ti	Titanium
TCDD	2,3,7,8-tetrachlorodibenzoparadioxine
Tl	Thallium
TOMPS	Toxic organic micro-pollutants
TP	Total particulates
TSP	Total suspended particulates
TSPM	Total suspended particulate matter
V	Vanadium
VOCs	Volatile organic compounds. Collective term for many molecular configurations of hydrogen and carbon atoms
Xe	Xenon
Zn	Zinc

A.3 UNITS

Ambient air quality is usually measured in units of concentration (i.e. the amount of substance per unit volume (μg/m^3)), or in a weight format which is related to a defined time scale or averaging period (e.g. mg/m^2/day)

Bq/m^3	Becquerels per cubic metre
^{o}C	degrees centigrade
^{o}C/km	degrees centigrade per kilometre
COHb	carboxyhaemoglobin
cm	centimetres
^{o}f	Fahrenheit
F/m^3	fibres per cubic metre
f/ml	fibre per millilitre
fibre/litre	Fibre per litre
fg/m^3	femtogrammes per cubic metre
gal	gallon
g/km	grammes per kilometre
g S/m^2	grammes of sulphur per metre square
g/m^3	grammes per cubic metre
g/mol	grammes per mol
g N/m^2	grammes of nitrogen per metre square
g N/m^2/yr	grammes of nitrogen per metre square per year
g/sec	grammes per seconds
GtC	gigatonnes of carbon
hr	hour
K	degrees Kelvin
kg	kilogrammes
kg day/mg	kilogrammes per day per milligramme
kg/hr	kilogrammes per hour
kg/kwh	kilogrammes per kilowatt hour
kg/t	kilogramme per tonne
kg/yr	kilogrammes per year
kg N or S /ha/yr	kilogrammes of nitrogen or sulphur per hectare per year
km	kilometre
km/hr	kilometres per hour
km^2	square kilometre
kph	kilometres per hour
kt	knots
kts	kilo tonnes
kt/a	kilotonnes per annum
kw/hr	kilowatt hours
J/^{o}C/kg	Joules per degrees centigrade per kilogramme
lb	pounds
l	litres
l/min	litres per minute
m	metre
mb	millibars

m/sec	metres per second
mm	millimetres
m^3	cubic metre
m^3/min	cubic metres per minute
m^3/sec	cubic metres per second
m^3/day	cubic metres per day
m^3/ha	cubic metres per hectare
mg/m^2/day	milligrammes per metre square per day
mg/m^3	milligrammes per cubic metre
mg/Nm^3	milligrammes per normalised cubic metre
min	minutes
mph	miles per hour
mt	million tonnes
MWe	megawatts of electricity
ng/m^3	nanogrammes per cubic metre
ng/Nm^3	nanogrammes per normalised cubic metre
ou	odour units
ou/sec	odour units per second
pg/m^3	picogrammes per cubic metre of air
ppb	parts per billion (by volume)
ppm	parts per million (by volume)
t	tonnes
T	time
t/a	tonnes per annum
t/ha	tonnes per hectare
t/km^2/a	tonnes per square kilometre per annum
ton/km^2	tons per square kilometre
ton/km^2/month	tons per square kilometre per month
t/m^3	tonnes per cubic metre
μg/s/m^2	microgrammes per second per square metre
μm	micron (1 millionth of a metre)
μg/dl	microgrammes per decilitre
μg/m^3	microgrammes per cubic metre
μg/(dm^2.d)	microgrammes per decimetre square per day
μg/sec	microgrammes per second
yr	year

A.4 ANNOTATIONS

Λ	washout coefficient
B_c	background concentration of the discharged pollutant for a particular district or area (mg/m^3)
C_d	actual pollutant emission concentration
C_h	corrected stack height
C_s	normalised pollutant emission concentration
C_m	maximum hourly ground level concentration (pphm)
C_{max}	3 minute maximum ground level concentration

C	the box, including at ground level, units for C are in $\mu g/m^3$
d	inside stack diameter (m)
D	discharge rate of the pollutant (g/sec)
D_o	dilution factor
d_{max}	maximum distance from source within which odour complaints are likely to occur
DTT	odour concentration (ou/m^3), where E is the product of the odour concentration and the volume discharge rate at the exit point
e	logarithm
E	odour mass flow rate (ou/sec)
E_e	collection efficiency
EL	emission loading
F	buoyancy flux (m^4/s^3)
g	acceleration of gravity (9.8 m/s^2)
G_d	guideline concentration of the discharged pollutant (mg/m^3)
h	height of source (m)
H	plume height above ground level (m)
ΔH	rise of the plume above the stack (m)
h_b	height of building (m)
h_{bx}	height of box (m)
k	constant
L	length (m)
L_{max}	maximum length of scales of a phenomena (km)
L_{min}	minimum length of scales of a phenomena (km)
M	momentum
$O_{2(s)}$	standard stack gas oxygen level
$O_{2(d)}$	discharge stack gas oxygen level
P	precipitation rate
P_i	pollution index
Q	release rate of gas or particulates less than 20 μm in size from source type(s); units for Q are usually g/sec
Q_a	pollution emission rate in mass per time (g/sec)
Q_h	heat release (MW)
Q_s	rate of pollutant emission (m^3/sec) reduce to STP
RfD	reference dose
T	time period over which assumption of uniform mixing in box holds valid, a typical period might be one hour; (secs)
T_a	air temperature (K)
T_s	stack gas temperature (K)
u	wind speed (m/sec)
u_c	uncorrected stack height
u_h	wind speed at stack top (m/sec)
u_p	wind speed at plume level (m/sec)
U_b	buoyancy plume rise
U_m	momentum plume rise
u_z	wind speed (m/sec) at vertical height z above ground level
u_a	wind speed (m/sec) at anemometer height above ground level
v	gas flow rate (m/sec)

v_v	gas volume flow rate (m^3/sec)
V_A	ventilation rate
V_B	blood volume
W_{inlet}	pollutant inlet weight
W_{outlet}	pollutant outlet weight
x	downwind dimension of box, chosen based on average wind speed and physical aspects of terrain; units for x are in metres
y	crosswind dimension of box, chosen based on average wind speed, source configuration and physical aspects of terrain; units for y are in metres
z	vertical dimension of box, chosen based on limiting inversion heights in area and physical aspects of terrain; units for z are in metres
x_d, y_d, z_d	distance in the x, y, z direction in metres
Z	depth of boundary layer
Z_1	vertical height above ground level (m)
Z_a	anemometer height above ground level (m)
χ	air pollution concentration in mass per volume, usually (g/m^3)
σ_y	the standard deviation of the concentration distribution in the crosswind direction (m), at the downwind distance x
σ_z	the standard deviation of the concentration distribution in the vertical direction (m), at the downwind distance x
π	mathematical constant pi (3.1215)
$d\vartheta/dz$	change of potential temperature with height (K/m) is the adiabatic lapse rate and equal to 0.0098 K/m
ΔT	stack gas temperature minus ambient air temperature (K)
ρ	atmospheric pressure (mb)
2.68×10^{-3}	constant (mb^{-1} m^{-1})
%	percentage
α	constant
τ	surface shearing stress

A.5 USEFUL CONTACT DETAILS

Detailed below are useful Internet addresses for national and international air pollution agencies. The list is not exhaustive, however, it does provide the principal addresses.

A.5.1 National environmental agencies

- Environment Canada: (www.ec.gc.ca)
- National Environmental Technology Centre, AEA Technology (www.aeat.co.uk/netcen)
- Scottish Environment Protection Agency: (www.sepa.org.uk)
- UK DETR: (www.detr.gov.uk)

- UK DETR The National Air Quality Information Archive: (www.aeat.co.uk/netcen/airqual/)
- UK DETR Air Quality and Environment Quality Information: (www.environment.detr.gov.uk/airq/aqinfo.html)
- UK DETR Air Quality Research Programme: (www.aeat.co.u/netcen/airqual/reports/research/index.html)
- UK Department of Health: (www.doh.gov.uk)
- UK Environment Agency (www.environment-agency.gov.uk)
- UK MAFF: (www.maff.gov.uk)
- USEPA: (www.epa.gov)

A.5.2 International institutions

- World Bank: (www.worldbank.org)
- World Health Organisation: (www.who.dk)
- European Environment Agency: (www.eea.eu.int)
- United Nations Framework Convention on Climate Change: (www.unfccc.de/resource/convkp/conv)

A.6 CONVERSION TABLES

Table A1 Conversion units

Pollutant	*WHO*	*EU*
	25°C and 1013 mb	20°C and 1013 mb
O_3	1 ppb = 1.96 μg/m^3	1 ppb = 2.00 μg/m^3
NO		1 ppb = 1.25 μg/m^3
NO_2	1 ppb = 1.88 μg/m^3	1 ppb = 1.91 μg/m^3
CO	1 ppm = 1.15 mg/m^3	1 ppm = 1.16 mg/m^3
SO_2	1 ppb = 2.62 μg/m^3	1 ppb = 2.66 μg/m^3
Benzene		1 ppb = 3.24 μg/m^3
	1 ppb = 3.2 μg/m^3	
1,3-butadiene	1 ppb = 2.2 μg/m^3	

A consistent system of units is necessary if concentrations of air pollutants in different countries are to be compared. Tables A1 and A2 provide a summary of units and conversion factors. For both gases and particles WHO (2000a) has adopted a mass per unit volume system, with concentrations generally expressed as microgrammes per cubic metre (μg/m^3). The volume of a mass of air varies with ambient temperature and atmospheric pressure and thus these conditions should be specified. In considering pollutants on a global scale this is clearly important. The alternative system, the volume mixing ratio, is applicable only to gases. In this system the concentration of gas is expressed as parts per billion (ppb), for example, and assuming ideal gas behaviour, does not depend upon the conditions of sampling because these will affect the air containing the pollutant and the pollutant

Table A2 Units to characterise air quality (from British Standard 6069, Part 1; NSCA, 1994)

Units for substances	*Quantity*	*Unit*	*Symbol*
Gases	Volume fraction of main constituents of air	per cent (by volume)	%
	Mass concentration of gaseous pollutants[a]	milligrammes per cubic metre	mg/m^3
		microgrammes per cubic metre	mg/m^3
Particles	Mass concentration of suspended matter	milligramme per cubic metre	mg/m^3
		microgrammes per cubic metre	mg/m^3
		nanogrammes per cubic metre	ng/m^3
	Size of particles	micrometre	µm
	Atmospheric dustfall[b] (deposit gauges)	gramme per square metre thirty days	g/m^2.30d
		milligramme per square metre thirty days	mg/m^2.30d
	Biological, microbiological and other suspended matter	reciprocal cubic metre	m^3
		reciprocal cubic decimetre	dm^3
Units for specifying the state of a gas			
	Celsius temperature	degree Celsius	°C
	Pressure	pascal	Pa
		kilopascal	kPa
	Relative humidity	per cent	%
Meteorological quantities			
	Wind speed	metre per second	m/sec
	Wind direction[c]	degree	°
	Precipitation intensity	millimetre per day	mm/d
	Irradiance	watt per square metre	W/m^2
Time			
	Time	second	sec
		minute	min
		hour	hr
		day	d
Miscellaneous			
	Geographical location (e.g. north (N))	degree	°
	Altitude	metre	m

a) *The units mg/m^3 and µg/m^3 express concentrations in mass volume and because the mass of a volume of gas depends on temperature and pressure, the m^3 can only be transformed to mass accurately, if ambient temperature and pressure are known (see above).*

b) *When deposit gauges are used, no account is taken of the volume of air from which the atmospheric dustfall is deposited; the duration of collection of the atmospheric dustfall should also be reported*

c) *Wind direction is conventionally reported as degrees measured clockwise from north as 0° to the north 360°*

itself to the same extent. A gas present at one part per million (ppm) thus occupies 1 cm^3 per m^3 of polluted air; is present as 1 molecule per 1 x 10^6 molecules and exerts a partial pressure of 1 x 10^{-6} atmospheres. The two systems are inter-convertible as under ideal conditions, 1 mole of gas occupies 22.4 litres at 273K and 13mb pressure, dry air Standard Temperature and Pressure Dry (STPD). The inter-conversion formula is:

$$mg/m^3 = ppm \times (molecular\ weight/molar\ volume) \quad (13.1)$$

$$\text{molar volume} = 22.4 \times T \times 1013/273 \times P \qquad (13.2)$$

Where:

T = absolute temperature (K)
P = atmospheric pressure (mb)

References

Aberdeen City Council (1999), Personal Communication from the Environmental Health Department, CO and NO_x Data.

Addink R., Govers H. A. J. and Olie K. (1998), Isomer Distribution of Polychlorinated Dibenzo-p-dioxins/dibenzofurans Formed During De Novo Synthesis on Incinerator Fly Ash, *Environmental Science and Technology*, 32, 1888-1893.

AEA Technology (1994), *Odour Measurement and Control - An Update*, Woodfield M. and Hall D. (Eds), Abingdon, Oxfordshire.

AEA Technology (1995), Air Pollution Abatement Review Group (APARG), Report on the Abatement of Toxic Organic Micro-pollutants (TOMPS), Workshop Report, May, National Environmental Technology Centre, Oxfordshire.

AEA Technology (1996), UK Smoke and Sulphur Dioxide Monitoring Networks - Summary Tables for April 1994 - March 1995.

AEA Technology (2000a), Air Quality Management and Capacity Building, Liaoning Integrated Environmental Programme, Inception Report to the European Union, CHN/B7-3000/97/4 –env/liep/d1, November, and reports from the Air Quality Management and Capacity Building Project, Liaoning Integrated Environment Programme.

AEA Technology (2000b), Information from Internet Site, www.aeat.co.uk.

Alcock R. E. and Jones K. C. (1996), Dioxins in the Environment: A Review of Trend Data, *Environmental Science and Technology*, 30(11), 3133-3143.

Alcock R. E., Gemmill R. and Jones K. C. (1998), Improvements to the UK PCDD/F and PCB Atmospheric Emission Inventory Following an Emissions Measurement Programme, *Chemosphere*, 38, 759 – 770.

Altshuller A. P., Lindhorst R. A., Nader J. S., Niemeyer L. E. and McFee W. W. (1983), *Acidic Deposition Phenomena and its Effects: Critical Assessment Review Papers, Volume 1: Atmospheric Sciences* and *Volume 2: Effects Sciences*, EPA-600/8=016a and EPA-600/8=016b, PB84-171644 and PB84-171651, Raleigh, NC, USA, North Carolina State University.

American Petroleum Institute (1987), Atmospheric Hydrocarbon Emissions from Marine Vessel Transfer Operations, API Publication 2514A, August.

American Petroleum Institute (1995), Emission Factors for Oil and Gas Production Operations, API Publication 4615, January.

Andres R. J. (1993), Sulphur Dioxide, Particle and Elemental Emissions from Mount Etna, Italy During July 1987, *Geolo. Rundsch*, 82, 1001.

Artis D. (1984), *Odour Nuisances and their Control*, Shaw & Sons, London.

Ashby E. and Anderson M. (1981), *The Politics of Clean Air*, Oxford University Press, Oxford.

Aspinwall & Company (1992), Air Quality Data from an Environmental Statement, Walford Manor, Baschurch, Shropshire.

Aspinwall & Company (1994), Environmental Statement for a Sewage Sludge Incinerator, Northern Ireland, Walford Manor, Baschurch, Shropshire.

Aspinwall Enviros (2000), Quantitative Risk Assessment Study of a Proposed Waste to Energy Plant, Hong Kong.

Association of London Authorities, London Boroughs Association and the South East Institute of Public Health (1995), Air Quality in London, Second Report of the London Air Quality Network, April.

Association of London Chief Environmental Health Officers and Greater London Council Scientific Services Branch (1985/86), Nitrogen Dioxide - Environmental Health Aspects of the 1984/85 London Diffusion Tube Survey, *London Environmental Supplement*, No 14.

Australian Government, National Health Medical Research Council (2000), Internet Site, www.environment.gov.au/epg/air toxics/iaq.

Bach W. (1972), *Atmospheric Pollution*, McGraw Hill Problems Series in Geography, McGraw Hill, New York.

Bailey J. and Bedborough D. (1979), *Sensory Measurement and Instrumental Analysis of Odours in Septic Sewage: Problems and Solutions*, The Institute of Water Pollution Control, 37-54.

Barnett J. L., Clayton P. and Davis B. J. (1987), Fugitive Emissions – a Directional Sampler for Particulates, Warren Spring Laboratory, Stevenage, Hertfordshire, August.

Barret K. and Berge E. (1996a), Estimated Dispersion of Acidifying Agents and of Near Surface Ozone, Part 1, EMEP MSC-W, Status Report, Transboundary Air Pollution in Europe, Det Norske Meteorologiske Institutt, The Norwegian Meteorological Institute, Research Report No. 32.

Barret K. and Berge E. (1996b), Estimated Dispersion of Acidifying Agents and of Near Surface Ozone, Part 2, EMEP MSC-W, Status Report, Transboundary Air Pollution in Europe, Det Norske Meteorologiske Institutt, The Norwegian Meteorological Institute, Research Report No. 32.

Barrowcliffe R. (1992), Air Quality Monitoring, *Environmental Policy and Practice*, 1(4), 3 – 26.

Bates K. J., *et al* (1990), Impact of Dust from Mineral Workings, Mineral Planning into the 1990's, County Planning Officers Society Committee No. 3 Conference, Loughborough University, 19-21 September 1990.

Bates T. S. *et al* (1992), Sulphur Emissions to the Atmosphere from Natural Sources, *Journal of Atmospheric Chemistry*, 14, 315 - 337.

Belfast City Council Environmental Health Department, (1993), Belfast City Council - Personal Communication.

Benson P. E. (1979), CALINE-3, A Versatile Dispersion Model for Predicting Air Pollution Levels Near Highways and Arterial Streets, Report No. FHWA/CA/TL-79/23.

Berry, R. W., Brown, V. M., Coward, S. K. D., Crump, D. R., Gavin, M., Grimes, C. P., Higham, D. F., Hull, A. V., Hunter, C. A., Jeffrey I. G., Lea, R. G., Llewellyn, J. W. and Raw G. J. (1996), Indoor Air Quality in Homes, The Building Research Establishment Indoor Environment Study, London Construction Research Communications.

Beychok M. R. (1979), Fundamentals of Stack Gas Dispersion, California.

Bisset R. (1991), Role of Monitoring and Auditing in EIA, First Annual Portuguese Seminar on EIA, Albufeira, April, Centro de Estudos de Planeamento a Gestao do Ambient.

Booker J. and Moorcroft S. (1988), Measurement and Identification of Odour Producing Compounds - Part 2 Odour Perception, *London Environmental Supplement*, 18, Autumn.

Bosanquet C. H., Carey W. F. and Halton E. M. (1950), Dust From Chimney Stacks, *Proceedings Institute of Mechanical Engineers*, 162, 355-367.

Bosanquet C. H. (1957), The Rise of a Hot Waste Gas Plume, *Journal Institute of the Fuel*, 30, 197, 322-328.

Bovee H. H. and Robinson R. J. (1961), Sodium Diphenylaminesulfonate as an Analytical Reagent for Ozone, *Analytical Chemistry*, 33, 1115.

BMT Fluid Mechanics (1999), Personal Communication with Mr S. Rowe, Teddington, London.

BP Research (1992), A Survey of Hydrocarbons and Acidic Gases in the Community around BP Chemicals and BP Oil at Grangemouth Between June 1991 and June 1992.

Bretschneider B. and Kurfurst J. (1987), *Air Pollution Control Technology*, Elsevier, Oxford.

Briggs D., Corvalen, C. and Nurminen, M. (1996), Linkage Methods for Environment and Health Analysis, WHO/UNEP/USEPA.

Briggs G. A. (1969), Plume Rise, Atomic Energy Commission Critical Review Series TID-25075, National Technical Information Service, 81.

Briggs G. A. (1975), Plume Rise Predictions, in Lectures on Air Pollution and Environmental Impact Analysis, *American Meteorological Society*, Boston, MA, 59-111.

Brimblecombe P. (1987), *The Big Smoke - A History of Air Pollution in London since Medieval Times*, Routledge, London.

British Coal Opencast Northern (1993), Environmental Review.

British Standards Institution (1969a), *Methods for the Measurement of Air Pollution,* Part 1, Deposit Gauges, BS 1747, HMSO, London.

British Standards Institution (1969b), *Methods for the Measurement of Air Pollution*, Part 2, Deposit Gauges, BS 1747, HMSO, London.

Broman D., Naf C. and Zebuhr Y. (1991), Long-term High and Low Volume Air Sampling of Polychlorinated Dibenzo-p-dioxins and Dibenzofurans and Polycyclid Aromatic Hydrocarbons along a Transect from Urban to Remote Areas on the Swedish Baltic Coast, *Environmental Science and Technology*, 25, 1841-1849.

Brooks K. and Schwar M. J. R. (1987), Dust Deposition and the Soiling of Glossy Surfaces, *Environmental Pollution*, 43, 129 - 141

Broughton G. F. J. *et al.* (1992), Air Quality in the UK, A Summary of Results from Instrumental Air Quality Monitoring Networks in 1990/91, LR883 (AP), Warren Spring Laboratory, Stevenage, Hertfordshire.

Buckland A. T. (1998), Validation of a Street Canyon Model in Two Cities, *Environmental Monitoring and Assessment*, 52(1-2), 255-267.

Buckland A. T. and Middleton D. R. (1999), Nomograms for Calculating Pollution within Street Canyons, *Atmospheric Environment*, 33, 1017-1036.

Building Research Establishment (1991), Radon: Guidance on Protective Measures for New Dwellings, Watford.

Cambridge Environmental Research Consultants Ltd. (2000), ADMS Technical Specification (Version 2.2).

Campbell G. W., Cox J., Downing C. E. H., Stedman J. R. and Stevenson K. (1992), A Survey of Nitrogen Dioxide Concentrations in the United Kingdom Using Diffusion Tubes: July to December 1991, WSL Report LR893, Warren Spring Laboratory, Stevenage, UK.

Canter L. (1985), *Environmental Impacts of Agricultural Production Activities*, Lewis Publishers Inc., Chelsea, Michigan, 169-209.

Canter L. W (1996), *Environmental Impact Assessment*, 2nd Edition, McGraw-Hill Book Company, New York.

Carruthers D. (1995), Atmospheric Dispersion Modelling System, Cambridge Environmental Research Consultants Ltd., National Society for Clean Air 1995 Spring Workshop, 29-30th March.

Carruthers D. J, Edmunds H. A., Lester A. E., McHugh C. A. and Singles R. J. (1998), Use and Validation of ADMS-Urban in Contrasting Urban and Industrial Locations, Proceedings of the Fifth Workshop on Harmonisation within Atmospheric Dispersion Modelling for Regulatory Purposes, Rhodes.

Catcott E. J. (1961), Effects of Air Pollution on Animals, in *Air Pollution*, World Health Organisation, Geneva.

Chalmers A. K. (1930), *The Health Of Glasgow 1818 - 1925 - An Outline*, Bell & Bain, Glasgow.

Chartered Institute of Water and Environmental Management (1997), *Monograph on Best Practice – Odour Control.*

China Daily (2000), New Plan to Reduce Desert Dust, 16th June, 20(6280).

Cirillo R. R., Tschanz J. F. and Camaioni J. E. (1975), An Evaluation of Strategies for Airport Pollution Control, Argonne, US Argonne National Laboratory, Report ANL/ES-45.

City of Edinburgh Council (1999), Review and Assessment of Air Quality in the City of Edinburgh, Stages 1 and 2.

City of Glasgow Council, (1999), Personal Communication from the Environmental Health Department, Air Quality Data.

Cleland D. I. and Kerzner H. (1986), *Engineering Team Management*, Van Nostrand Reinhold Company, New York.

Cline J. D. and Bates T. S. (1983), Dimethylsulphide in Equatorial Pacific, *Geophysical Research Letters* 10, 949 - 952.

Clyde Analytical (1993), Contaminants in the Ambient Air at Grangemouth During Several Periods Between August 1992 and May 1993, Report 8340W/FALKTRAF.

Cohrssen J. J and Corvello V. T. (1989), Risk Analysis: Guide to the Principles and Methods for Analysing Health and Environmental Risks, United States Council on Environmental Quality, Executive Office of the President, NTIS Springfield, VA, USA.

Cooper C. D. and Alley F. C. (1986), *Air Pollution Control: A Design Approach*, PWS Engineering, Boston.

Corbett J. J. *et al*, (1999), Global Nitrogen and Sulfur Inventories for Oceangoing Ships, *Journal of Geophysical Research*, 104, No. D3, 3457-3470, February.

Cordah Limited (1997), First Phase Air Quality Review Report for Falkirk Council, Aberdeen, Scotland.

Cordah Limited (1999), VOC Emission Inventory Review of Nigg Oil Terminal, Talisman Energy (UK) Ltd, Aberdeen, Scotland.

Cordah Limited (2000a), Emission Inventory of Falkirk Council, Aberdeen, Scotland.

Cordah Limited (2000b), Air Monitoring Study of Stirling Council, Aberdeen, Scotland.

Cordah Limited (2000c), Third Phase Air Quality Review Report for Falkirk Council, Aberdeen, Scotland.

Cordah Limited (2000d), Stirling Council First Stage Review and Assessment, Aberdeen, Scotland, February.

Cordah Limited (2000e), Stirling Council Second Stage Review and Assessment, Aberdeen, Scotland, May.

Cordah Limited (2001), Notes on Inspection and Auditing Techniques, Aberdeen, Scotland.

Council for European Communities (1980), Directive on Air Quality Limit Values and Guidelines for Sulphur Dioxide and Suspended Particulates, OJ No. L229, 30 August 1980.

Council for European Communities (1982), Council Directive on a Limit Value for Lead in Air, Brussels, OJ L378, 31.12.82, pp 15-18.

Cumnock Academy (1990), Climatological Station C6574 Weather Information 1973 – 1989, Cumnock, Ayrshire.

Cumnock & Doon Valley District Council (1996), Air Quality Data - Smoke, Personal Communication, April.

Davidson-Bryant W. F. (1949), The Dispersion and Spreading of Gases and Dust from Chimneys, *Trans. Conf. on Ind. Wastes*, 14th Annual Meeting Industrial Hygiene Foundation, America, pp. 38-55.

de Nevers N. H., Neligan R. E. and Slater H. H. (1977), Air Quality Management, Pollution Control Strategies, Modelling and Evaluation, in Stern A. C. (Ed.), *Air Pollution 5, Air Quality Management*, pp.3 – 40.

Department of the Environment Northern Ireland (1990), Air Quality, Statistical Bulletin (90)1, Supplement of Digest of Environmental Protection and Water Statistics No 12, 1989.

Department of the Environment (1992), *Environmental Effects of Surface Mineral Workings*, HMSO, London.

Department of the Environment (1993), *Urban Air Quality in the United Kingdom*, HMSO, London.

Department of the Environment (1994a), Critical Loads Advisory Group, Critical Loads of Acidity in the United Kingdom, Summary Report, February.

Department of the Environment (1994b), *Expert Panel on Air Quality Standards – Carbon Monoxide*, HMSO, London.

Department of the Environment, Minerals Division (1995a), *The Environmental Effects of Dust from Surface Mineral Workings*, Volumes 1 and 2, HMSO, London.

Department of the Environment (1995b), *Expert Panel on Air Quality Standards - Particles*, HMSO, London.

Department of the Environment (1995c), *Air Quality A to Z*, Air Quality Division, June, Met Office, Bracknell.

Department of the Environment (1997a), *The United Kingdom National Air Quality Strategy*, HMSO, London.

Department of the Environment (1997b), Air Quality Data from Internet Site, www.detr.gov.

Department of the Environment (1997c), *Air Quality in the UK: 1995*, AEA Technology, Abingdon, Oxfordshire.

Department of the Environment, Transport and the Regions (1998), Review of the United Kingdom National Air Quality Strategy, A Consultation Document, HMSO, London.

Department of the Environment, Transport and the Regions (1999a), *Design Manual for Roads and Bridges*, Volume 11, Section 3, Part 1 Air Quality, HMSO, London.

Department of the Environment, Transport and the Regions (1999b), *An Economic Analysis of the National Air Quality Strategy Objectives*, HMSO, London.

Department of the Environment, Transport and the Regions (2000a), *Draft Guidance Review and Assessment: Estimating Emissions*, (LAQM.TG2(00).

Department of the Environment, Transport and the Regions (2000b), Air Quality Data and Information from Internet Site, www.detr.gov.uk.

Department of the Environment, Transport and the Regions (2000c), *Review and Assessment: Monitoring Air Quality*, (LAQM.TG1(00)).

Department of the Environment, Transport and the Regions (2000d), Climate Change Programme from Internet Site, www.environment.detr.gov.uk/climate change.

Department of the Environment, Transport and the Regions, the Scottish Executive, the National Assembly for Wales and the Department of the Environment in Northern Ireland (2000e), *The Air Quality Strategy for England, Scotland, Wales and Northern Ireland*, HMSO, London, January.

Department of the Environment, Transport and the Regions, the Scottish Executive, and the National Assembly for Wales (2000f), *Review and Assessment: Selection and Use of Dispersion Models*, LAQM.TG3(00).

Department of the Environment, Transport and the Regions, the Scottish Executive, and the National Assembly for Wales (2000g), *Review and Assessment: Pollutant Specific Guidance*, LAQM.TG4(00).

Department of Health (1998a), Committee on the Medical Effects of Air Pollutants Reports.

Department of Health (1998b), *Quantification of the Effects of Air Pollution on Health in the United Kingdom*, HMSO, London.

Department of Health (1999), *Do Particulates from Opencast Coal Mining Impair Children's Respiratory Health?*, HMSO, London.

Department of Health (2000), Internet Site, www.doh.gov.uk.

Derwent R. G. and Middleton D. R. (1996), An Empirical Function for the Ratio NO_2:NO_x, *Clean Air*, 26(3/4), 57 – 60.

Dorney R. S. and Dorney L. C. (1989), *The Professional Practice of Environmental Management*, Springer-Verlag, New York.

Duarte-Davidson R., Clayton P., Davis B., Halsall C., Jones K. C. and Jones P. (1994), PCDDs and PCDFs in British Urban Air and Deposition, *Environmental Science and Pollution Research*, 1, 262-270.

Eggleston H. S. (2000), Emission Inventories, Lecture given to LEIP, Shenyang, People's Republic of China, February.

Eggleston H. S. and McInnes G. (1987), Methods for the Compilation of UK Air Pollutant Emission Inventories, LR 634(AP)M, Warren Spring Laboratory, Stevenage, Hertfordshire.

Eitzer B. D. and Hites R. A. (1989), Polychlorinated Dibenzo-p-dioxins and furans in Ambient Atmosphere of Bloomington, Indiana, *Environmental Science and Technology*, 23, 1389-1395.

Elsom D. (1987), *Atmospheric Pollution*, Basil Blackwell Ltd., Oxford.

Eltimate Ltd (2000), Scottish Environmental Protection Agency, Public Register.

EMEP/CORINAIR (1999), *Atmospheric Emission Inventory Guidebook*, 2nd Edn, Internet Site, http://reports.eea.eu.int/EMEPCORINAIR/en.

EMEP (2000), Information from Internet Site, http://projects.dnmi.no/~emep.

ENDS (1999), Setback for EMAS as Momentum Builds Behind ISO14001, *ENDS*, 291, 11.

Energy From Waste Ltd. (1999), Environmental Statement for a Waste to Energy Plant.

Enterprise Ayrshire (1996), Environmental Statement of a Proposed Chipboard Manufacturing Plant, Barony Colliery.

Environment Agency (1997), *Guidance for Operators and Inspectors of IPC Processes, Best Practicable Environmental Option Assessments for Integrated Pollution Control, Volume 1: Principles and Methodology* and *Volume 2: Technical Data (for Consultation)*, HMSO, London.

Environment Agency (1999), A Baseline Soil Survey in the South Dudley Area, July.

Environment Agency (2000), Report into an Air Pollution Episode – Sulphur Dioxide, September 2nd 1998, Midlands and South Yorkshire.

Environment Canada (1999), Emission Inventory Data, Internet Site, http://www.ec.gc.ca, August.

Environment Canada (2000), Internet Site, http://www.ec.gc.ca.

Environmental Analysis Co-operative (1999), *Emissions and Your Licence to Operate – A Guide for Assessing Releases to the Environment*, Institute of Chemical Engineers, Rugby, England.

Environmental Resources Management Ltd. (2000a), An Assessment of Dioxin Emissions in Hong Kong, Final Report, March.

Environmental Resources Management Ltd. (2000b), Air Quality Assessment, Integra South West Energy Recovery Facility, May.

Enviros Ltd. (1999), Widening of the Tolo/Fanling Highway between the Island House Interchange and Fanling (Hong Kong), People's Republic of China

European Environment Agency (1997), *Air Pollution in Europe*.

European Environment Agency (1998a), *Europe's Environment: Statistical Compendium for the Second Assessment*.

European Environment Agency (1998b), *Europe's Environment, The Second Assessment*.

European Environment Agency (2000a), Internet Site, www.eea.eu.int.

European Environment Agency (2000b), Guidelines for Defining and Documenting Data on Costs of Possible Environmental Protection Measures, Internet Site, ww.eea.eu.int, environmental themes.

Faith W. L. and Atkisson A. A. (Eds) (1972), *Air Pollution*, Wiley-Interscience, John Wiley and Sons Inc, New York.

Fangmark I., Stromberg B., Berge N. and Rappe C. (1995), The Influence of Fly Ash Load and Particle Size on the Formation of PCDD, PCDF, PCBz and PCB in a Pilot Incinerator, *Waste Management and Research,* 13, 259-272.

Ferman M. A., Wolff, G. T. and Kelly N.A. (1981), The nature and sources of haze in the Shenandoah Valley/Blue Ridge Mountains area, *Journal of the Air Pollution Control Association*, 31, 1074.

Fiedler H. (1993), Formation and Sources of PCDD/PCDF, *Organohalogen Compounds*, 11, 221 - 228.

File R. (1988), Dutch Smell Comes to London, *London Environmental Bulletin*, Autumn, 4(4), 18 – 19.

Fortin C. and Caldbick D. (1997), Are Dioxins and Furan Predominantly Anthropogenic? *Organohalogen Compounds*, 32, 417 – 429.

Fuller J. G. (1977), *The Poison that Fell from the Sky*, Randon House, New York.

Georgii H-W. (1986), *Atmospheric Pollutants in Forest Areas - Their Deposition and Interception*, D Reidel Publishing Company, Dordrecht.

Gibson N. (1994), Identification and Quantification of Offensive Odour Emissions, in Woodfield M. and Hall D. (Eds), *Odour Measurement and Control – An Update*, AEA Technology, Abingdon, Oxfordshire, August.

Gillham C. A., Couling S., Leech P. K., Eggleston H. S. and Irwin J. G. (1994), UK Emissions of Air Pollutants 1970 - 1991 (Including Methodology Update), LR961, Department of the Environment.

Glasgow Scientific Services (1998), Recommendations on the Emissions Limits for Waste Water Treatment Works, Colston Laboratory.

Goodwin J. W. L., Salway A. G., Eggleston H. S., Murrells T. P. and Berry J. E., (1999), UK National Atmospheric Emissions Inventory (NAEI), January.

Graham S. (1927), *The Gentle Art of Trampling*, Ernest Benn Ltd, London.

Grant G. A., Katz M. and Haines R. L. (1951), A Modified Iodine Pentoxide Method for the Determination of Carbon Monoxide, *Canadian Journal of Technology*, 29, 43.

Grennfelt P. and Thornelof E. (Eds) (1992), Critical Loads for Nitrogen – A Workshop Report, Nord 41, Nordic Council of Ministers, Copenhagen.

Griffing G. W. (1980), Relationships Between the Prevailing Visibility, Nephelometer Scattering Coefficient, and Sunphotometer Turbidity Coefficient, *Atmospheric Environment*, 14, 577 - 584.

Gruderian R. (1977) *Air Pollution Phytotoxicity of Acidic Gases and its Significance in Air Pollution Control,* Ecological Studies, 22, Springer-Verlag, Berlin.

Hagenmaier H., Kraft M., Brunner H. and Haag R. (1987), Catalytic Effects of Flyash from Waste Incineration Facilities on the Formation and Decomposition of PCDDs and PCDFs, *Environmental Science and Technology*, 21, 1080 - 1084.

Hall, D. J. and Kukadia, V. (1994), Approaches to the Calculation of Discharge Heights for Odour Control, *Clean Air*, 24, 2.

Hall D. J., Spanton A. M., Dunkerley F., Bennett M. and Griffiths R. F. (2000a), An Inter-comparison of the AERMOD, ADMS and ISC Dispersion Models for Regulatory Applications, Environment Agency.

Hall D. J., Spanton A. M., Dunkerley F., Bennett M. and Griffiths R. F. (2000b), A Review of Dispersion Model Inter-comparison Studies Using ISC, R91, AERMOD and ADMS, Environment Agency.

Hamilton E. M. and Jarvis W. D. (1963), *The Identification of Atmospheric Dust by Use of the Microscope*, Monograph, Central Electricity Generating Board.

Hampton E., Harrop D. O. and MacDonald C. (1993), The Changing Face of Glasgow's Air Quality, *Clean Air*, 22(4), 233 - 237.

Harrad S. and Jones K. (1992), Dioxins at Large, *Chemistry in Britain*, 1110-1112.

Harrop D. O. (1986), Tackling Air Pollution Problems with Computer Models, *London Environmental Bulletin*, 3/4 (4/1), 11-12.

Harrop D. O. and Daunton R. (1987), Air Quality and Heathrow Results of Air Monitoring at Cranford and East Bedfont During 1984 – 1985, *London Environmental Supplement*, 15, 1-22.

Harrop D. O. and Pollard S. J. T. (1998), Quantitative Risk Assessment for Incineration: Is it Appropriate in the UK, *Journal of the Chartered Institution of Water and Environmental Management*, 12, 48 - 53.

Harrop D. O. (1999), Air Quality Assessment, in Petts J. (Ed.*), Handbook of Environmental Assessment, Volume 1 Environmental Impact Assessment: Process, Methods and Potential*, Blackwell Science, Oxford, pp. 252-272.

Harrop D. O. and Nixon J. A. (1999), *Environmental Assessment in Practice*, Routledge, London.

Harter P. (1986), Acidic Deposition – Materials and Health Effects, IEA Coal Research, London, ICTIS/TR 36, December.

Harvey D. and Obasaju E. (1999), Accuracy of Techniques for Modelling the Effects of Building Downwash on the Dispersion of Emissions to Atmosphere, *Clean Air*, 29(2), 42 – 45.

Health & Safety Executive (1982), Environmental Hygiene Series, Atmospheric Pollution in Car Parks, Guidance Note EH33, August.

Health & Safety Executive (1984), Guidance Note EH43, August.

Health & Safety Executive (HSE) (2000), *Occupational Exposure Limits*, HMSO, London.

Heister E., Bruckman P., Bohm R., Eynck P., Mulder A. and Ristow W. (1995), *Organohalogen Compounds*, 24, 147-152.

Her Majesty's Inspectorate of Pollution (1993a), *Sampling Facility Requirements for the Monitoring of Particulates in Gaseous Releases to Atmosphere, Technical Guidance Note (Monitoring) M1*, HMSO, London.

Her Majesty's Inspectorate of Pollution (1993b), *Technical Guidance Note (Dispersion) D1*, HMSO, London.

Her Majesty's Inspectorate of Pollution, (1995a) ADMS-Urban User Guide Version 1.53, November 1999, CERC Ltd. Validation of the UK-ADMS Dispersion Model and Assessment of its Performance Relative to R91 and ISC Using Archived LIDAR Data, HMIP Commissioned Research. DoE Report Number. DoE/HMIPIRR/95/022.

Her Majesty's Inspectorate of Pollution, (1995b), *Standards for IPC Monitoring: Part 1 – Standards, Organisations and the Measurement Infrastructure; Technical Guidance Note (Monitoring) M3*, HMSO, London.

Hesketh H. E. (1974), *Understanding and Controlling Air Pollution*, Ann Arbor Science, Michigan.

Hester R. E. (Ed.) (1987), *Understanding Our Environment*, The Royal Society of Chemistry, London.

Hewitt, C. N. and Harrison, R. M. (1986), Monitoring, in Hester R. E. (Ed.), *Understanding our Environment*, The Royal Society of Chemistry, London.

Hickman A. J. And Colwill D. M. (1982), The Estimation of Air Pollution Concentrations from Road Traffic, TRRL Laboratory Report 1052, TRRL Assessment Division, Crowthorne, UK.

Hoekstra E. J., de Weerd H., de Leer E. W. B. and Brinkman U. A. Th. (1999), Natural Formation of Chlorinated Phenols, Dibenzo-p-dioxins in Soil of a Douglas Fir Forest, *Environmental Science and Technology*, 33, 2543 – 2549.

Holgate, M. W. (1979), *A Perspective of Environmental Pollution,* Cambridge University Press, London.

Holland J. Z. (1953), A Meteorological Survey of the Oak Ridge Area, Atomic Energy Comm., Report ORO-99, Washington, D.C, p. 540.

Hong Kong Environmental Protection Department (1997), Technical Memorandum on Environmental Impact Assessment Process.

Hong Kong Labour Department (1995), Reference Note on Occupational Exposure Limits for Chemical Substances in the Work Environment, Occupational Health Division, Hong Kong.

Hunt G. T. and Maisel B. E. (1990), Atmospheric PCDDs/PCDFs in Wintertime in a North-eastern US Urban Coastal Environment, *Chemosphere*, 20, 1455-1462.

Industrial Science Division, Department of Economic Development (1988), Nitrogen Dioxide Survey at Belfast Harbour using Passive Sampling Tubes.

Industrial Science Division, Department of Economic Development (1990), A Nitrogen Dioxide Survey at Belfast Harbour Estate Using passive Diffusion Samplers.

Industrial Science Centre (1991), A Survey of Nitrogen Dioxide Levels in the Atmosphere of Areas Adjacent to Kilroot Power Station.

Innes J. L. (1987), Air Pollution and Forestry, *Bulletin 70*, Forestry Commission, HMSO, London.

Institute of Environment and Health (1997), Indoor Air Quality in the Home.

Institute of Terrestrial Ecology (1999), Maps of Ozone and Trends at Scottish Sites 1986 – 97, Bush, Penicuik.

Intergovernmental Panel on Climate Change (1996), *The Science of Climate Change 1995, Summary for Policy Makers*, Cambridge University Press, Cambridge.

International Chamber of Commerce (1991), Business Charter for Sustainable Development, London.

International Organisation for Standardisation (1996), *Environmental Management Systems – Specification Guidance for Use*, International Organisation for Standardisation, Geneva, Switzerland.

Janssen L. H. J. M. (1988), A Classification of NO_2 Oxidation Rates in Power Plant Plumes Based on Atmospheric Conditions, *Atmospheric Environment,* 22, 43-53.

Japan Environmental Management Association for Industry (1997), Private Communication.

Japanese Government (2000a), State of Japan's Environment at a Glance - Air Pollution, Environment Agency.

Japanese Government (2000b), Air Quality Data, Internet Site, www.jin.japan.org/stat/stats.

Jordon B. C. (1977), An Assessment of Potential Air Quality Impact of General Aviation, Research Triangle Park, NC, Environmental Protection Agency Office of Air Quality Planning and Standards.

Keddie A. (1982), The Quantification of the Emissions and Dispersion of Odours from Sewage-Treatment Works, *Water Pollution Control*, 266-279.

Keeling C. D. and Whorf T. P. (1998), Atmospheric CO_2 Concentrations – Mauna Loa Observatory, Hawaii 1958 – 1997, Carbon Dioxide Information Analysis Center, Oak Ridge National Laboratory, Tennessee.

Kemp K. (Ed.) (1993), Danish Air Quality Monitoring Network. Technical Description, Report LMP-4/93. National Environmental Research Institute. Roskilde, Denmark.

Konig J., Theisen W. J., Gunther K. H., Liebl K. H. and Buchen M. (1993), Ambient Air Levels of Polychlorinated Dibenzofuran and Dibenzo-p-dioxins at Different Sites in Hessen, Germany, *Chemosphere*, 26, 851-861.

Krause F., Bach W. and Koomey J. (1990), *Energy Policy in the Greenhouse*, Earthscan Publications Ltd., London.

Kuwaiti Data Archive (KuDA) (2000), Atmospheric and Meteorological Data from the Kuwait Oil Fires Taken from Ground, National Centre for Atmospheric Research, Boulder, CO, USA.

Laxen D. P. H. (1985), Nitrogen Dioxide – An Air Quality Problem in London?, *London Environmental Bulletin*, 3(2), 10 – 12.

Laxen D. P. H. (1989), Winter Smogs Return to London, *London Environmental Bulletin*, 5(2), 7 - 8.

Leaderer B.P. and Stolwijk J.A. (1979), Optical Properties of Urban Aerosol and their Relation to Chemical Composition, Presented at the New York Academy of Science Symposium on Aerosols: Anthropogenic and Natural Sources and Transport, Jan. 9-12.

Lee D. S. and McMullen T. (1994), The Effects of Air Pollution on Materials, Proceedings of a Workshop Held in London, July 27 – 28, AEA Technology, Abingdon, Oxfordshire.

Leeds University (2000), Incineration of Municipal Waste with Energy Recovery, The Department of Fuel and Energy, September.

Lefohn A. S., Husar J. D. and Husar R. B. (1999), Estimating Historical Anthropogenic Global Sulphur Emission Patterns for the Period 1850-1990, *Atmospheric Environment*, 33, 3435-3444.

Leithe W. (1971), *The Analysis of Air Pollutants*, Ann Arbor Science, Michigan,

Ligon W. V. S. B. and Dorn R. J. (1989), Chlorodibenzofuran and Chlorodibenzo-p-dioxin Levels in Chilean Mummies Dated About 2800 Years Before the Present, *Environmental Science and Technology*, 23, 1286 – 1290.

London Research Centre (1998), Atmospheric Emissions Inventories, Glasgow, Middlesborough and West Yorkshire (Leeds, Bradford and Kirkless), October.

London Research Centre (2000), Internet Site, www.london-research.gov.uk.

Lowles, I. and ApSimon, H. (1996), The Contribution of Sulphur Dioxide Emissions from Ships to Coastal Acidification, *International Journal of Environmental Studies*, 51, 21 - 34.

Mansfield T. (1987), Soiling of London's Buildings, *London Environmental Bulletin*, 14(2), Spring, 6 – 7.

McCrae I. S., Hamilton R. S., Revitt R. M. and Harrop D. O. (1988), Modelling the Dispersion of Vehicle-Emitted Pollutants, Presentation at 81st Annual Meeting of Air Pollution Control Association, Dallas Texas, June, 1 - 15.

McHugh, C.A, Carruthers, D. J and Edmunds, H.A. (1997), ADMS-Urban: An Air Quality Management System for Traffic, Domestic and Industrial Pollution, *International Journal of Environmental Pollution*, 7, Nos 5 - 7.

Meteorological Office (1995), Wind Speed and Direction Data for Prestwick Airport, Bracknell, Berkshire.

Meteorological Office (1996a), Rainfall Data for Dalblair 1961 - 1990.

Meteorological Office (1996b), Total Precipitation and Evapotranspiration Data for Cumnock, Bracknell Berkshire.

Meterological Office (2000), Meteorological Data for Prestwick, Kinross, Turnhouse and Dyce Airports.

Middleton D. (1996), Physical Models of Air Pollution for Air Quality Reviews, *Clean Air*, 26(2), 28 - 36.

Middleton D. R. (1999), Development of AEOLIUS for Street Canyon Screening, *Clean Air*, 29(6), 156-161.

Middleton D. R., Butler J. D. and Colwill D. M. (1979), Guassian Plume Dispersion Model Applicable to a Complex Motorway Interchange, *Atmospheric Environment*, 13, 1039-1049.

Miller M.E., Canfield N.L., Ritter T.A. and Weaver C.R. (1972), Visibility changes in Ohio, Kentucky, and Tennessee from 1962 to 1969, *Monthly Weather Review*, 100, 67-71.

Ministerio de Planificacion (2000), Nacional y Politica Economica (MIDEPLAN), Costa Rica, Internet Site, www.mideplan.go.cr/sides/ambiental

Ministerio del Ambiente y de Los Recursos Naturales Renovables (MARNR) (1997), Boletin de Calidad del Aire en las Ciudades de Puerto Ordaz, San Cristobal y Valencia, Caracas, Venezuela.

Ministry of Agriculture, Fisheries and Food and the Welsh Office Agriculture Office (1992), Code of Good Agriculture Practice for the Protection of Air, London.

Moncrieff R. W. (1967), *The Chemical Senses*, 3rd Edn, Leonard Hill, London.

Moore D. J. (1973), Factors Influencing Plume Rise and Ground Level Concentrations, Chimney Design Symposium, Edinburgh University.

Moorcroft J. S. (1985), Airborne Asbestos Fibres – Identification and Measurement, *London Environmental Supplement*, 11, Spring.

Moorcroft J. S. and Eyre S. (1989), Assessment of Dust Nuisance Measurement and Guidelines, *London Environmental Supplement*, 19, Autumn.

Moses H. and Strom G. H. (1961), A Comparison of Observed Plume Rises with Values Obtained from Well-known Formulas, *Journal of the Air Pollution Control Association*, 11(10), 455-466.

Muir D. M. (Ed.) (1992), *Dust and Fume Control*, Revised Edn, Institute of Chemical Engineers, Rugby, England.

Munn R.E. (1973), Secular Increases in Summer Haziness in the Atlantic Provinces, *Atmosphere*, 11, 156-161.

Murley L. (1993), *Clean Air Around the World*, 2nd Edn, IUAPPA, Brighton.

Murley L. (1995), *Clean Air Around the World*, 3rd Edn, IUAPPA, Brighton.

Murray Fenton Edon Liddiard Vince BMT (2000), Study on the Economic, Legal, Environmental and Practical Implications of a European Union System to Reduce Ship Emissions of SO_2 and NO_x, European Commission Contract B4-3040/98/000839/MAR/B1, August.

Murrells T.P. (2000), UK Road Transport Emission Projections, The Assumptions Used and Results of the 1997 National Atmospheric Emissions Inventory Base Projections, AEA Technology Internet Site, www.aeat.co.uk/netcen/airqual.

National Atmospheric Emission Inventory (1999), Internet Site, www.detr.gov.uk.

National Environmental Monitoring Centre of the Peoples Republic of China (2000), Air Pollution Index in Major Chinese Cities (June - July).

National Material Exposure Programme (1999), Personal Communication.

National Society for Clean Air and Environmental Protection (1994), *Pollution Handbook*, Brighton.

National Society for Clean Air and Environmental Protection (1995), *Indoor Air Pollution*, Brighton.

National Society for Clean Air and Environmental Protection (1996), Asthma and Air Quality, Fact Sheet 15.

National Society for Clean Air and Environmental Protection (2000a), *Pollution Handbook*, Brighton.

National Society for Clean Air and Environmental Protection (2000b), Air Quality Action Plans: Interim Guidance for Local Authorities, Brighton, England.

National Swedish Environmental Protection Board (NSEPB) (1988), Dioxins - A Program for Research and Action.

Newton M. and Harrop D. O. (1987), Exposure to High Carbon Monoxide Concentrations in a Car Wash, *London Environmental Bulletin*, 4(2), Spring, 15-16.

New Zealand Government (2000), Air Quality Data, Internet Site, http//agdb.niwa.crinz/aqdb/.

Noll K. and Duncan J. (1973), *Industrial Air Pollution Control*, Ann Arbor Science, Michigan.

Occupational Safety & Health Administration (2000), United States Environmental Protection Agency, Occupational Health Standards.

Organisation for Economic Co-operation and Development (1964), Methods of Measuring Air Pollution, Paris.

Official Journal (1994), Framework Directive on Ambient Air Quality Assessment and Management (96/62/EC).

Oke T. R. (1978), *Boundary Layer Climates*, Methuen & Co Ltd, London, p.372.

O'Neill P. (1985), *Environmental Chemistry*, George Allen & Unwin, London.

Ortolano L. (1985), Estimating Air Quality Impacts, *Environmental Impact Review*, 5 (1), March, 9-35.

Occupational Safety & Health Administration (2000), United States.

Palmer D. G. (1974), *Introduction to Air Pollution*, New Educational Press Ltd., Lewes, Sussex.

Palmgren F., Kemp K. and Manscher O.H. (1992), The Danish Air Quality Programme (LMP II), Annual Data Report 1991. NERI Technical Report No. 60. National Environmental Research Institute. Roskilde, Denmark.

Parker A. (1978), *Industrial Air Pollution Handbook*, McGraw-Hill Book Company (UK) Limited, London, p.658.

Pasquill F. and Smith F. B. (1982), *Atmospheric Diffusion*, 3rd Edn, Ellis Horwood, Chichester.

Pasternach (1989), *The Risk of Environmental and Human Health Hazards*, John Wiley & Sons, New York.

People's Republic of China State Environmental Protection Agency (1996), National Air Quality Standards.

Petersen W. B. (1980), Users Guide for HIWAY-2, Report EPA 600/8-018, Environmental Protection Agency, NC27711.

Photochemical Oxidants Research Group (1997), Ozone in the United Kingdom, The Fourth Report of the Photochemical Oxidants Review Group.

Pope C. A. and Dockery D. W. (1992), Acute Health Effects of PM_{10} Pollution on Symptomatic and Asymptomatic Children, *American Review of Respiratory Disease*, 145, 1123 – 1128.

Pope C. A., Dockery D. W. And Schwartz J. (1995), Review of Epidemiological Evidence of Health Effects of Particulate Air Pollution, *Inhalation Toxicology*, 7, 1-18.

Purdy D. R. and Williams D. H. (1977), *Fibre Identification Using Optical Microscopy*, Electricity Council, London.

Quality of Urban Air Review Group (1996), *Airborne Particulate Matter in the UK*, Department of the Environment, London.

Ralph M., Barrett F. and Upton S. (1982), Wind Tunnel Study of the Inlet Efficiency of the WSL Suspended Particle Sampler, Stevenage, Warren Spring Laboratory, Report LR 420(AP).

Rappe C. (1992), Sources of Exposure, Environmental Levels and Exposure Assessment of PCDDs and PCDFs, *Chemosphere*, 27, 211 – 226.

Rappe C. (1993), Environmental Concentrations and Ecotoxicological Effects of PCDDs, PCDFs and Related Compounds, Extended Abstract Dioxin '93, *Organohalogen Compounds*, 12, 163-170.

Rappe C., Andersson R., Boinner M., Cooper K., Fiedler H., Lau C. and Howell F. (1997), PCDDs and PCDFs in Lake Sediments from a Rural Area in the USA, *Organohalogen Compounds*, 32, 88 – 93.

Rickard A. and Ashmore M. R. (1996), The Size Fractionation and Ionic Composition of Airborne Particulates in the London Borough of Greenwich, *Clean Air*, 26(2), 37 – 42.

Robins P. C. and Clark E. R. (1987), Concentrations of Fluoride in the Atmosphere and Herbage in the Environs of the Bedfordshire Brickworks between October 1984 and September 1986, Aston University, Birmingham.

Robinson E. and Valente R. J. (1982), Atmospheric Turbidity over the United States, 1948-1978. Research Publication GMR-3474, Env #92, General Motors Corp.

Rosenberg N. J. (1974), *Microclimate: The Biological Environment*, John Wiley & Sons, Chichester.

Rotty R. M. and Marland G. (1986), Production of CO_2 from Fossil Fuel Burning by Fuel Type, Report NDP-006, Carbon Dioxide Information Analysis Center, Oak Ridge National Laboratory, Tennessee.

Royal Commission on Environmental Protection (1993), Seventeenth Report, *Incineration of Waste*, HMSO, London.

Royal Commission on Environmental Protection (2000), Twenty Second Report, *Energy – The Changing Climate*, HMSO, London.

Royal Meteorological Society (1995), Atmospheric Dispersion Modelling: Guidelines on the Justification of Choice and Use of Models, and the Communication and Reporting of Results, Policy Statement, May.

Ruch W. E. (1970), *Quantitative Analysis of Gaseous Pollutants*, Ann Arbor Science, Michigan.

Saltzman B. E. (1954), Colorimetric Microdetermination of Nitrogen Dioxide in the Atmosphere, *Analytical Chemistry*, 26, 1949.

Sanderson H. P. (1963), Grit and Dust, Air Pollution Notes, *Journal of the Air Pollution Control Association*, 13, 461.

Schinder L. (1998), The Indonesian Fires and SE Asean Haze 1997/98, Review, Damages, Causes and Necessary Steps, Paper presented at the Asia-Pacific Regional Workshop on Transboundary Atmospheric Pollution, 27-28. May 1998, Pan Pacific Hotel Singapore.

Schulz T. J. and van Harreveld A. P. (1996), International Moves Towards Standardisation of Odour Measurement using Olfactometry, 18th Biennial Conference on the IAWQ, Singapore, 23-28 June.

Schwartz J. and Zeger S. (1990), Passive Smoking, Air Pollution and Acute Respiratory Symptoms in a Diary Study of Student Nurses, *American Review of Respiratory Diseases*, 36, 62-67.

Scorer R. S. (1998), Modelling of Air Pollution – Its Uses and Limitations, *Clean Air*, 28(3), 102-104.

Scottish Coal (1996a), Environmental Statement for Spireslack OCCS.

Scottish Coal (1996b), Personal Communication, November.

Scottish Environment Protection Agency (1999a), IPC Public Register for Longannet Power Station.

Scottish Environment Protection Agency (1999b), Circular on the Ozone Layer.

Scottish Environment Protection Agency (1999c), Circular on Climate Change, UK Impacts Programme of Climate Change in Scotland (1998).

Scottish Environment Protection Agency (2000), State of the Environment Air Quality Report.

Scottish Hydro-Electric (1997), Air Dispersion Modelling Study of Peterhead Power Station, Aberdeenshire.

Scottish Office (1994), SN1 1(94).

Scottish Office (1998), An Assessment of Air Quality Monitoring in Scotland, Central Research Unit.

Segal H. M. and Yamartino R. (1981), The Influence of Aircraft Operations on Air Quality at Airports, *Journal of the Air Pollution Control Association*, 31, 846-50.

Shenyang Urban Planning Project (SUPP) (2000), Mission Report, EU-China Liaoning Integrated Environment Programme LIEP.

Shy C. M. and Finklea J. F. (1973), Air Pollution Effects Community Health, *Environmental Science and Technology*, 7, 204 - 208.

Simms K. L., Wilkinson S. and Bethan S. (2000), Odour Nuisance and Dispersion Modelling - An Objective Approach to a Subjective Problem, CERC, Cambridge.

Singapore Government (2000), Environment Department, Internet Site, www.gov.sg/env/info.

Sloane C. S. (1984), Meteorologically Adjusted Air Quality Trends: Visibility, *Atmospheric Environment*, 18, 1217.

Smith K. R. (1996), Indoor Air Pollution in Developing Countries: Growing Evidence of its Role in the Global Disease Burden, in Ikeda K. and Iwata T. (Eds), *Indoor Air '96*, published by the Organising Committee of the 7th International Conference on Indoor Air Quality and Climate, SEEC ISHIBASHI Inc., Japan.

Smith R. G. and Diamond P. (1952), The Microdetermination of Ozone, *Ind. Hyg. Quart.*, 13, 235.

Smith R. I., Metcalf S. E., Coyle M., Finnegan D., Whyatt J. D., Pitcairn C. E. R., Cape J. N. and Fowler D. (1996), Sulphur Deposition in Scotland, The Scottish Office.

Smith W. H. (1981), *Forest Stress Symtomatic Foliar Damage Caused by Air Contaminants in Air Pollution and Forests - Interactions Between Air Contaminants and Forest Ecosystems*, Springer - Verlag, New York.

Stanger (2000), Design Manual for Roads and Bridges (DMRB), Model from Stanger Internet Site, (www.stanger.co.uk/airqual/modelhlp.).

State Environmental Protection Administration (1993), Emission Standards for Odor Pollutants, People's Republic of China (GB 14554-93).

State Environmental Protection Administration (1998), Selected Environmental Standards of the People's Republic of China (1979 – 1997), December.

Stedman J. R. (2000), Secondary Air Pollutants, *Clean Air*, 30(3), 86 – 87.

Stern A. C. (1976), *Air Pollution,* Academic Press, New York.

Stevenson K. J. and Bush T. (1996), UK Nitrogen Dioxide Survey 1994, AEA Technology, January.

Stratospheric Ozone Review Group (1996), Review Report, Department of the Environment, Transport and the Regions.

Sturges W. T. and Harrison R. M. (1989), *Atmospheric Environment*, 23(9), 1987-1996.

Sugita K., Asada S., Yokochi T., Ono M. and Okazawa (1993), Polychlorinated Dibenzo-p-dioxins, Dibenzofurans, Co-planar PCBs and Mono-ortho PCBs in Urban Air. Extended Abstract Dioxin '93, *Organohalogen Compounds*, 12, 127-130.

Sustainable Shenyang Project Office (1998), Shenyang Environmental Profile – Shenyang Municipal Government, April,CPR/96/321/A/01/99.

Swedish Ministry of Agriculture (1982), *Acidification Today and Tomorrow*, Environment '82 Committee.

Szepesi D. J. (1989), *Compendium of Regulatory Air Quality Simulation Models*, A Kadémiai Kiadó, Budapest.

Szepesi D. J. and Fekete K. (1987), Background Levels of Air and Precipitation Quality for Europe, *Atmospheric Environment*, 21, 1623 – 1630.

Taucher J. A., Buckland S. J., Lister A. R. and Porter L. J. (1992), Levels of Polychlorinated Dibenzo-p-dioxins and Polychlorinated Dibenzofurans in Ambient Air in Sydney Australia, *Chemosphere*, 25, 1361 - 1365.

Taylor H. J., Ashmore M. R. and Bell J. N. B. (1988), *Air Pollution Injury to Vegetation*, Institute of Environmental Health Officers, London.

Tchobanoglous G., Theisen H. and Vigil S. A. (1993), *Integrated Solid Waste Management – Engineering Principles and Management Issues*, McGraw-Hill International Editions, Civil Engineering Series, New York.

Territory Development Department South East New Territories Development Office, Hong Kong (1992), Tseung Kwan O Development Sewage Treatment and Disposal, Environmental Assessment.

Thain W. (1980), *Monitoring Toxic Gases in the Atmosphere for Hygiene and Pollution Control*, Pergamon Press, Oxford.

Thomas M. D. (1961), Effects of Air Pollution on Plants, in *Air Pollution*, World Health Organisation, Geneva.

Times Newspaper (1881).

Timmis, R. (1995), Modelling Lecture Notes, Her Majesty's Inspectorate of Pollution, London.

Treshow M. (1984), *Air Pollution and Plant Life*, John Wiley & Sons, Chichester.

Trijonis J. C. (1982), Existing and Natural Background Levels of Visibility and Fine Particles in the Rural East, *Atmospheric Environment*, 16, 2431 (1982).

Troyanowsky C. (Ed.) (1985), *Air Pollution and Plants*, VCH Verlagsgesellschaft.

Tsyro, S. and Berge, E. (1997), The Contribution of Ship Emission from the North Sea and the North-East Atlantic Ocean to Acidification in Europe, EMEP/MSC-W, July.

Turner D. B. (1970), Workbook of Atmospheric Dispersion Estimates, Air Resources Field Research Office, Environmental Science Services Administration, Environmental Protection Agency, Offices of Air Programs, Research Triangle Park, North Carolina.

Turner B. (1979), Atmospheric Dispersion Modelling - A Critical Review, *Journal of the Air Pollution Control Association*, 29 (5), 502 - 519.

Turner B. (1994), *Workbook of Atmospheric Dispersion Estimates - An Introduction to Dispersion Modelling*, 2nd Edn, Lewis Publishers, Boca Raton.

Turner S. M. and Liss P. S. (1985), Measurement of Various Sulphur Gases in a Coastal Marine Environment, *Journal of Atmospheric Chemistry*, 2, 223 - 232.

Tysklind M., Fangmark I., Marklund S., Lindskog A., Thaning L. and Rappe C. (1993), Atmospheric Transport and Transformation of Polychlorinated Dibenzo-p-dioxins and Dibenzofurans, *Environmental Science and Technology*, 27, 2190 - 2197.

United Nations Economic Commission for Europe (1995), *Strategies and Policies for Air Pollution Abatement, Convention on Long-Range Transboundary Air Pollution*, New York and Geneva.

United States Department of Health, Education and Welfare (1970), *Air Quality Criteria for Carbon Monoxide*, Washington, D.C.

United States Environmental Protection Agency (1971), National Primary and Secondary Ambient Air Quality Standards, *Federal Register*, 36, 3825, 22384.

United States Environmental Protection Agency (1980), Guidelines and Methodology used in the Preparation of Health Assessment Chapters of the Consent Decree Water Quality Criteria, *Federal Register*, 45, 79347 – 79357.

United States Environmental Protection Agency (1984), Method T02, Method for the Determination of Volatile Organic Compounds in Ambient Air by Carbon Molecular Sieve Adsorption and Gas Chromatography/Mass Spectrometry, April.

United States Environmental Protection Agency (1986), Superfund Public Health Evaluation Manual, EPA/540/1-86/60, October, Washington DC.

United States Environmental Protection Agency (1987a), *Industrial Source Complex (ISC) Dispersion Model User's Guide* - Second Edn. (Revised) Volumes 1 and 2, EPA-45014-88-002A and EPA-450/4-88-002B, December.

United States Environmental Protection Agency (1987b), Assessment of Municipal Waste Combustor Emissions under the Clean Air Act, Advance Notice of Rulemaking, 52 FR 25399, Washington, 7 July 1987.

United States Environmental Protection Agency (1988), Method T014, Determination of Volatile Organic Compounds (VOCs) in Ambient Air using SUMMA Passivated Canister Sampling and Gas Chromatographic Analysis, June.

United States Environmental Protection Agency (1989), Risk Assessment Guidance for Superfund, 1, Human Health Evaluation Manual (Part A), Interim Final, Washington DC.

United States Environmental Protection Agency (1994), Indoor Air Pollution – An Introduction for Health Professionals, US Government Printing Office Publication No. 1994-523-217/81322.

United States Environmental Protection Agency (1995a), Compilation of Air Pollution Emission Factors, AP-42.

United States Environmental Protection Agency (1995b), Air Chief, United States Environmental Protection Agency, Washington DC.

United States Environmental Protection Agency (1998) Compilation of Air Pollution Emission Factors, AP-42. USEPA Internet Site, www.epa.gov.

United States Environmental Protection Agency (2000), Internet Site, www.epa.gov.

Utah State Department for the Environment (2000), Internet Site, www.eq.state.ut.us.

Valentin F. H. H. (1980), General Principles of Odour Control, in Valentin F. H. H. and North A. A. (Eds), *Odour Control – A Concise Guide*, Warren Spring Laboratory, Stevenage, Hertfordshire.

Valentin F. H. H. and North A, A. (1980), *Odour Control - A Concise Guide*, Warren Spring Laboratory, Stevenage, Hertfordshire.

van Gemert L. J. And Nettenbreijer A. H. (1977), Compilation of Odour Threshold Values in Air and Water, National Institute for Water Supply, Voorburg and the Central Institute for Nutrition and Food Research TNO, Zeist, Netherlands.

Warren Spring Laboratory (1966), National Survey of Smoke and Sulphur Dioxide, Instruction Manual, Stevenage, Hertfordshire.

Warren Spring Laboratory (1980), *Odour Measurement and Control*, Stevenage, Hertfordshire.

Warren Spring Laboratory (1993), Quarterly Air Quality Report for Department of the Environment Automatic Monitoring Networks, October-December 1992.

Water Research Centre (1999), Odour Emission Rates from Sewage Treatment Works, PT1048, March 1995 (as revised 1999).

Weber E. (1982), *Air Pollution: Assessment Methodology and Modelling*, Vol. 2, Plenum Press, New York.

Weiss R. E., Waggonner A. P., Charlson R. J. and Ahlquist N. C. (1977), Sulfate Aerosol: Its Geographical Extent in the Midwestern and Southern United States, *Science*, 197, 997-998.

Wevers M. R., De Fre R., Van Cleuvenbergen R. and Rymen T. (1993), Concentrations of PCDDs and PCDFs in Ambient Air at Selected Locations in Flanders, Extended Abstract Dioxin '93, *Organohalogen Compounds*, 12, 123-126.

Wolff G. T., Kelly N. A. and Ferman M. A. (1982), Source Regions of Summertime Ozone and Haze Episodes in the Eastern United States, Water Air Pollution.

Woodfield M. and Hall D. (1994), *Odour Measurement and Control – An Update*, AEA Technology, Abingdon, Oxfordshire.

Woodfield N. (2000), Declaring An Air Quality Management Area – Is it an Easy Business?, *Clean Air*, 30(4), 112 – 114.

World Bank (2000), Internet Site, www.worldbank.org/wbi/cleanair.

World Health Organisation (1979a), Carbon Monoxide, *Environmental Health Criteria*, 13, Geneva.

World Health Organisation (1979b), Sulfur Oxides and Suspended Particulate Matter, *Environmental Health Criteria*, 8, Geneva.

World Health Organisation (1982a), *Rapid Assessment of Sources of Air, Water and Land Pollution*, Geneva.

World Health Organisation (1982b), Indoor Air Pollutants: Exposure and Health Effects, EURO Reports and Studies 78, June, Copenhagen.

World Health Organisation (1986), Asbestos and Other Natural Mineral Fibres, *Environmental Health Criteria*, 33, Geneva.

World Health Organisation (1987), *Air Quality Guidelines for Europe*, WHO Regional Publications, European Series, 23.

World Health Organisation (1994), Updating and Revision of the Air Quality Guidelines for Europe – Inorganic Air Pollutants, EUR/ICP/EHAZ 94 05/MT04, Regional Office for Europe, Copenhagen.

World Health Organisation European Centre for Environment and Health (1995a), Concern for Europe's Tomorrow.

World Health Organisation (1995b), Update and Revision of the Air Quality Guidelines for Europe, Meeting of the Working Group 'Classical' Air Pollutants Bilthoven, The Netherlands 11 - 14 October 1994, WHO Regional Office for Europe, Denmark.

World Health Organisation (1998), Healthy Cities Air Management Information System, AMIS 2.0.CD ROM, Geneva.

World Health Organisation (2000a), Guidelines for Air Quality, WHO/SDE/OEH/00.02, Geneva.

World Health Organisation (2000b), Internet Site, www.who.dk.

World Health Organisation/ASMA (2000), Air Quality Information, Internet Site, www.enviromalaysia.com.my/who_asma_amis.htm.

Wright R. (1994), Odour Abatement, in Woodfield M. and Hall D. (Eds), *Odour Measurement and Control – An Update*, AEA Technology, Abingdon, Oxfordshire.

Wright R. and Woodfield M. (1994), Appendix C, in Woodfield M. and Hall D. (Eds), *Odour Measurement and Control – An Update*, AEA Technology, Abingdon, Oxfordshire.

Yamartino R. J., Smith D. G., Bremers A., Heinbold D., Lamich D. and Taylor B. (1980), Impact of Aircraft Emissions on Air Quality in the Vicinity of Airports, 1, Washington Federal Aviation Administration, FAA-EE-80-09A.

Yin F., Grosjean D. and Seinfield J. H. (1990), Photooxidation of Dimethyl Sulphide and Dimethyl Disulphide, I, Mechanism Development, *Journal of Atmospheric Chemistry,* 11, 365 - 399.

Young R. A. and Cross F. L. (1980), *Operation & Maintenance for Particulate Control Equipment*, Ann Arbor Science, Michigan.

Country index

Subject index